3rd International Conference on Chemical and Food Engineering 2016

AA004448

MATEC Web of Conferences
Volume 62 (2016)

Tokyo, Japan
8-9 April 2016

Editors:

C. J. Park
H. Morita
H. H. Huang
M. R. Khan
C. K. Cheng
R. L. De Leon

ISBN: 978-1-5108-2600-7

Printed from e-media with permission by:

Curran Associates, Inc.
57 Morehouse Lane
Red Hook, NY 12571

Some format issues inherent in the e-media version may also appear in this print version.

This work is licensed under a Creative Commons Attribution license:
http://creativecommons.org/licenses/by/2.0/
You are free to:
Share – copy and redistribute the material in any medium or format.
Adapt – remix, transform, and build upon the material for any purpose, even commercial.
The licensor cannot revoke these freedoms as long as you follow the license terms.
Under the following terms:
You must give appropriate credit, provide a link to the license, and indicate if changes were made. You may do so in any reasonable manner, but not in any way that suggests the licensor endorses you or your use. The copyright is retained by the corresponding authors.

Printed by Curran Associates, Inc. (2016)

For additional information, please contact EDP Sciences – Web of Conferences at the address below.

EDP Sciences – Web of Conferences
17, Avenue du Hoggar
Parc d'Activité de Courtabœuf
BP 112
F-91944 Les Ulis Cedex A
France

Phone: +33 (0) 1 69 18 75 75
Fax: +33 (0) 1 69 28 84 91

contact-edps@webofconferences.org

Additional copies of this publication are available from:

Curran Associates, Inc.
57 Morehouse Lane
Red Hook, NY 12571 USA
Phone: 845-758-0400
Fax: 845-758-2633
Email: curran@proceedings.com
Web: www.proceedings.com

TABLE OF CONTENTS

Committees

Conference General Chair

Prof. Miwako Hosoda, SEISA University, Japan

Technical Program Co-Chairs

Prof. Chan Jin Park, Incheon National University, Republic of Korea
Prof. Hideki Nakagome, Graduate School of Engineering, Chiba University, Japan
Associate Prof. Chi-wai Kan, The Hong Kong Polytechnic University, Hong Kong

Contact Chair

Ms Eve Lee, Asia-Pacific Chemical, Biological & Environmental Engineering Society

International Technical Committees

Prof. Chan Jin Park, Incheon National University, Republic of Korea

Assoc. Prof. Mohammad Reza Alizadeh, Rice Research Institute of Iran, Iran

Assoc. Prof. Chi-wai Kan, The Hong Kong Polytechnic University, Hong Kong

Prof. Hideki Nakagome, Graduate School of Engineering, Chiba University, Japan

Assoc. Prof. Maria Lourdes P. Dalida, University of the Philippines, Philippines

Assoc. Prof. Hui Hsin Huang Aletheia University, Taipei, Taiwan

Assoc. Prof. Maksudur. R. Khan, University Malaysia Pahang, Malaysia

Assoc. Prof. Chin K. Cheng, Universiti Malaysia Pahang, Malaysia

Assoc. Prof. Rizalinda Leon, University of the Philippines, Philippines

Assoc. Prof. T. Wajima, Chiba University, Japan

Prof. Ashok Patel, Ghent University, Belgium

Prof. Tony J. Fang National Chung Hsing University, Taichung, Taiwan

Dr. Jutatip Poubol Kasetsart, University, Thailand

Dr. Aree Innun Kasetsart University, Thailand

Dr. Pharima Phiriyangkul, Kasetsart University, Thailand

Dr. Jing-Jing Zhang University of Illinois Urbana-Champaign, USA

Preface

2016 3rd International Conference on Chemical and Food Engineering has been successfully held in Tokyo, Japan during April 08-09, 2016, which was organized by CBEES.

The goal of the conference is to provide a forum for the discussion of new developments, recent progress, and innovations in the chemical and food engineering. It addresses all aspects of the related fields. The emphasis is on current and future challenges in research and development in both academia and industry. It emphasizes long-term, fundamental research aimed at discovering novel phenomena, processes, and tools.

The proceedings present a selection of 35 papers accepted in the conference from universities, research institutes and industries. All of the papers were subjected to peer-review by conference committee members and international reviewers. The papers selected for publishing in the proceedings depended on their quality and their relevancy to the conference. The proceedings tend to present to the readers the recent advances in the field of chemical and food engineering and various related areas. The proceedings mainly cover 7 specific areas, namely, Material Science, Food Science, Biochemistry, Chemistry Engineering, Environment & Chemistry, Environment & Bioscience, Environment & Building.

We are grateful to all the authors for submissions to this conference and for sharing their research fruits and valuable experiences they've acquired. And we would like to express our special thanks to the conference chair, Prof. Miwako Hosoda and the program chair, Prof. Hideki Nakagome, and all the reviewers and conference committee members.

At last, we are very delighted to welcome you to ICCFE next year.

Editors,
Prof. Chan Jin Park and Assoc. Prof. Maksudur. R. Khan

April 2016

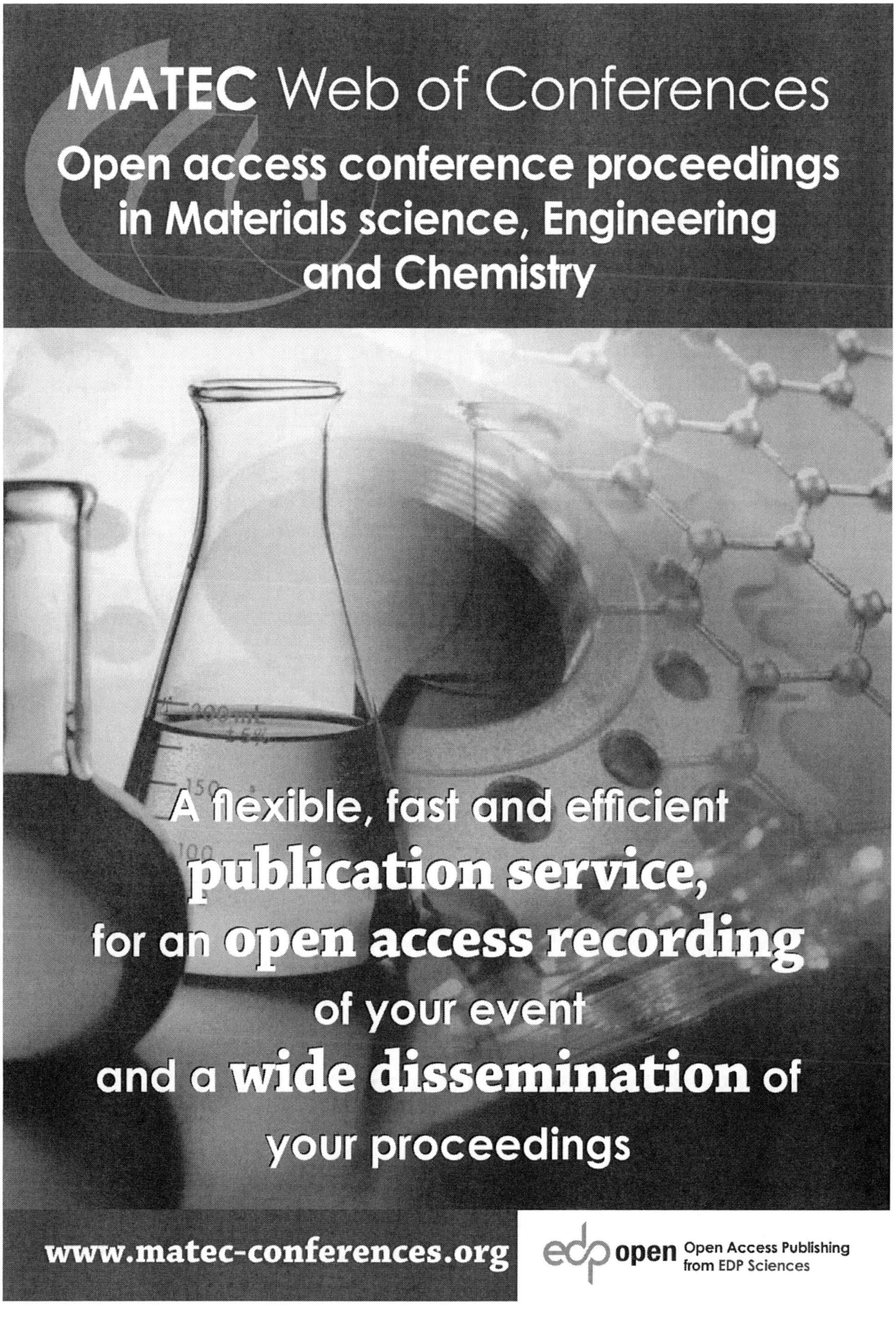
MATEC Web of Conferences
Open access conference proceedings in Materials science, Engineering and Chemistry
A flexible, fast and efficient publication service, for an open access recording of your event and a wide dissemination of your proceedings
www.matec-conferences.org
edp open Open Access Publishing from EDP Sciences

Aims & scope

MATEC Web of Conferences is an open-access journal devoted to the publishing of conference proceedings dealing with Materials science, Engineering and Chemistry.

The journal aims to make conference proceedings visible, citable and widely usable by the scientific community. It offers outstanding services for the publication and dissemination of contents in order to ensure long-term archiving and maximum exposure of scientific events.

Coverage includes all engineering disciplines: civil, naval, mechanical, chemical, and electrical engineering as well as nanotechnology and metrology. The journal also deals with the physical-chemical characterization of materials, their implementation, their resistance in their environment, etc. Others subdisciplines of chemistry, such as analytical chemistry, petrochemistry, organic chemistry, etc., and even pharmacology, are also welcome.

Why publish with MATEC Web of Conferences?

- **Open access:** conference papers available worldwide.
- **Fast publication:** less than 3 months.
- **Total flexibility:** no page restrictions, free unlimited use of colour in the online version, any kind of scientific document welcomed (traditional articles, posters, abstracts and slideshows).
- **High visibility:** all documents indexed in main databases and easily found online.
- **Fair author rights:** CC-BY Creative Commons license.
- **Range of additional publishing options:** printed proceedings, CD-ROMs, copy-editing, etc.

Contact: Isaline Augusto
matec@webofconferences.org

www.matec-conferences.org

edp open Open Access Publishing from EDP Sciences

DOI: 10.1051/matecconf/20166201001

Evaluation on the Collection Efficiency and Performance of the Sound Pressure Machine Equipped with a HEPA Filter Eliminating Asbestos Particles

Wha Me Park[1,a], Rew Jae Myung[1], Key Young Kim[1]

[1]*Institute of Environmental & Industrial Medicine(IEIM), Han-yang University, Korea*

Abstract. When the one-pass collection efficiency of each size of particles of the sound pressure machine equipped with a glass fiber HEPA filter to get rid of friable asbestos at the asbestos elimination field was evaluated, the collection efficiency of those with the size of 0.3um was examined to be 98.91%. That of the particles of 0.5um size was proved to be 99.21% on average, which is a little bit higher than that of 0.3um size. The 1.0um particles showed 100% of efficiency, and the collection efficiency of each size had statistically meaningful differences.

1 Introduction

Asbestos is a mineral fiber found in nature. It is a fire-resistant building material and comes in such different forms as asbestos tile and slate, etc. It is also widely used as a heat-resistant industrial good in household products, electric goods and jets. Use of anti-scattering products for high-risk asbestos work classified under the most stringent Class H in the UK in terms of carcinogenicity and hazard are advised in order to keep asbestos particles from scattering indoors [1, 2]. Specifications for performance, test standards and evaluation of anti-scattering products are still much too loose compared to those adopted outside of Korea, which limits the efforts to measure performance of the products. Most of them are not properly graded and sound pressure machines used for removing asbestos can rarely be applied indoors due to their high filter differential pressure, use of high electricity to deliver high air flow and high note [3]. Hence, an indicator that grades anti-scattering products based on their removal efficiency, performance evaluation and controls high-risk groups is needed.

2 Research Methods

2.1 Specification and performance of high-performance air filter (HEPA filter)

2.1.1 Korea

There is no certified specification or test standards for HEPA filters sold in the domestic market as of yet. HEPA filter manufacturers and marketers that have in-house test lab may require all inspection before sales. Others commission private test labs for an inspection if an outside certification is required. Two KS for domestic HEPA filter are KS A 4812: 2001 high-performance air filter for radioactive aerosol by Korean Agency for Technology and Standards (KATS) and KS B 6740: 1995 air filter performance test method for cleanroom by KATS [4, 5].

KS A 4812: 2001 standard is radioactive aerosol with a view to remove nuclear facilities used by the powerful about to be in the fire flame is required, and air filter And allow efficiency, evaluation with a particle DOP (dioctyl phthalate) 0.15 μm [6].

KS B 6740: 1995 standards are only distributed in the form of particles 0.3 μm DOP standard is to the Particle Collection with a rate test [7].

2.1.2 England

British is embracing national standards to conform to the standard HEPA filter (EN 1822) of the European Union [8]. The European Union have specifications for the HEPA filter of a unified one. Matters concerning a total of five HEPA filters, divided by the specifications and type classification, test methods, measurement equipment, test methods is shown in each standard. The EN 1822 standard test aerosol dos, dop, paraffin oil, and aerosols, size is writing MPPS(Most Penetrating Particle Size). Sampling equipment is used to play Optical counter or Condensation nucleus counter.

National standard f HEPA filter of England 1) High efficiency air filters(EPA, HEPA and ULPA). C lassification, performance testing marking(BS EN 1822-1: 2009), 2) High efficiency air filters(EPA, HEPA and ULPA). Aerosol production, measuring equipment, particle counting statistics(BS EN 1822-2: 2009), 3) High

[a] Corresponding author: wmpark2045@hanmail.net

© The Authors, published by EDP Sciences. This is an open access article distributed under the terms of the Creative Commons Attribution License 4.0 (http://creativecommons.org/licenses/by/4.0/).

efficiency air filters(EPA, HEPA and ULPA). Testing flat sheet filter media(BS EN 1822-4: 2009) 4) High efficiency air filters(EPA, HEPA and ULPA). Determining leakage of filter elements(scan method)(BS EN 1822-4: 2009) 5) High efficiency air filters(EPA, HEPA and ULPA). Determining the efficiency of filter elements (scan method)(BS EN 1822-5: 2009)

2.1.3 United States

There are many HEPA filter standards for using the United States. Each rule all developed for different history of the methods are used has been flouting respectively. The specification used in KS standards, testing institutions and many standards a descendant of HEPA filter of the United States.

In 1956, how dop MIL-STD-282<Filter Unit, Protective Clothing, Gas-mask Components and Related Products> to use was published, Heat the liquid into this standard the test aerosol DOP, superimposes steam and concentrate again make the particles an average of 0.3 μm.

Published the ANSI/UL 586 <Air Filter Units> in 1959.

In 1975 and 1976 ANSI N 45.8 Committee < American Society of Mechanical Engineers Committee on Nuclear Air and Gas Treatment> ASME/ANSI N510 and ASME N509published. ASME/ANSI N510-2007 is the latest edition.

Between 1992 and 1993 Institute of Environmental Sciences and Technology the HEAP/ULPA filter test standards-IEST-RP-CC007.1: 1992 <Testing ULPA Filters>, IEST-RP-CC001.3:1993 <HEAP and ULPA Filters>, IEST-RP-CC006.2:1993 <Testing Cleanrooms>- publication series against the brothers.

In 2005 published the IEST-RP_CC001.4 <HEPA and ULPA Filters>, IEST-RP-CC04.2 <HEAP and ULPA Filter Tests>.

2.2 Specification and performance of sound pressure machines

2.2.1 Korea

Specification of sound pressure machines is based on approval certification that requires(KOSHA Guide H-53-2009) [9-11] 1) HEPA filter installation (higher than 99.97% in DOP removal efficiency), 2) installation of pre-treatment filter in HEPA filter front, 3) installation of filter differential pressure gauge, 4) successful leak test, 5) installation of air blower in filter rear and 6) installation of equipment preventing scattering during mobility. Sound pressure machines should be loaded with HEPA filter (99.97% in removal efficiency) and pass leak test (99.95% in removal efficiency) based on approval certification standards. (Table 1), [12]

Sound pressure of an sale in the country to a survey of performance requirements is shown below. Sound pressure flow rate, which is the most important when using a machine sound pressure If I had not displayed in a 42 % at its maximum. Sound pressure often not given leak testing is a 42 % and sound pressure on a bi ban song for the inside flap if they are not installed, and a 58 %. Flow controller 29 % if they are not installed. (Table 2)

Table 1. Producing sound pressure machine in domestic capacity.

Rate(m³/h)	Number
500 ~ 1,000	3
1,000 ~ 2,000	4
2,000 ~ 3,000	4
3,000	1

Table 2. Domestic performance installation requirements for producing sound pressure machi.

	Yes	No
Sound pressure flap bibansong installation inside the machine	3	4
Intromitter gasket(sea wave filters and between cases)	7	0
Flow bean or damper setup	5	2
Sound pressure meter upward pressure on an installation	6	1(Nonresponse)
Alarm installations for the sound pressure machine	5	2
See flow in the external sound pressure machine	4	3
See jejonyeon in the sound pressure machine	7	0
Sound pressure machine displayed by manufacturer name or mark	7	0
Test whether the sound pressure machine	4	3

2.2.2 England

Performance of Sound Pressure in England for national standards-BS 8520-2: 2009 Equipment used in the controlled removal of asbestos-containing materials- Part 2: Negative pressure units-Specification [13-15].

The purpose of this standard HEPA filter to be used, remove asbestos containing materials equipped with portable sound pressure to specify the performance.

Be easy to clean the outer case, apartment area that is signed three times made of plastic welding, head end when moved to the flaps, prevent reflux filter cover, may be able to protect the front wheels and handles, HEPA filter, air inlet Ministry of the wheel, A manometer on the sound-pressure machine specifications show when I was stuck in traffic filters HEPA should be included.

The leakage test specifications Household and similar electrical appliance. Safety. Particular requirements for wet and dry vacuum cleaners including power brush for commercial use (BE EN 60335-2.69: 2009) in Annex a

Carrying yellow dusts shows. According to this standard, should more than 99.7% efficiency HEPA filter and mounted on the sound pressure machine, Maximum leakage should be allowed only 0.005 % from the leakage tests. Leakage test the same grades H a vacuum cleaner as applicable.

2.2.3 United States

a) OSHA class I of working with asbestos rigid an entire engineering standard [11].

OSHA law, used asbestos services in the dismantling and removal work, sound pressure specified for information regarding the structure of a particular point about not be, and sound pressure, using a specified.

- enclosure internal sound pressure measurements, - 5Pa(0.5mm H2O) to do maintenance.
- of differential pressure measurement.
- flow of air, ease, from worker to perform.
- Install a smoke test to detect the flow and air leakage.

b) American National Standards Institute and American Industrial Hygiene Association and portable ventilation system specifications.

American National Standards Institute and the American Industrial Hygiene Association to jointly published by mobile portable some matters were about standards in ventilation systems in the specification of Ventilation SystemInclude.

2.3 Evaluation of commercial sound pressure machines' one-pass dust collection efficiency

Sound pressure machines commonly used to remove asbestos were used for the evaluation and dust collection efficiency of glass fiber HEPA mounted in sound pressure machines was evaluated. (Figure 1)

Figure 1. Commercial sound pressure machine for removing asbestos.

Affiliations of authors should be typed in italic 8-point Arial. They should be preceded by a numerical superscript corresponding to the same superscript after the name of the author concerned. Please ensure that affiliations are as full and complete as possible and include the country. (Figure 2)

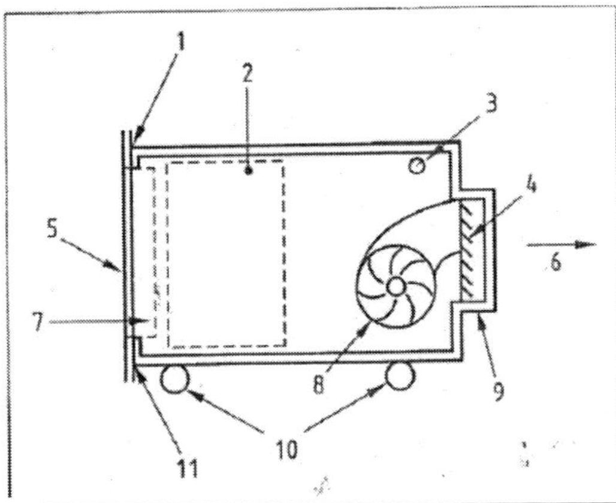

1. Flange 2. HEPA filter 3. Manometer gauge
4. non-return flaps 5. Transit cover 6. Airflow direction
7. Pre-filter 8. Fan/motor 9. Exhaust air spigot
10. casters 11. Flange

Figure 2. Sound pressure machine configuration.

2.4 Measurement of one-pass dust collection efficiency

Airborne Particle Counter (Rion, KC-01C) was used to evaluate dust collection efficiency of all sound pressure machines. Airborne Particle Counter applied light scattering method to measure particle density and measured particles 0.3μm, 0.5μm and 1μm in diameter. Flow Meter sampled 0.5L/min of flux and sample volume was 1L or 2.83L. Particle density from sound pressure machine's inlet and outlet was alternately measured after air volume stabilized. (Figure 3)

$$\eta = (1 - C_0/C_i)X100$$

η: Dust collection efficiency(%)
C_0: Particle density of outlet(ea/L)
C_i: Particle density of air inlet(ea/L)

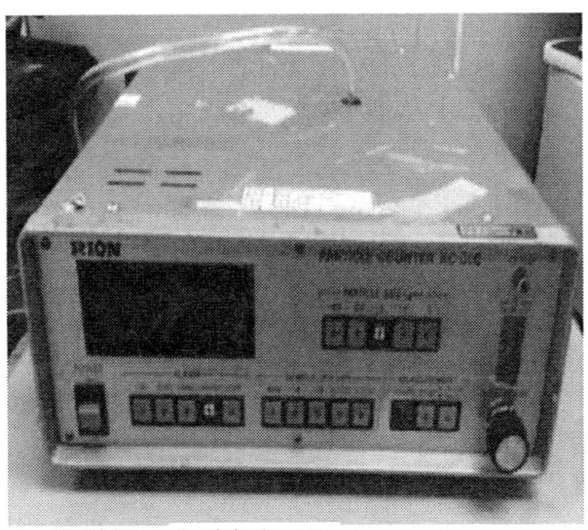

Figure 3. Airborne Particle Counter.

2.5 Evaluation of dust collection efficiency of asbestos-removing sound pressure machines by particle size

The following table presents evaluation of one-pass dust collection efficiency of a sound pressure machine mounted with glass fiber HEPA filter. Efficiency is evaluated by particle size in a commercially-used sound pressure machine. (Table 3)

Table 3. Dust collection efficiency of asbestos-removing sound pressure machine mounted with glass fiber HEPA filter by particle size.

Filter Type	Particle Size (μm)	N	Removal Efficiency (%)				
			Mean	Std. Dev	Min	Max	P-value
Fiber Glass	0.30	10	98.91	0.110	98.72	99.07	<0.05
	0.50	10	99.21	0.735	97.71	100	
	1.0	10	100	0	100	100	

3 Study Results

Evaluation of a one-pass dust collection efficiency of a commercial sound pressure machine mounted with glass fiber HEPA filter by particle size indicates dust collection efficiency of 0.3 um in particle size to average at 98.91%. That of 0.5 um and 1.0 um in particle size showed a dust collection efficiency of 99.21% on average and 100%.

The dust collection efficiency test is based on KS B 6740 using mono-dispersed 0.3 um DOP particle. Performance evaluation requires particles 0.3um in size to show a dust collection efficiency exceeding 99.97%. Dust collection efficiency for each size in the test shows a statistically significant difference.

Acknowledgements

This study was funded by the Korea Ministry of Environment(MOE) as "the Environmental Health Action Program" [11].

References

1. Tsai, P.P., Gibson, H.S., and Gibson, P., Different electrostatic methods for making electret filters. Journal of Electrostatics, 54, 333-341(2002)
2. Young Man Roh, Mi Hee Seok, Kang Yoon Kim, Tae Jae Cho, Sang Hyo Sim, Hyun Cheol Ha., Study on "Infrastructure Standards to Remove Asbestos", Occupational Safety & Health Research Institute(2008)
3. Tae Hyung Kim, Hyun Cheol Ha, Kang Yoon Kim, Jong Cheol Kim, Yoo Jin Jung., "Study on Asbestos Removal Guidelines in Buildings to Prevent Scattering, Ministry of Environmen"t(2010)
4. Health and Safety Executive. Approved Code of Practice and guidance, Work with materials containing asbestos, Control of Asbestos Regulations. (2006)
5. Korea Air Cleaning Association. High-performance air filter test method(2009)
6. Korean Agency for Technology and Standards. High-performance air filter for radioactive aerosol, KS A 4812(2001)
7. Korean Agency for Technology and Standards. Air filter performance test method for clean room, KS B 6740(1995)
8. British Standards Institution. High efficiency air filters, Aerosol production, measuring equipment, particle counting statistics, BS EN 1822-2 (2009)
9. Health and Safety Executive. Approved Code of Practice and guidance, The management of asbestos in non-domestic premises(2006)
10. Health and Safety Laboratory. Investigation into the effectiveness of modular hygience unit(2001)
11. Occupational Safety and Health Administration, "Safety and Health Regulations for Construction" (1926.1101). Retrieved on(2012-01-10)
12. Korea Occupational Safety and Health Agency. Recommended guidelines for demolition and removal procedures when working with asbestos, KOSHA Guide H-70-2012(2012)
13. British Standards Institution. Controlled wetting of asbestos-containing materials-Specification, Equipment used in the controlled removal of asbestos-containing materials-Part 1, BS 8520-1 (2009)
14. British Standards Institution. Negative pressure unit-Specification, Equipment used in the controlled removal of asbestos-containing materials-Part 2, BS 8520-2(2009)
15. British Standards Institution. Operation, Cleaning and maintenance of Class H vacuum cleaners-Code of practice, Equipment used in the controlled removal of asbestos-containing materials-Part 3, BS 8520-3(2009)

Water Diffusion Modelling of CFB Fly Ash Thermoset Composite

Ralph P. Villa[1,a], Herbie T. Bella[1], Julian S. De Lara[1] and Romelando R. Juanatas[1]

[1]*Department of Chemical Engineering, University of the Philippines-Diliman, Quezon City 1101 Philippines*

Abstract. The shift in coal-fired power plants from pulverized coal (PC) boiler technology into the greener circulating fluidized bed (CFB) boiler technology resulted into a major deviation in the properties of the waste fly ash generated making it less suitable for its previous application as additives for construction materials. A new market for CFB fly ash had to be found for it not to end up as a zero value by-product. Using CFB fly ash as filler for thermoset composites is a new and remarkable application. Only a few studies, however, have been done to characterize the properties of this new material. Further experimentation and analysis may be costly and time-consuming since common procedures are material destructive. A computer-aided modeling of the composite's water sorption behavior was done. The effect of particle loading, size and shape were considered. These properties were varied and the resulting overall diffusivities were compared to previous experimental studies. The comparison of the model and experimental diffusivity values showed satisfactory results. This model may then provide a cheaper and more time-efficient method for the characterization of the water sorption properties of CFB fly ash thermoset composites. In the future, this may lead to further studies on its application as a green material.

1 Introduction

The shift of the technology used in coal-fired power plants from pulverized coal (PC) boiler technology into circulating fluidized bed (CFB) boiler technology had caused major changes into how the fly ash generated from the plants have been utilized over the years. Fly ash particles from coal-fired power plants were usually sold to cement companies and were used as an additive or filler to their building materials when the PC boiler technology was still in use. Now that the shift was made into CFB boiler technology due to its more promising energy conversion and it being more environment-friendly because of less emissions, the fly ash generated are of different physical and chemical properties which caused it to be unfit for its previous applications [1]. Cement companies used to be the market for fly ash residues from power plants but with this change, a new market had to be found and established.

The proceedings of the 15th International Conference on Fluidized Bed Combustion presented how CFB fly ash could be helpful in construction usages but high chlorine content, the lack of the existence of free lime in the ash and the need for ash pretreatment had caused limitations to its applications in the construction industry. The non-uniformity in the size and the shape of the fly ash particles and the high sulfur content also proves CFB fly ash inadequate for use as a cement additive [2]. Given that all coal-fired power plants will eventually shift to the new technology and that the fly ash they produce will not anymore be used by the cement industry in the Philippines for their products, the problem on how to dispose of or utilize these particles arises. Putting large amounts into landfill has become the most common method for handling CFB ash. Incineration and/or thermal degradation for the treatment of CFB fly ash also prove to be very expensive and therefore uneconomical. With the current trends in treating CFB fly ash and while putting economics into consideration, it is imperative to focus on developments through which these fly ash particles could be put to use. One possible use of the CFB fly ash is as fillers to polymer composites.

Studies to determine the properties of CFB fly ash particle-filled polymer composites or any new material can be done in the laboratory by making samples. This is feasible but can also be expensive given that several tests done to determine a material's properties are usually destructive and so the necessity of creating more and more samples for the results to be considered reproducible. An engineering breakthrough would be to create a nondestructive test (NDT) for material testing. One NDT technique that can be explored is the development of a predictive model for material property prediction.

The effect of particle loading, temperature and surface treatment on the moisture absorption of CFB fly ash reinforced polyester composite was recently studied by Onishi (2015) [3]. The effect of particle-filler shape on the moisture behavior through a different composite was also recently studied by Pajarito et.al. (2014) [4]. Modeling water diffusion through polymer composites are usually done by Finite Element Method (FEM) as shown in various literature [5-7]. In some, non-isothermal

[a] Corresponding author: ralph.villa@coe.upd.edu.ph

© The Authors, published by EDP Sciences. This is an open access article distributed under the terms of the Creative Commons Attribution License 4.0 (http://creativecommons.org/licenses/by/4.0/).

diffusion is also also studied using FEM as shown in the work of Yoon et.al, (2007) [8].

In this study, the isothermal diffusion of water through a CFB fly ash thermoset composite will be modeled using FEM. The effect of the filler fraction, filler shape and size distribution on the diffusion behavior will also be studied. A predictive model for the prediction of effectivity diffusivity of the composite will be formulated via regression analysis.

2 Methodology

2.1 Simulations on pure resin

To run the diffusion simulations, the student edition of the software ABAQUS was used. To simulate mass diffusion, ABAQUS requires the equilibrium moisture content and diffusivity of the material [9]. The pure properties of the resin were obtained from Onishi 2015 experimental results [3]. A sample, as seen in Figure 1, was drawn in the ABAQUS interface. The sample had a length of 2 mm, and a height of 0.04 mm. The size used was based on the size of the samples in the actual experiment. The length of 2 mm represents the thickness of the samples wherein the diffusion is assumed to occur in. The whole height of the sample was not used as it was assumed that a small slice is sufficient enough to model the diffusion that is assumed to occur uniformly throughout its length.

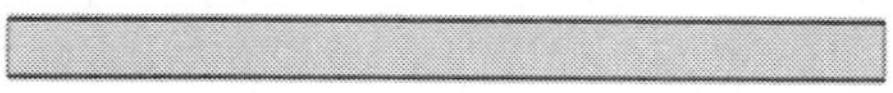

Figure 1. Resin Sample as seen on ABAQUS

As for the boundary conditions, both left and right sides of the sample were assumed to have a mass concentration of 1 with the water flowing through the x-axis during the diffusion process.

The time period for the simulation was set at 25020 minutes, the same time period used in the actual experiment.

A triangular mesh was used in the FEM Analysis (Figure 2). The fineness of the mesh was limited by the software as it was just a trial version for academic purposes. The finest mesh possible was used.

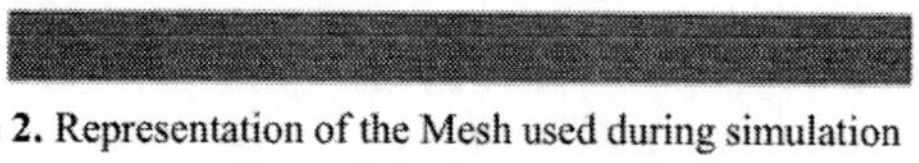

Figure 2. Representation of the Mesh used during simulation

Results of the simulation were then compared to the experimental results as well as the analytical solution to the Fickian Model of Diffusion.

2.2 Obtaining filler diffusion properties

To completely model the diffusion of water through the composite, the sorption properties of the filler must also be specified. However, no such properties were obtained from experimental data.

In order to obtain these properties, parametric studies were conducted by varying the diffusion and equilibrium moisture content one-at-a-time.

Figure 3. Composite RVE as seen in ABAQUS

Samples were made of the same size as the one for the pure resin. The average particle size of the fly ash used in the experiment were used in constructing the RVE. Fillers were arranged in a uniform, periodic manner. Fly ash volume concentrations of 4.82% and 23.28% were used. Figure 3 shows a sample used in the parametric study.

2.3 Simulation run

For the simulation run, a 3-factor, 2-level full factorial (2^3) design of experiment was used [10]. The factors considered were volume fraction, filler shape, and particle size.

For this research, a two-dimensional drawing was used. The volume fraction was defined as the ratio of the area occupied by the filler and the area occupied by the resin. The filler shape was quantified by defining the ratio between its vertical length and horizontal length. Meanwhile, the particle size is defined by the horizontal length of the filler.

The following table summarizes each run.

Table 1. Summary of Simulation Runs

Run	Volume Fraction	Shape	Particle Size (μm)
1	0.04815	0.8	15.3
2	0.04815	0.8	33.3
3	0.04815	1	15.3
4	0.04815	1	33.3
5	0.23285	0.8	15.3
6	0.23285	0.8	33.3
7	0.23285	1	15.3
8	0.23285	1	33.3

As with the previous procedures, the fillers were drawn in a uniform periodic manner, and the sample had a length of 2 mm and a height of 0.04 mm.

The time period for each run were based on the actual time period used in the experiment, in all cases, 25020 minutes.

For the boundary conditions, the left and right side of the filler were set to have a mass concentration of 1 during the diffusion process.

Two responses are expected for each run, the effective diffusivity of the composite and the equilibrium moisture content of each.

2.4 Statistical analysis

Results of the diffusion simulations were then input in Design Expert, which conducts the regression analysis and ANOVA or analysis of variance. ANOVA is done to determine which of the three factors considered significantly affected the response variables.

Validation of the model was done by comparing its results to the experimental results.

DOI: 10.1051/matecconf/20166201002

3 Results and Discussions

3.1 Determination of resin and composite diffusion parameters from experimental data

The Fickian model of diffusion, equation (1) [11], was fitted to experimental results obtained from Onishi (2015) [3] to determine the diffusion parameters, diffusivity D and saturated moisture content, M_∞ of the pure resin and of the composite. These values are to be used for the validation of the predictive model for the prediction of the effective diffusivity of the composite as well as input in the FEM modeling of diffusion.

$$\frac{M_t}{M_\infty} = 1 - \frac{8}{\pi^2}\sum_{n=0}^{\infty}\frac{1}{(2n+1)^2}\exp\left(\frac{-(2n+1)^2\pi^2 D_x t}{l^2}\right) \quad (1)$$

Where:
M_t is the moisture uptake at time t
M_∞ is the moisture uptake at $t = \infty$
l is the specimen thickness

Equation (1) is obtained by:

$$M_t = \int_0^l c(x,t)\,dx \quad (2)$$

Where c(x,t) is the solution of the Fick's 2nd law [11]:

$$\frac{\partial c}{\partial t} = D\frac{\partial^2 c}{dx^2} \quad (3)$$

Where:
c is the moisture concentration,
t is the time
D is the diffusivity through the thickness
x represents the thickness dimension

Subject to the following boundary conditions

$$c(0 < x < l) = c_0$$
$$c(0,t) = c_1$$
$$c(l,t) = c_1$$

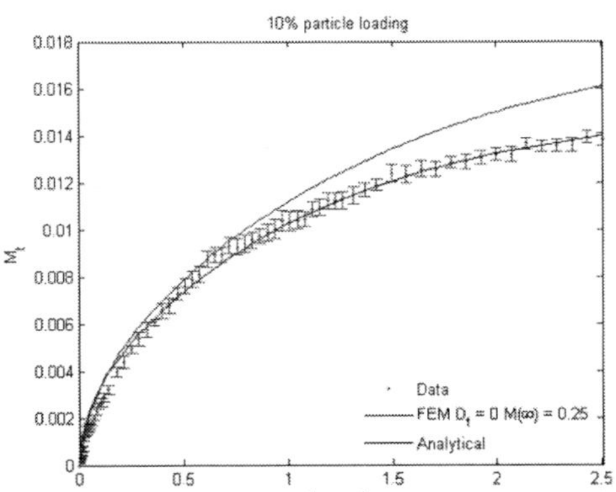

Figure 4 Sample Experimental results fitted with Fickian and FEM Model

Fitting the experimental data to the Fickian Model was performed using the *lsqcurvefit* command of MATLAB. Figure 4 shows the sample fitting of moisture content-time experimental data with the fickian model for a particle loading of ten weight percent. The comparison of experiment with the FEM model is only shown in the same figure (Figure 4).

The graphs above indicate that Fickian diffusion is a valid mechanism to describe the diffusion of water through a thermoset CFB fly ash composite.

The diffusion parameters were also determined at different filler loading conditions as shown in Table 2.

Table 2. Experimental Resin and Composite Diffusion Parameters

particle loading [vol fraction]	D [µm^2/min]	M_∞ [wt. Fraction]
0	47.69364997	0.010742137
0.048151333	42.90478368	0.015245192
0.102189781	17.21908693	0.02444381
0.163265306	2.288066141	0.084700255
0.232848233	3.862657302	0.090250862

3.2 Predictive model for effective composite diffusivity

A 2^3 factorial design matrix (Table 3) consisting of particle loading, particle shape and particle size as factors and the composite diffusion parameters D and M_∞ as the response was implemented to develop a predictive model to determine the composite diffusion parameters from only the filler properties.

Table 3. 2^3 Factorial Design Matrix

Run	Volume Fraction (φ)	Shape (S)	Particle Size (D_p)	M_infinity	D_effective
1	0.04815	0.8	15.3	0.03007	28.249
2	0.04815	0.8	33.3	0.01658882	26.479
3	0.04815	1	15.3	0.02585	28.922
4	0.04815	1	33.3	0.01610	25.275
5	0.23285	0.8	15.3	0.09839	2.8751
6	0.23285	0.8	33.3	0.03470	9.4096
7	0.23285	1	15.3	0.10334	3.4984
8	0.23285	1	33.3	0.05163	3.2123

The three factors mentioned above was used to vary the representative volume element element (RVE) in the FEM analysis of diffusion of water through the composite. FEM results give the curves corresponding to (1). Equation (1) is then used to determine the corresponding diffusion parameters (also shown in Table 3).

Using the *Design Expert* software, regression analysis of the design matrix yield the following empirical models for the prediction of the diffusion parameters of thermoset CFB fly ash composite as a function of particle properties:

$$D_{effective} = 33.09312 - 121.72837\phi \quad (4)$$

$$M_\infty = 0.00862016 + 0.60678\phi + 0.000021972S + 0.013861\phi S \quad (5)$$

The models above show that only particle loading significantly affects the diffusivity of the composite while the moisture content is affected by particle loading and particle size.

DOI: 10.1051/matecconf/20166201002

3.3 Validation of models

Equations (4) and (5) were validated by comparing predicted values with those obtained from experiment (Table 2). The validation results are shown below in Figure 5 and Figure 6.

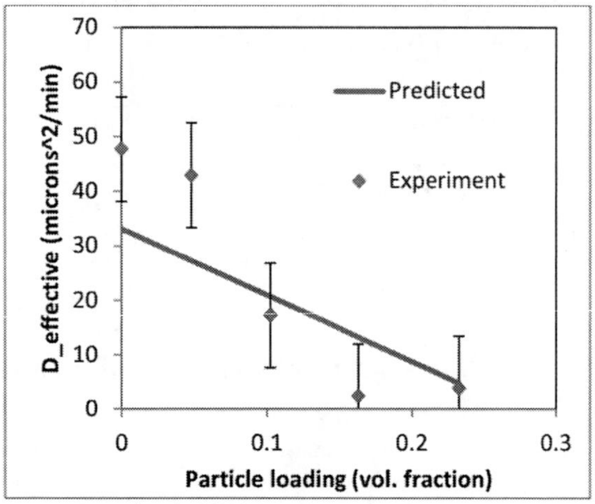

Figure 5 Predicted Diffusivity versus Experiment

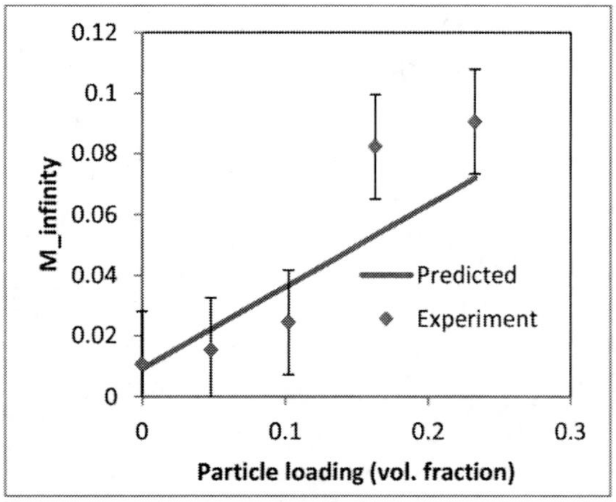

Figure 6 Predicted Saturation Moisture versus Experiment

4 Conclusion

A predictive model for the water sorption behavior of a CFB Fly Ash Thermoset composite was developed using Finite Element Method (FEM) and regression and analysis. The model was validated with experimental data and it was shown to fit satisfactorily.

From the model, it is also seen that only the volume fraction of the CFB fly ash in the composite had a significant effect in characterizing the sorption behavior of the composite. The equilibrium moisture content of the sample was also found to be affected mainly by the volume fraction of the CFB fly ash composite. An interaction between the volume fraction and the particle size also had a significant effect in the moisture content. Smaller particle size with the same volume fraction had caused a slight increase in the moisture content of the sample.

Results suggest that material properties of composites e.g. diffusivity can be easily determined by just knowing the filler's loading content and using an appropriate model. This validates the claim that a nondestructive test for material properties determination is a viable alternative to conventional destructive experimental techniques.

References

1. J. Utt and R. Giglio, *Industrial Fluidization South Africa* (2011)
2. R. E. Conn, K. Sellakumar, and A.E. Bland, *Proceedings of the 15th International Conference on Fluidized Bed Combustion* (1999)
3. S. E. Onishi, Int. J. Chem. Eng. & App., **6** (2015)
4. B. B. Pajarito, M. Kubouchi, and S. Aoki, J. Mat. Sci, **37** (2014)
5. J. Yang, Q. Yang, L. Ma, and W. Liu, Front. Mech. Eng. China, **5** (2010).
6. Y. Joliff, L. Belec, and J.F. Chailan, Comp. Struc., **97** (2013)
7. L. Canal and V. Michaud, Comp. Struc., **111** (2014)
8. S. Yoon, B. Han, and Z. Wang, JEP, **129** (2007)
9. ABAQUS 6.13, Analysis User's Manual, Sec 6.9
10. D. Montgomery, *Design and Analysis of Experiments* (John Wiley & Sons, Inc., 2013)
11. J. Crank, *The Mathematics of Diffusion* (Clarendon Press, England, 1975)

MATEC Web of Conferences **62**, 01003 (2016) DOI: 10.1051/matecconf/20166201003
ICCFE 2016

Green Composites Using Lignocellulosic Waste and Cellulosic Fibers from Corn Husks

Terence P. Tumolva[1,a], Danilo S. Enguero Jr.[1], Tracy Joy C. Laus[1], Benedict A. Requejo[1]

[1]*University of the Philippines Diliman, Quezon City, Philippines*

Abstract. This study explores the feasibility of using lignocellulosic waste and cellulosic fibers from corn husks in the production of green composites, with orthophthalic unsaturated polyester (ortho-UP) resin as a matrix. Lignocellulose was extracted from corn husk fibers by alkali treatment using 1M NaOH, and the dried lignocellulose extract was characterized using FTIR spectroscopy. Composites containing varying weight fractions of lignocellulose, treated fibers and ortho-UP were fabricated, and the tensile and flexural strengths and moduli were measured. Based on the results, it was observed that the composite containing 15wt% fiber possesses the highest tensile modulus, while the one with 20wt% lignocellulose showed the highest flexural modulus. The composites were also subjected to scanning electron microscopy to examine the fracture surfaces of the composites. Furthermore, the water sorption behavior of the composites was also studied, and it was observed that all the composites obey Fickian diffusion.

1 Introduction

Green composites that are partially or totally derived from biodegradable plant materials are viable alternatives to petroleum-based composites. Commercially used natural fibers are typically derived from agro-fiber plants such as abaca, kenaf and hemp; however, there is also an increasing trend in the use of agricultural residues such as sugarcane bagasse, rice hull and coconut coir as potential fiber sources.

Corn husk is a lignocellulosic material that currently has no established commercial value and is often simply considered as agricultural waste. In the Philippines, corn husk is very abundant and widely available since corn is the second most important crop of the country. To utilize this material as fiber reinforcement in the production of green composites, chemical treatments such as alkali must be employed to separate the cellulose from its other components such as lignin and hemicellulose, which are later disposed as lignocellulosic waste. Yin, et al. have reported that the corn-derived lignin in the waste may also be used to produce biocomposites [1]; however, purification of lignin can only be done with additional processing, which translates to additional energy and cost. In this study, it is proposed that the separated lignocellulosic wastes be blended raw into a polymer matrix as an extender rather than merely disposing them, and the resulting composite properties be evaluated.

2 Experimental

Corn husks were sourced from a local market in Quezon City, Philippines. For the lignocellulose extraction, 1M NaOH and 10 wt% aqueous H_2SO_4 were prepared. For the fabrication of the composites samples, orthophthalic polyester (ortho-UP) resin, R10-103 from Polymer Products (Phil) Inc., was used as polymer matrix with methylethyl ketone peroxide (MEKP) as hardener. A liquid release wax was also used to facilitate the easy removal of the composites from the curing molds.

[a] Corresponding author: terence.tumolva@coe.upd.edu.ph

2.1 Alkali treatment

The corn husks were first sun-dried for 12 hours and then crushed into short fibers using a mill. The fibers were then immersed in 1M NaOH, using a ratio of 1 g fiber to 15 mL of NaOH solution. The mixture was then heated at 170°C for 90 minutes (see Figure 1), and, afterwards, allowed to cool down for 30 minutes before filtering.

Figure 1. Alkali treatment set-up.

© The Authors, published by EDP Sciences. This is an open access article distributed under the terms of the Creative Commons Attribution License 4.0 (http://creativecommons.org/licenses/by/4.0/).

DOI: 10.1051/matecconf/20166201003

The treated pulp was washed, dried, and set aside, while the filtrate was gradually acidified using 10wt% H_2SO_4 to pH 5. The precipitation of lignocellulose would then begin as indicated by the change in the color of the liquid from dark brown to a lighter murky brown color. The solution was then heated in a water bath at 60°C for 30 minutes, and then further acidified down to pH 3. The lignocellulose precipitates were allowed to settle for at least 1-2 hours and then separated by vacuum filtration for 6 to 8 hours; afterwards, the filter cake was sun-dried and pulverized.

2.2 Composite fabrication

Composite samples with varying ratios of lignocellulosic waste (LW), treated fiber and resin were prepared (see Table 1). Each component was weighed in an analytical balance and blended using a hand mixer. The resin mixture was then placed in a vacuum desiccator to remove air bubbles.

Table 1. Component weight fractions of fabricated composites.

Sample	LW wt%	Fiber wt %	Resin wt%	MEKP wt%
Blank	0	0	98.5	1.5
10 % LW	10	0	88.5	1.5
15 % LW	15	0	83.5	1.5
20 % LW	20	0	78.5	1.5
15 % Fiber	0	15	83.5	1.5
1:1 (F:LW)	10	10	78.5	1.5
2:1 (F:LW)	10	5	83.5	1.5
1:2 (F:LW)	5	10	83.5	1.5

For the specimen preparation, the mixture was then poured into steel molds with dimensions that are in accordance to ASTM standards D638 and D790 for tensile and flexural tests, respectively, and placed in a thermal press set at 55°C for 4-5 hours. The resulting specimens are shown in Figure 2 below.

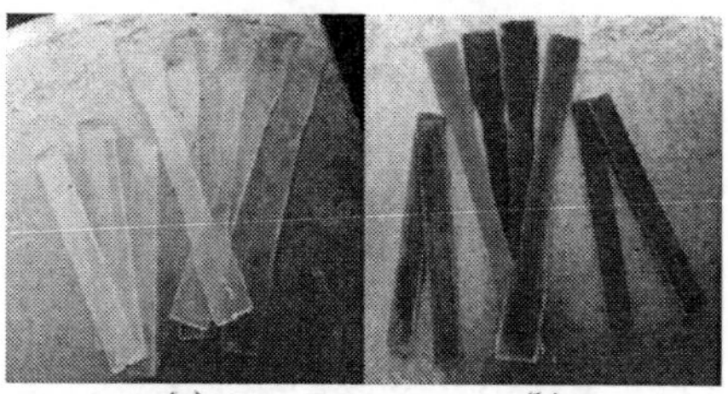

(a) (b)

Figure 2. Fabricated specimens of (a) neat ortho-UP and (b) with corn husk fibers and lignocellulosic wastes.

2.3 Tests and analyses

The LW powder was subjected to Fourier transform infrared spectroscopy (FTIR) to verify the presence of lignin in the samples. Tensile and flexural tests were conducted using a universal testing machine to determine the mechanical properties of the samples. The fracture surfaces of the test specimens were then observed using a scanning electron microscope (SEM). Lastly, water sorption tests were done by immersing the samples in water at ambient temperature for 72 hours to study the water sorption behavior of the composites.

3 Results and discussion

3.1 FTIR spectroscopy

Figure 3 shows the FTIR spectrum generated by the dried pulverized lignocellulose extract from the corn husk. The spectrum was compared to the FTIR spectra of various lignin samples in the study reported by Lisperguer, et al. [2]. All spectra show a broad band between wavenumbers 3400-3500 cm^{-1} with similar peaks. These strong bands are attributed to the hydroxyl groups present in the phenolic and aliphatic structures of lignin. Hence, the presence of lignin in the samples could be confirmed based on the similarities in characteristic peaks.

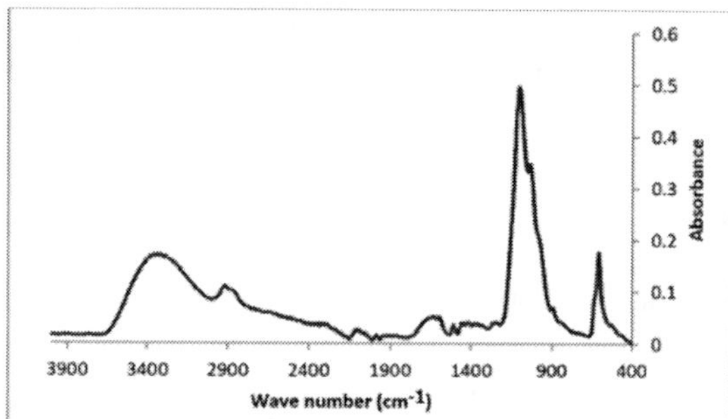

Figure 3. FTIR spectrum of dried lignocellulose powder.

3.2 Tensile tests

Figure 4 shows the tensile moduli of the corn husk composite samples. Among the samples made, it can be observed that the composite containing 15% fiber has the highest modulus, and the only one that is higher than that of a neat ortho-UP resin. It can also be observed that the composites containing both lignocellulosic waste and fibers have higher moduli than those of composites containing lignocellulosic waste only. This is because the rigid and somewhat brittle cellulosic fibers improve the composites' resistance to elongation during loading [3].

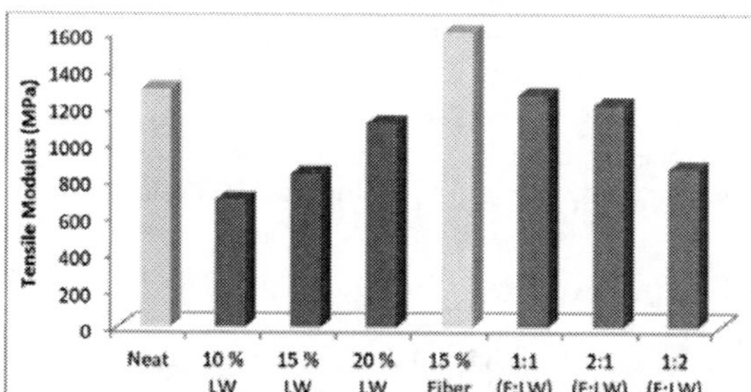

Figure 4. Tensile moduli of the corn husk green composites at varying fractions of fiber and lignocellulosic waste.

Figure 5 shows the tensile strengths of the different corn husk composites. It can be observed that the strengths of the corn husk fiber-reinforced composites are significantly lower than that of the neat resin, which is consistent with the results reported by Jaya Chitra and Vasanthakumari in 2012 [4]. This is an indication of the weak adhesion between the corn husk fibers and the ortho-UP matrix. Also, the results reveal that the simple mixing of the ortho-UP and LW produces a blend that is heterogeneous and has weaker tensile strength; this may be improved by initiating cross-linking between the lignin with the ortho-UP or by further polymerizing the lignin during the blending process.

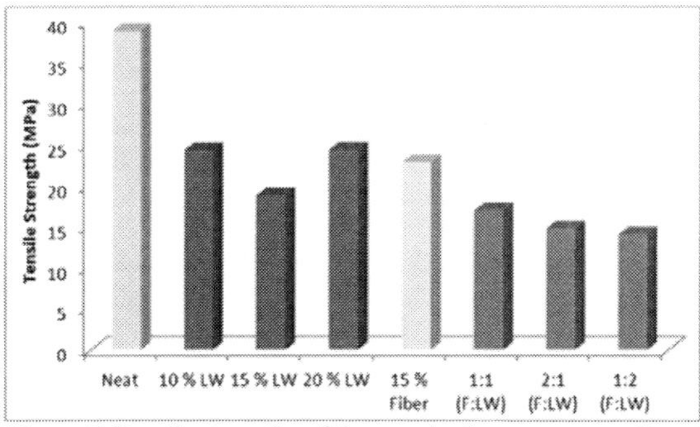

Figure 5. Tensile strengths of the corn husk green composites at varying fractions of fiber and lignocellulosic waste.

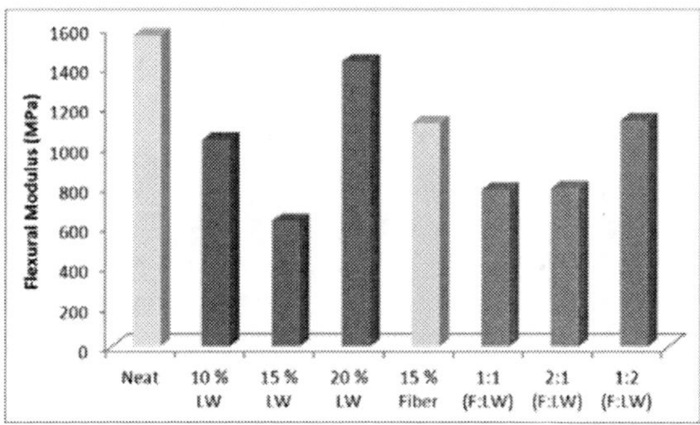

Figure 6. Flexural moduli of the corn husk green composites at varying fractions of fiber and lignocellulosic waste.

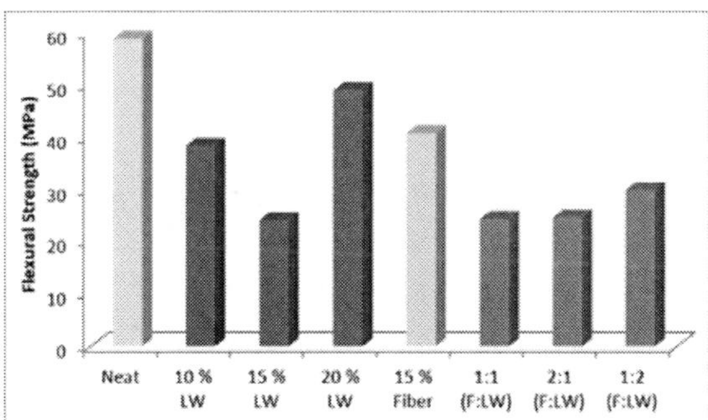

Figure 7. Flexural strengths of the corn husk green composites at varying fractions of fiber and lignocellulosic waste.

3.3 Flexural tests

Figures 6 and 7 show the flexural moduli and strengths of the corn husk green composites, respectively. The composite with 20% lignocellulose exhibits the highest modulus, comparable to that of the neat resin. From the results, it can be observed that the corn husk composites containing only the lignocellulosic waste have higher flexural strength and moduli than those containing corn husk fibers. As observed from the figures, there is always a sudden decrease in modulus and strength upon the introduction of a small amount of filler into the neat resin. This is because of the competing effects of introducing heterogeneity into the material and the plasticizing effect of the added lignocellulosic waste (due to the presence of hemicellulose). This means that increasing the amount of lignocellulosic waste can possibly overcome the negative effects of the introducing heterogeneity due to the increased presence of the fibers.

3.4 Fracture surface morphology

From the SEM images of the fracture surfaces for the different corn husk green composites, it can be observed that fracture surface becomes rougher as the amount of lignocellulose is increased, which can be attributed to the heterogeneous nature of the composite. From the comparison of Figures 8, 9 and 10, it can be observed that the lignocellulosic particles become more apparently visible under SEM for the corn husk LW/ortho-UP blends with increasing lignocellulosic waste content.

Figure 8. SEM image of the fracture surface for the 10% corn husk LW/ortho-UP blend.

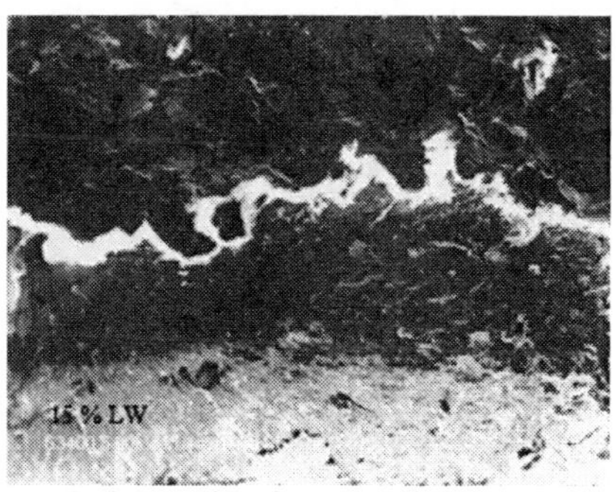

Figure 9. SEM image of the fracture surface for the 15% corn husk LW/ortho-UP blend.

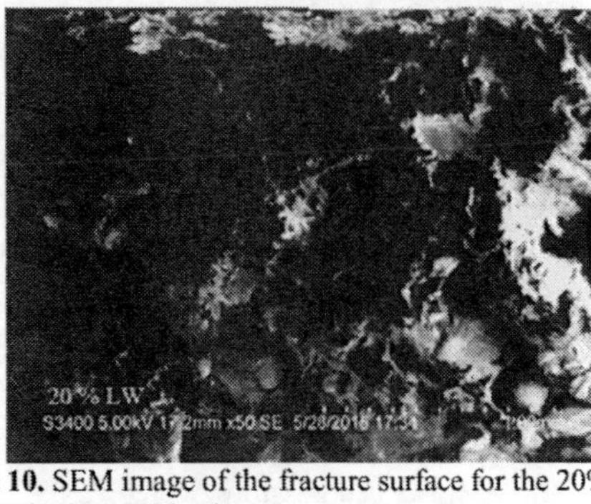

Figure 10. SEM image of the fracture surface for the 20% corn husk LW/ortho-UP blend.

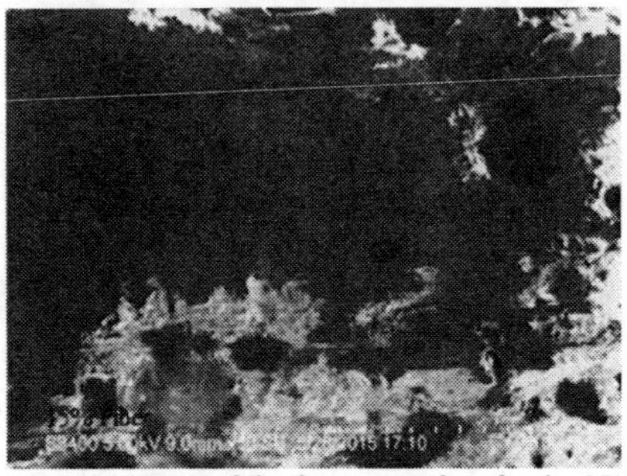

Figure 11. SEM image of the fracture surface for the 15% corn husk fiber green composite.

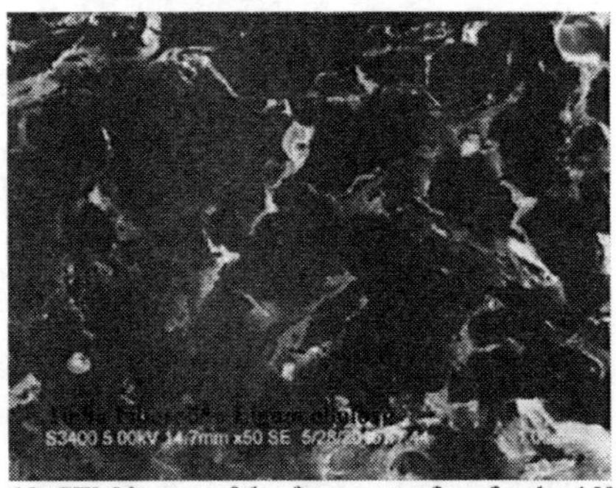

Figure 12. SEM image of the fracture surface for the 10% fiber/5% LW corn husk green composite.

Figure 13. SEM image of the fracture surface for the 10% fiber/10% LW corn husk green composite.

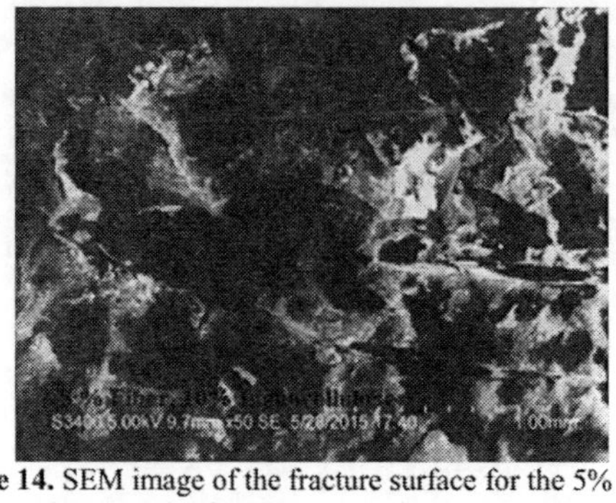

Figure 14. SEM image of the fracture surface for the 5% fiber/10% LW corn husk green composite.

The same trend is observed from the SEM images for the fracture surfaces of the corn husk fiber-reinforced composites, as shown in Figures 11 to 14. As the amount of lignocellulosic waste is increased, the fracture surface becomes rougher. The presence of voids that can be observed from the SEM images of these fiber-containing composites are due to fiber pull-out during testing.

3.5 Water sorption

Water retention is observed for the prepared green composites that has been immersed in water at ambient temperature for 72 hours, with the total mass of the wet specimens measured at different points in time using an analytical balance. Figure 15 shows the change in the weight of the composites over time. It can be observed that the addition of the hydrophilic cellulose fibers and the heterogeneity brought by the introduction of the lignocellulosic waste have increased the hydrophilicity of the polymer composite. Additionally, the presence of the corn husk fibers also increases the water sorption of the composite because of the lumen they inherently possess [5]. This may be remedied by improving the blending procedure and by using a more effective fiber treatment.

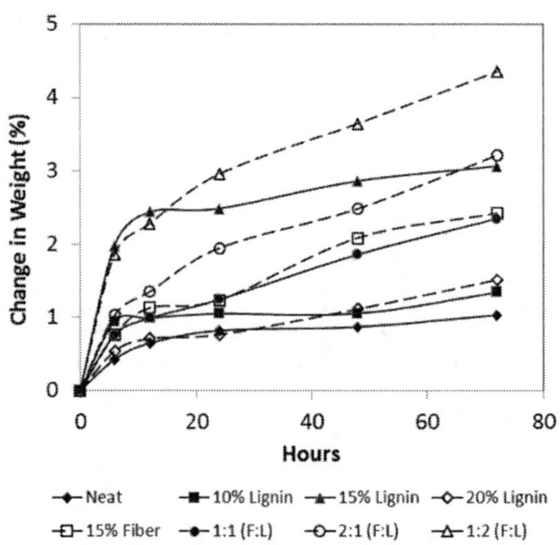

Figure 15. Percent weight change with time due to water sorption of the different corn husk green composites.

Figure 16 shows the plot of the change in weight plotted against the square root of time. A linear trend can be observed for all the samples since the equilibrium moisture contents of the composites are not yet reached after 72 hours; however, the results are still consistent with the expected Fickian diffusion behavior of the water sorption of green composites [6].

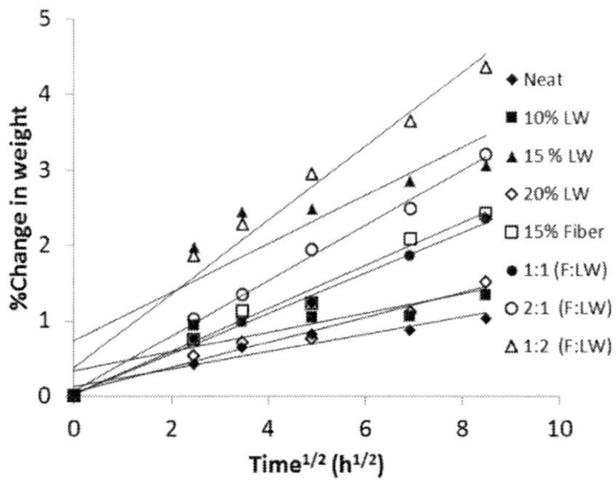

Figure 16. Percent weight change due to water sorption of the different corn husk green composites against $t^{1/2}$.

Figure 17 shows the total change in weight of the samples after 72 hours. The change in weight ranges from 1.03% to 4.36%. As expected, the corn husk fiber-filled composites absorbed the most amount of water due to the lumen and the hydrophilic nature of cellulose. The fiber-matrix interface also contributes to the hydrophilicity [7]. On the other hand, the lignocellulosic waste/ortho-UP blends generally have a lower weight change because the lignin is less hydrophilic than cellulose, especially after it has been subjected to alkali treatment (which reduces the number of active hydroxyl groups).

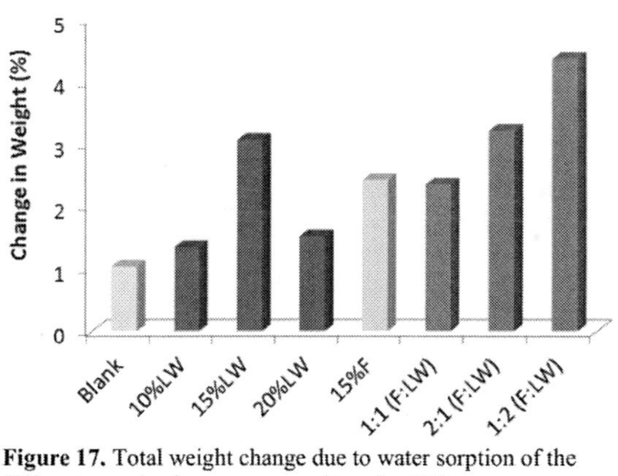

Figure 17. Total weight change due to water sorption of the different corn husk green composites after 72 hours.

The increased water sorption can also be caused by the presence of voids in the composites resulting from the failure of fiber-matrix adhesion. This is because alkali treatment only increases surface roughness to improve mechanical interlocking and does not chemically alter the fibers [8]. Lastly, neat ortho-UP resin exhibits the lowest weight change from the water immersion due to its inherent hydrophobicity.

4 Conclusions and Recommendations

Green composites using corn husk fibers and ligno-cellulosic waste and unsaturated polyester resin can be produced. Composites with 15% fibers exhibited the highest tensile properties. Also, the addition of fibers improves the tensile properties of the composite. Among the prepared green composites, the one with 20% lignocellulosic waste shows the highest flexural strength and modulus. Among the corn husk composites containing both fibers and lignocellulosic waste, the one with 1:1 ratio of LW/fiber has the highest tensile properties, while the one with a 1:2 LW/fiber ratio has the highest flexural strength and modulus. Overall, the addition of the recycled corn husk fillers shows a slight decrease in strength but a retention of moduli compared to that of a neat ortho-UP resin.

The corn husk composites obey Fickian diffusion, and the addition of fillers causes an increase in the amount of water that can be absorbed. Lignin-containing composites generally exhibited lower water sorption than fiber-containing composites.

For future studies, it is recommended to use a wider range of lignocellulosic waste weight fractions and more variations in the fiber-to-lignocellulose ratio. Also, other properties of the composites such as wet mechanical properties and thermal properties must still be evaluated. This study may also be extended to the use of other polymeric resins. Lastly, optimization studies may be conducted and the interaction of various parameters still needs to be explored.

References

1. Q. Yin, W. Yang, C. Sun, M. Di, BioResources **7**, 4 (2012)
2. J. Lisperguer, P. Perez, S. Urizar, J. Chil. Chem. Soc. **54**, 4 (2009)
3. D. Ray, B.K. Sarkar, A.K. Rana, N.R. Bose, Bull. Mater. Sci. **24**, 2 (2001)
4. N. Jaya Chitra, R. Vasanthakumari, Int. J. Sci. Eng. Res. **3**, 7 (2012)
5. T. Tumolva, M. Kubouchi, S. Aoki, T. Sakai, *Proceedings of JCOM-38*, 310-313 (2009)
6. M. Kubouchi, T. Tumolva, Y. Shimamura, *Biofiber-Reinforced Thermoset Composites,* in *Polymer Composites* **3**, 213-237 (2013)
7. T. Tumolva, M. Kubouchi, S. Aoki, T. Sakai, Materials Science and Technology of Japan **48**, 87-94 (2011)
8. R. Lopez, V.M. Poblano, A. Licea-Claverie, M. Avalos, A. Alvarez-Castillo, V.M. Castaño, Adv. Composite Mater. **9**, 2 (2000)

The Effect of Computer-Assisted Instruction for Elementary School Students in Food Safety and Sanitation Knowledge

Yi-Horng Lai[1,a]

[1]*Oriental Institute of Technology (Taiwan), New Taipei City, Taiwan*

Abstract. The purpose of this study is to introduce the graphic presentation food safety and sanitation learning system with parent participation in element school's health and physical education curriculum. The students were divided into four groups: control group, control group with parent participation, learning system group, and learning system group with parent participation. There were three extra variables in this study: learning system, and parent participation. The research data (three exams scores) was obtained before the course, in the middle of the course, and at the end of the course. The results indicate that, first, the estimate of slope of learning system is significantly correlated with parent participation; second, male elementary school students and female elementary school students were similar in the growth rate between each time points; third, the relationship between the initial exam score and the following two exam scores were not significant. Based on the results, it can be concluded that the use of learning system and parent participation were helpful for elementary school students to acquire food safety and sanitation knowledge in the health and physical education curriculum.

1 Introduction

Schools play an important role in students' health promotion and disease prevention. Since elementary school students tend to form their health knowledge, attitudes and behaviour through school education, it is important for schools to deliver effective health program. [1].

In 2000, Taiwan's Ministry of Education implemented the Nine-Year Integrated Course, and "Health and Sports" is one of the main fields of study. The integration of information technology is a new learning tool for food safety and sanitation, which is an important part of health education.

Computer tailored food safety and sanitation education may be more effective than traditional food safety and sanitation education because messages are tailored to individual behaviour, needs and beliefs. Therefore, the messages are more capable of targeting at individuals and may have stronger motivational effects. Computer tailored food safety and sanitation education has been studied for different dietary behaviours, in different target populations, and in different settings. In recent years, studies [2-4] that assessed the effects of interactive technology in food safety and sanitation education were based on the behaviour change theory. Computer tailored food safety and sanitation education is more likely to be read, remembered, and experienced as it is more personally relevant compared to standard materials. Furthermore, the computer tailored food safety and sanitation education also appears to have a greater effect in motivating people to change their diet, their fat intake in particular, although at present no definite conclusions can be drawn.

The purpose of this study is to explore the effects of the graphic presentation food safety and sanitation education informatics network (Computer Assist Instruction; CAI) with parent participation in elementary school's health and physical education curriculum.

1.1 The Information technology and food safety and sanitation education

Computer tailored food safety and sanitation education is an innovative and promising tool to motivate people to make healthy dietary changes. It provides respondents with individualized feedback about their dietary behaviours, motivations, attitudes, norms, and skills and mimics the process of person-to-person dietary counselling. The available evidence indicates that computer tailored food safety and sanitation education is more effective in motivating people to make dietary changes than the traditional food safety and sanitation information, especially for reduction of dietary fat. The effectiveness of computer tailoring has been attributed to the fact that individualized feedback commands greater attention, is processed more intensively, contains less redundant information, and is appreciated better than more general intervention materials. Interactive technology offers good opportunities for the application of computer tailored food safety and sanitation education,

[a] Corresponding author: FL006@mail.oit.edu.tw

© The Authors, published by EDP Sciences. This is an open access article distributed under the terms of the Creative Commons Attribution License 4.0 (http://creativecommons.org/licenses/by/4.0/).

and some studies of web-based computer tailoring have shown promising results. [5].

The results of Oenema, Brug, and Lechner's study [6] indicated that interactive, web-based computer-tailored food safety and sanitation education can lead to changes in determinants of behaviour. Food safety and sanitation educators are encouraged to explore the opportunities and challenges of these new technologies to enhance their work [7]. Brug, Steenhuis, Assema, and Vries's study [2] pointed out that computer-tailored food safety and sanitation information is a promising means of stimulating people to change their diet toward dietary recommendations.

Bensley, Anderson, Brusk, Mercer, and Rivas's study [8] claimed that Internet food safety and sanitation education was a viable alternative to traditional food safety and sanitation education for increasing fruit and vegetable consumption in some women, infants, and children clients. Besides, food safety and sanitation education through the telehealth service resulted in positive effects on the risk factors for metabolic syndrome, nutrient intake, and dietary habits [3]. In Tyrovolasa, Tountasb, Polychronopoulosa, and Panagiotakos's study [4], active food safety and sanitation policy and enhancement of food safety and sanitation services within the public health care system can contribute to improved health and quality of life among older populations. The food guidance system provides the basis for the food guidance presented in the American Red Cross food safety and sanitation course, "Better Eating for Better Health", and in "Dietary Guidelines and Your Diet", which is a series of bulletins developed by USDA to help consumers use the Dietary Guidelines [9].

1.2 The parent participation and food safety and sanitation education

Niemeiera, Hektnerb, and Enger's study [10] mentioned that weigh-related health interventions with parent participation could more effectively reduce body mass indexes of child and adolescent participants. In addition, longer interventions that include parent participation appear to have greater success. Suggestions for future research and related interventions are provided. Parents who participated in the intervention increased the scores on the food safety and sanitation knowledge test, and there also was a significant association between degree of family involvement, higher grain servings, and lower cholesterol intake [11].

2 Methods

The sample of this study was students of an elementary school in Taiwan. The participants came from four classes, and they were assigned in four groups respectively. The four groups were: control group, control group with parent participation, learning system group, and learning system group with parent participation. The research framework is shown in Figure 1.

The Latent Growth Model of this study is illustrated in Figure 2. Learning system group was the group (class) that applied the graphic presentation food safety and sanitation education system in the four-week food safety and sanitation education.

Learning system group with parent participation was the group (class) that applied the graphic presentation food safety and sanitation education system in the four-week food safety and sanitation education. Each parent was given a guide book and had the access to the feedback system for this graphic presentation food safety and sanitation education program. Students were required to do homework with the graphic presentation food safety and sanitation education system with their parents.

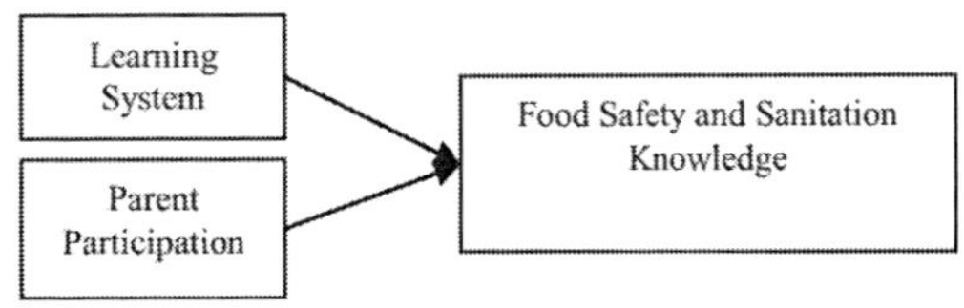

Figure 1. Research framework of this study.

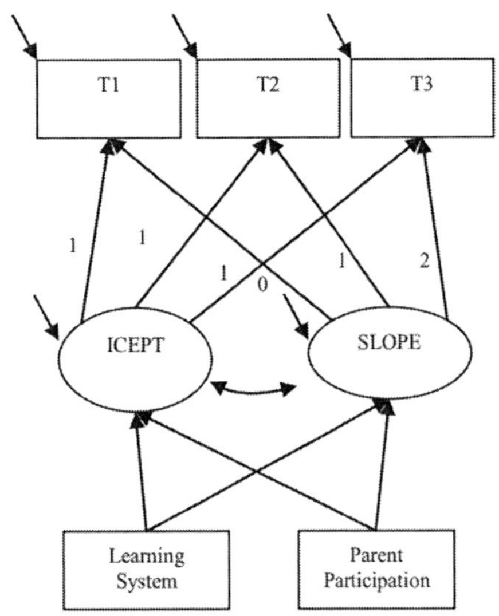

Figure 2. The latent growth model of the effect of food safety and sanitation education system for elementary school students in food safety and sanitation knowledge.

Control group was the group (class) that did the four-week food safety and sanitation education by traditional teaching method. This group did not involve the graphic presentation food safety and sanitation education system. Control group with parent participation was the group (class) that did the four-week food safety and sanitation education by the traditional teaching method. Each parent was given a guide book and the access to the feedback system. Students were required to do food safety and sanitation education homework by the traditional method with their parents.

The 3 exam scores (T1, T2, and T3) were obtained before the course, in the middle of the course, and at the end of the course. The questions on the three exams were based on the learning materials of the four-week food safety and sanitation education of health and physical education curriculum [12].

2.1 Research data

124 students of Linong elementary school participated in this study. The data of this study was collected by Y.Y. Chu for the research: The Food Safety and Sanitation Education Information System (Research Project ID is NSC93-2516-S-034-001). This research was completed on July, 31, 2005. Data in this present study was obtained from Survey Research Data Archive (SRDA) [12] (Table 1).

There were four groups in this study: control group (learning system=0, parent participation=0), control group with parent participation (learning system=0, parent participation=1), learning system group (learning system=1, parent participation=0), and learning system group with parent participation (learning system=1, parent participation=1). The 3 exam scores on food safety and sanitation knowledge (T1, T2, & T3) were obtained at 3 time points before and after the learning system was used.

There were three extra variables in this study: learning system, and parent participation, and they were all binary variables. Learning System=1 means using the learning system; Learning System=0 means not using the learning system. Parent Participation=1 means learning with parent participation, and Parent Participation=0 means learning without parent participation.

There were a total of 59 female students (47.58%) and 65 male students (52.42%). There were 17 male students and 14 female students in the control group; there were 16 male students and 15 female students in the control group with parent participation; there were 17 male students and 14 female students in the learning system group; and there were 16 male students and 15 female students in the learning system group with parent participation. Each group all had 31 students (Table 1).

Table 1. Data Summarize of the research.

		N	%
Gender	Female	59	47.58
	Male	65	52.42
Group	Control Group	31	25.00
	Control Group with Parent Participation	31	25.00
	Learning System Group	31	25.00
	Learning System Group with Parent Participation	31	25.00
Total		124	100.00

2.2 Data analysis

The data was analysed with latent growth modelling by using the Mplus 7 software and R 3.2.3. Latent growth modelling is a statistical technique used in the structural equation modelling framework to estimate growth trajectory. It is a longitudinal analysis technique to estimate growth over a period of time.

Latent Growth Model represents repeated measures of dependent variables as a function of time and other measures. The relative standing of an individual at a specific time point is modelled as a function of an underlying process, the parameter values of which vary randomly across individuals. Latent Growth Curve Methodology can be used to investigate systematic change, or growth, and inter individual variability in this change. A special topic of interest is the correlation among the growth parameters, the so-called initial status and growth rate, as well as their relation with time varying and time invariant covariates [13].

3 Results

The relationship between time points and the average of exam scores are shown in Table 2.

Table 2. Data summarize of the scores in three time points.

	N	Mean	S.D.	Low	High
Time 1	124	11.63	1.07	9.00	14.00
Time 2	124	13.87	1.86	9.00	17.00
Time 3	124	13.80	1.91	10.00	17.00

As for the model fit information of this study, Chi-Square Test of model fit was 57.296 (df=4), RMSEA was 0.328, and CFI was 0.626. The estimate of the effect of learning system, and parent participation were found in Table 3 and Table 4. The intercepts of the estimate of use of learning system, and parent participation were not different. The 4 groups (control group, control group with parent participation, learning system group, and learning system group with parent participation) in this study were similar at the initial time point.

The estimates of fixed and random parameter for latent growth curve of this study were displayed in Table 3. It can be found that the estimate of intercept mean and slope mean were 3.060 (p-value <0.001) and 1.777 (p-value <0.001). But the estimate of relationship between intercept and slope was -0.004, P-Value was 0.972. The relationship between intercept and slope was not significant. It means that the scores were higher than the previous time. Besides, no matter what the initial score is, students could all perform better in the following exams.

Table 3. Fixed and random parameter estimates for latent growth curve of this study.

		Estimate	S.E.	P-Value
Intercept	Mean	3.060	0.166	<0.001
	Variance	0.134	0.208	0.519
Slope	Mean	1.777	0.098	0.007
	Variance	0.029	0.090	0.750

Table 4 shows that the relationship between the estimate of slope of learning system and parent participation was significant (P-value<0.05). The estimate of slope of factor of learning system was 0.385, and the estimate of slope of factor of parent participation was 0.429. The growth rate of each exam time was positively correlated with the use of learning system and parent participation. The score of elementary school students that used food safety and sanitation education information system was higher than those who did not use food safety and sanitation education information system by 0.385 points in each growth time. The score of elementary school students with participating parents was higher than those without participating parents by 0.429 points in each growth time. Male elementary school students and female elementary school students were similar in slope (the growth rate between each time).

Table 4. The effect of learning system, and parent participation.

		Estimate	S.E.	P-Value
Intercept	Learning System	0.136	0.168	0.419
	Parent Participation	0.183	0.169	0.279
Slope	Learning System	0.385	0.100	<0.001
	Parent Participation	0.429	0.100	<0.001

4 Conclusions

In recent years, computer applications have emerged as a viable means of gathering and disseminating food safety and sanitation information. Both stand-alone and on-line applications are being used to provide information on in the food safety and sanitation education for the public, paraprofessionals, and professionals. While the use of on-line communication applications such as multimedia and electronic discussion groups are just emerging as important tools among food safety and sanitation educators, the exponential growth of the Internet and the World Wide Web (WWW) are making these technologies more and more accessible. Food safety and sanitation educators are encouraged to explore the opportunities and cope with challenges of these new technologies to enhance their work.

The use of learning system was helpful for elementary school students to gain more food safety and sanitation knowledge, and this result was consistent with Brug, Oenema, and Campbell [5], Oenema, Brug, and Lechner [6], Brug, Steenhuis, Assema, and Vries [2], Park, Yang, and Kim [3], Tyrovolasa, Tountasb, Polychronopoulosa, and Panagiotakos [4]. Parent participation was helpful for elementary school students to learn more food safety and sanitation knowledge, and this result was the same as Niemeiera, Hektnerb, and Enger [10] and Hopper, Munoz, Gruber, MacConnie, Schonfeldt, and Shunk [11]. Besides, the initial score did not affect students' subsequent performances.

Finally, the learning of food safety and sanitation knowledge will be cultivated by both of the using of CAI (such as graphic presentation food safety and sanitation education system) and parent participation. Teachers could enhance the result of health education with computer learning system and parent support. CAI could be used as assistant teaching materials of food safety and sanitation education. Students could study with CAI on internet by themselves. CAI could reduce the time and expense of resource actual-food practicing. Furthermore CAI can also improve learning efficiency of school-children to achieve balance dietary behaviour.

It can be concluded that the acquisition of food safety and sanitation knowledge can be cultivated by both CAI (such as graphic presentation food safety and sanitation education system) and parent participation. Teachers can enhance the effectiveness of health education with computer learning system and parent support. CAI can supplement teaching materials of food safety, while parents' involvement can help students to achieve better in the learning system. To sum up, this paper has found that the use of CAI with parent participation can effectively improve the food safety and sanitation education in elementary school.

Acknowledgments

This study is based in part on data from the Survey Research Data Archive (SRDA) provided by the Academia Sinica. The interpretation and conclusions contained herein do not represent those of Survey Research Data Archive (SRDA) or Academia Sinica.

References

1. V.A. Ubbes, R.R. Cottrell, J.A. Ausherman, J.M. Black, P.C. Wilson, & J. Snider, Professional Preparation of Elementary Teachers in Ohio: Status of K-6 Health Education. Journal of School Health **69(1)**, 17-21 (1999)
2. J. Brug, I. Steenhuis, P.V. Assema, & H.D. Vries, The Impact of a Computer-Tailored Food safety and sanitation Intervention. Preventive Medicine **25**, 236-242 (1996)
3. S.Y. Park, Y.J. Yang, & Y. Kim, Effects of Food safety and sanitation Education Using a Ubiquitous Healthcare (u-Health) Service on Metabolic Syndrome in Male Workers. Korean Journal of Nutrition **44(3)**, 231-242 (2011)
4. S. Tyrovolasa, Y. Tountasb, E. Polychronopoulosa, & D. Panagiotakos, a Parametric Model of the Role of Nutritional Services within the Health Care System, in Relation to Cardiovascular Disease Risk among Older Individuals. International Journal of Cardiology **155(1)**, 110-114 (2012)
5. J. Brug, A. Oenema, & M. Campbell, Past, Present, and Future of Computer-Tailored Nutrition Education. The American Journal of Clinical Food safety and sanitation **77(4)**, 1028-1034 (2003)
6. A. Oenema, J. Brug, & L. Lechner, Web-Based Tailored Nutrition Education: Results of a Randomized Controlled Trial. Health Education Research **16(6)**, 647-660 (2001)
7. K.M. Kolasa, & M.G. Miller, New Developments in Nutrition Education Using Computer Technology. Journal of Food safety and sanitation Education **28(1)**, 7-14 (1996)
8. R.J. Bensley, J.V. Anderson, J.J. Brusk, N. Mercer, & J. Rivas, Impact of Internet vs Traditional Special Supplemental Nutrition Program for Women, Infants, and Children Nutrition Education on Fruit and Vegetable Intake. Journal of the American Dietetic Association **111(5)**, 749-755 (2011)

9. F.J. Cronin, A.M. Shaw, S.M. Krebs-Smith, P.M. Marsland, & L. Light, Developing a food guidance system to implement the dietary guidelines. Journal of Nutrition Education **19(6)**, 281-302 (1987)
10. B.S. Niemeiera, J.M. Hektnerb, , & K.B. Enger, Parentparticipation in weight-related health interventions for children and adolescents: A systematic review and meta-analysis. Preventive Medicine **55(1)**, 3-13 (2012)
11. C.A. Hopper, K.D. Munoz, M.B. Gruber, S. MacConnie, B.S. Schonfeldt, & T. Shunk, A School-Based Cardiovascular Exercise and Nutrition Program With Parent Participation: An Evaluation Study. Children's Health Care **25(3)**, 221-235 (1996)
12. Y.Y. Chu, The Graphic Presentation Food safety and sanitation Education Informatics Network. The Survey Research Data Archive. Retrieved July 18, 2012 from https://srda.sinica.edu.tw/search/gensciitem/1025 (2007)
13. K.J. Preacher, A.J. Wichman, R.C. MacCallum, & N.E. Briggs, Latent Growth Curve Modelling. Structural Equation Modelling. A Multidisciplinary Journal **19(1)**, 152-155 (2012)

Blumea Balsamifera (Sambong) Tea as a Therapeutic Drink for Calcium Oxalate Stones

Charlimagne M. Montealegre[1,a], Rizalinda L. De Leon[1]

[1]*Department of Chemical Engineering, College of Engineering, University of the Philippines, 1101 Diliman Quezon City*

Abstract. Calcium oxalate stones are among the most prevalent type of kidney stones. Methods of treatment range from surgical removal, shockwave lithotripsy, medical and natural treatment. In the Philippines, consuming tea preparation of *Blumea balsamifera* is believed to aid in the treatment of kidney stones. The effect of *Blumea balsamifera* tea in the nucleation of calcium oxalate crystals was studied at 0, 0.5 and 1.0 mg tea preparation per mL of the crystallization solution by UV-Vis spectrophotomery. Induction time decreased with increasing concentration of *Blumea balsamifera* tea. Further analysis showed that this decrease in induction time is due to a decrease in surface free energy from 20.95 mJ/m^2 to 19.48 mJ/m^2 and 17.22 mJ/m^2 at 0.5 and 1.0 mg/mL, respectively. This decrease in induction time correspond to an increased nucleation rate thus, a large number of smaller crystals are formed. This explains the therapeutic benefits of consuming *Blumea balsamifera* tea as stone formers would form smaller stones that are easier to eliminate by urination.

1 Introduction

Plants serve a variety of applications including food and as naturopathic agents. Its phytochemicals contain medicinal properties which not only improves but also augments health. There exists a large market for naturopathic plants consumed as food supplement capsules, in its raw form, or preparations as juices and tea.

The benefits of consuming herbal teas and decoction of leaves and other plant parts are widely recognized for the management and treatment of various ailments including kidney stones. This method of treatment does not only increase urinary volume but also provides phytochemicals which serve various purposes. The study by Rodgers et al showed that though urine volume (together with 12 other parameters) significantly influences stone formation, they key difference between stone formers and non-stone formers is the speciation of calcium ions in the urine [1]. Speciation is influenced primarily by an individual's metabolism and also by the presence of crystal growth modifiers acting as promoters and inhibitors of stone formation.

The crystal growth modifying property of various compounds are available in the literature. Some of these are inorganic salts such as citrate, magnesium and pyrophosphate [2]. These compounds were able to inhibit the crystallization of calcium oxalate by different means. Citrate forms a complex with calcium ions while pyrophosphate, to a certain extent, is incorporated into the forming crystal [3]. Biomolecules such as Tamm-horsfall protein, Uropontin, glycosaminoglycans, Bovine serum albumin, chondrotoin sulfate etc. also affects the crystallization process by modifying the habit of the crystal. Bovine serum albumin binds the {1 2 -1} and {0 2 1} face of the growing crystal changing the shape of the calcium oxalate crystal from hexagonal to diamond-shaped [4]. On the other hand, chondrotoin sulfate binds the (0 1 0) face resulting to elongated hexagonal crystals [4].

Compounds from plant such as quercetin, hyperoside [5] and crude plant extracts were also reported to inhibit the formation of calcium oxalate crystals. An in vivo study by Zhu et al. showed a significant decrease in crystal deposits for quercetin and hyperoside treated rats [5]. The extract of Khella or *Ammi visnagia* is used as an herbal remedy in Egypt, Middle East around Mediterranean and in South Africa. Extract from the seeds of Khella were shown to decrease the surface free energy and shift the morphology of calcium oxalate crystals from the monohydrate to calcium oxalate dihydrate [6]. A similar shift in morphology was shown by the extract of *Bergenia ciliata*, and was shown to be more effective in inhibiting nucleation and aggregation of COM crystals when compared with cystone [7]. Alok et al listed a total of 82 plants and their parts traditionally regarded to be herbal drugs used in the treatment of kidney stone, gall stone and urinary stone [8].

In the Philippines, *Blumea balsamifera* is recommended by the National Kidney Transplant Institute as an herbal remedy for kidney stones. Early studies found a statistically significant dissolution between the control and stones treated with urine from individuals taking *B. balsamifera* tablet [9] but the effect

[a] Corresponding author: charlimagne.montealegre@coe.upd.edu.ph

© The Authors, published by EDP Sciences. This is an open access article distributed under the terms of the Creative Commons Attribution License 4.0 (http://creativecommons.org/licenses/by/4.0/).

DOI: 10.1051/matecconf/20166202002

on the crystallization of calcium oxalate stones was unknown.

Crystallization is a complex process involving nucleation, crystal growth, and aggregation. Dissolution complicates the process further since these processes may occur simultaneously. In classical nucleation theory, the rate of nucleation is modeled in Arrhenius form

$$J = A\exp\left(-\frac{\Delta G}{kT}\right) \quad (1)$$

A thermodynamic assessment of the Free energy of formation of a stable nuclei (ΔG) shows that the free energy of nucleation is a function of surface free energy (γ) and supersaturation (s) shown as:

$$\Delta G = \frac{16}{3}\pi\gamma^3 \frac{\upsilon^2}{k^2T^2(\ln s)^2} \quad (2)$$

The rate of nucleation is difficult to measure due to the variations on the size of a nuclei which may vary from ten to several thousand molecules [10]. A more measurable parameter is induction time. Induction time is inversely proportional to the rate of nucleation. Thus converting to an equality using a proportionality constant k',

$$t_{ind} = \frac{k'}{J} = \frac{k'}{A}\exp\left(\frac{16}{3}\pi\gamma^3 \frac{\upsilon^2}{k^3T^3(\ln s)^2}\right) \quad (3)$$

This may be linearized by taking the natural logarithm of the equation.

$$\ln t_{ind} = \ln k'' + \frac{16}{3}\pi\gamma^3 \frac{\upsilon^2}{k^3T^3(\ln s)^2} \quad (4)$$

Equation 4 shows that a plot of the natural logarithm of the induction time and the reciprocal of the square of the natural logarithm of supersaturation will give a straight line and surface free energy may be calculate from the slope of this line.

This study aims to provide evidence on the effect of *Blumea balsamifera* tea on the crystallization of calcium oxalate through the induction time.

2 Materials and Methodology

Reactive crystallization is performed by mixing equal concentrations of calcium and oxalate solutions. Calcium solution is prepared from analytical grade calcium chloride dihydrate and Oxalate solution, from Sodium oxalate. These solutions are diluted using deionized water.

Blumea balsamifera decoction was prepared by boiling Golden Spoon™ all natural sambong tea in water for 5 minutes. The decoction was filtered through cellulose filter paper, cooled to room temperature and diluted to proper volume.

Induction time is the time difference between the mixing of a supersaturated solution and the detection of crystals in the solution. This parameter is inversely proportional to nucleation rate. Induction time is measured by monitoring the absorbance of the supersaturated solution using Perkin-Elmer Lambda 850 at 560 nm as optimized by a previous study [6]. An algorithm is used to systematically determine the induction time. An approximation of induction time is made by visual inspection of the spectra. The first derivative plot reflects the rate of change of the absorbance. When the changes in absorbance are gradual the first derivative plot is constant and an abrupt change is determined by a change in the slope of the first derivative plot. As the slope of the first derivative plot changes, the second derivative plot would also exhibit a sudden change or a jump discontinuity. The maximum of the second derivative plot is determined numerically to determine the time corresponding to the sudden change in the first derivative plot and consequently the induction time. Due to noise in the data, Savitzky-Golay smoothening is used.

To determine the effect of the decoction on the rate of nucleation, the degree of supersaturation is varied from 10 to 20 with increments of 2.5 The degree of supersaturation is based on the solubility of Calcium oxalate monohydrate at 25°C calculated using equation 5 [11]:

$$s = \sqrt{\frac{[Ca^{2+}][C_2O_4^{\ 2-}]}{K_{sp,25^oC}}} \quad (5)$$

The level of decoction was changed from a blank, 0.5 and 1.0 mg *Blumea balsamifera* tea preparation per mL of the crystallization solution.

3 Results and Discussion

The effect of *Blumea balsamifera* tea on the crystallization of calcium oxalate crystals is studied by determining the effect on induction time. Figure 1 shows absorbance-time plot at different levels of extract.

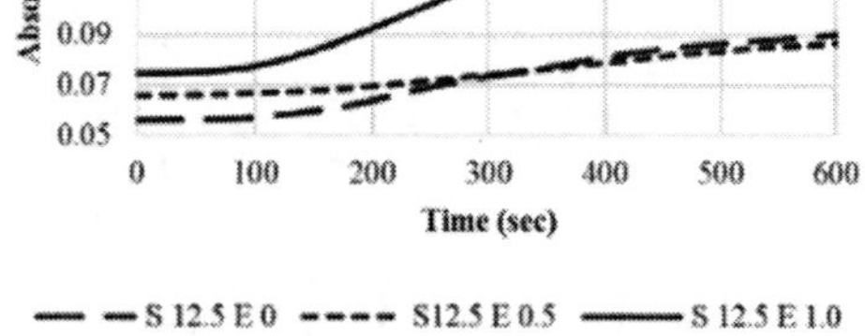

Figure 1. Absorbance versus time plot for a supersaturation of 12.5 and varying decoction concentrations shows an increase in induction time with increased concentration

The induction time is defined to be the point in time where there is a sudden increase in the absorbance from the initial absorbance measurements. This sudden increase is due to the formation of a stable absorbing specie, i.e. crystal formation. The absorbance increases continuously as the crystals grow. In some systems the absorbance decrease due to flocculation of the crystals formed. Figure 1, also shows that with an increase in extract concentration, the induction time decreases, i.e. stable crystals are formed faster and nucleation rate increases.

Induction time is plotted with respect to the reciprocal of the square of the natural logarithm of the supersaturation according to equation 4. These plots are summarized in Figure 2A for 0 mg/mL, 2B 0.5 mg/mL and Figure 2C for 1.0 mg/mL decoction. The surface free energy is computed from the slope of the plots and are summarized in the Table 1.

Table 1. Summary of slope, Pearson correlation factor and calculated Surface free energy.

Parameter	Slope	RSQ	γ (mJ/m^2)
Blank	26.901	0.644	20.95
0.5 mg/mL	21.618	0.642	19.48
1.0 mg/mL	14.929	0.147	17.22

The surface free energy of the system decreases upon the addition of the extract, consistent with previously published studies [6]. There was a 7.03% decrease in the surface free energy at an extract concentration of 0.5 mg/mL and a 17.82% decrease observed with the higher extract concentration of 1.0 mg/mL. The calculated surface free energy from the induction time studies was compared with a theoretical surface free energy using the proposed correlation of Mersmann [10].

$$\gamma = 0.414kT\left(\frac{\rho_C N}{M}\right)^{2/3} \ln\frac{C_S}{C_L} \qquad (6)$$

The proposed correlation of Mersmann enables the theoretical determination of surface free energy from the Boltzmann constant k, Avogadro's number N, molar mass of the crystal M, crystal density ρ_C, solute concentration in the solid phase C_S and solute concentration in the liquid phase C_L. However, the experimentally determined surface free energy differ from the Mersmann correlation predicted value by 71.82%. The correlation of Mersmann is under the simplification that surface free energy is only a function of relative supersaturation and material properties. In reality, the surface free energy varies across each crystal face, crystal edge, and within the surface as considered by more recent crystal growth theories such as the Burton-Cabrera-Frank theory, herring etc.

The critical radius of nuclei formation and change in critical Gibb's free energy were calculated and shown in Figures 3 and 4. Both the critical Gibb's free energy and critical nuclei radius decreased with the addition of the extract. At 0.5 mg/mL of extract, critical radius decreased by 7.03% while at 1.0 mg/mL the percentage of reduction was higher at 17.82%, across all degrees of supersaturation. Thus, a smaller nuclei size is required to form a stable crystal for growth.

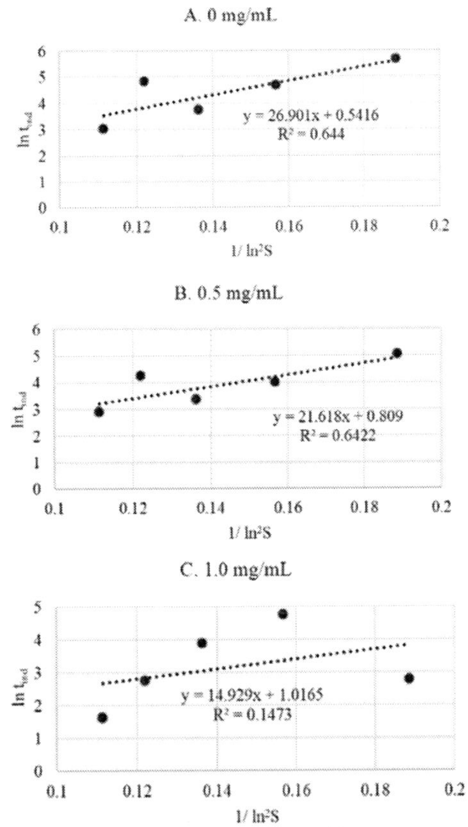

Figure 2. Variation of the induction time with respect to supersaturation at A 0, B 0.5 and C 1.0 mg/mL *Blumea balsamifera* tea preparation

The critical Gibbs' free energy decreased by 19.64% and 44.50% at 0.5 and 1.0 mg/mL of extract, respectively. The decrease in the critical Gibbs' free energy for the nucleation process shows that in the presence of the extract the formation of stable nuclei is easier. Nucleation is therefore promoted by the addition of the extract.

Addition of the extract decreased the induction time, free surface energy, decreased the critical Gibb's free energy and critical radius of a stable nuclei. All of these results support the observation that nucleation is promoted forming a large number of crystals that are smaller in size.

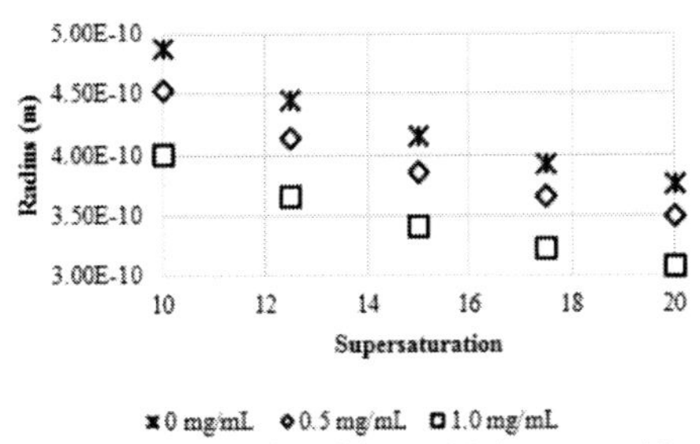

Figure 3. The critical radius of the nuclei decreases with increasing degree of supersaturation and with an increasing decoction concentration

 DOI: 10.1051/matecconf/20166202002

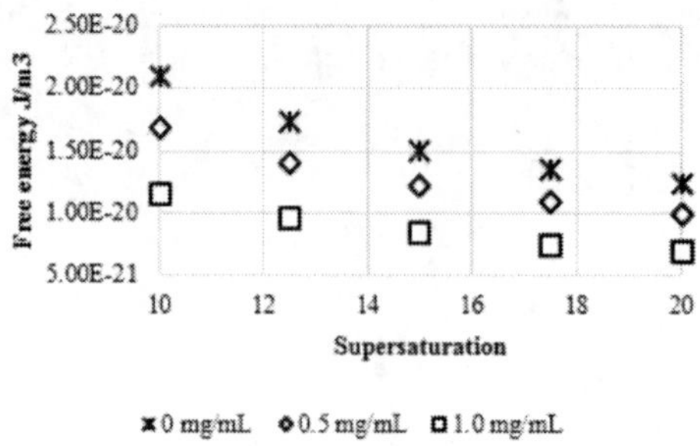

Figure 4. Critical Gibbs' free energy decreased as concentration is increased

4 Conclusion

A decoction of *Blumea balsamifera* tea preparation was shown to reduce the induction time thus increasing the nucleation rate. The surface free energy decreases with the addition of the extracts from 20.95 mJ/m^2 without the extract to 19.49 mJ/m^2 at 0.5 mg/mL and 17.21 mJ/m^2 at 1.0 mg/mL of *Blumea balsamifera* tea. This corresponds to a 7.03% and 17.82% reduction in surface free energy at 0.5 and 1.0 mg/mL, respectively. The calculated critical nuclei radius also decreased by 7.03% at 0.5 mg/mL and 17.82% at 1.0 mg/mL. Critical Gibbs free energy also decreased with increasing extract concentration showing that the nucleation process is favored. An increased nucleation rate implies an increased number of crystals that are smaller in size. This effect of the extract explains the therapeutic effect of consuming *Blumea balsamifera* tea as a therapeutic drink as stone formers would form smaller stones that are easier to eliminate by urination.

5 Compliance with Ethical Standards

The authors declare that they have no conflict of interest. This article does not contain any studies with human participants or animals performed by any of the authors.

References

[1] A.L. Rodgers, S. Allie-Hamdulay, G.E. Jackson, I. Durbach, J. Cryst. Growth 382 (2013) 67–74.
[2] M. Gupta, S. Bhayana, S.K. Sikka, Int. J. Res. Pharm. Chem. 1 (2011) 793–798.
[3] J.A. Muñoz, M. López-Mesas, M. Valiente, Urology 80 (2012) 1163.e13–8.
[4] S. Farmanesh, S. Ramamoorthy, J. Chung, J.R. Asplin, P. Karande, J.D. Rimer, J. Am. Chem. Soc. 136 (2014) 367–76.
[5] W. Zhu, Y.-F. Xu, Y. Feng, B. Peng, J.-P. Che, M. Liu, J.-H. Zheng, Urolithiasis (2014).
[6] E. a. Abdel-Aal, S. Daosukho, H. El-Shall, J. Cryst. Growth 311 (2009) 2673–2681.
[7] S. Saha, R.J. Verma, Arab J. Urol. 11 (2013) 187–192.
[8] S. Alok, S.K. Jain, A. Verma, M. Kumar, M. Sabharwal, Asian Pacific J. Trop. Dis. 3 (2013) 496–504.
[9] F. Rico, Philipp. J. Urol. 2 (1992) 9–13.
[10] J.W. Mullin, Crystallization, Elsevier, 2001.
[11] M. Öner, A. Khan, S.R. Khan, in:, Miner. Scales Depos., Elsevier, 2015, pp. 393–416.

Optimization for Dwarf Banana with Microwave Low Temperature Carbon Dioxide Flash Puffing Process by Response Surface Methodology and Factor Analysis

Bao-Yu Li[1,a], Kai-Dan Yin[1]

[1]*Guangdong AIB Polytechnic College, Guangzhou, Guangdong, 510507, China*

Abstract. A three factors quadratic regression rotation combination design was adopted to optimize the technical conditions of microwave low temperature carbon dioxide flash puffing (MLTCDFP) process for dwarf banana in the single factor's foundation. This paper analyzed the effect of puffing temperature, puffing pressure, vacuum drying temperature and the interaction of the three factors on color, crispness, and the effect of hardness and water content. Based on the experimental data, the quadratic regression model of four indexes was deduced, then variables were analyzed with response surface methodology (RSM). The weights of four evaluation indexes was determined by factor analysis. The regression square was obtained by the comprehensive score as follow: $Y=0.955+0.109X_1+0.16X_3-0.09X_1^2+0.015X_2^2-0.15X_3^2-0.018X_1X_2-0.056X_1X_3+0.016X_2X_3$ (R^2=0.837).Through factor analysis, the range of the optimal technical conditions of MLTCDFP for dwarf banana obtained. The result indicated the four indexes were affected significantly by the puffing temperature and vacuum drying temperature, and the interactions of the three factors did not affect the product obviously. The optimal technical parameters were as follows: puffing temperature 91～95℃, puffing pressure 0.11～0.19MPa, vacuum temperature 81～85℃.

1. Introduction

Dwarf banana is one of the most important tropical fruit in South of China, before lychee, pineapple, coconut [1]. Dwarf banana with rich nutrients and active ingredients, can regulate the gastrointestinal motility, beauty, enhance immunity, protecting cardiovascular, anti-cancer and other effects, highly favored by consumers [2]. Dwarf banana is a climacteric fruit species can easily degenerate during storage. Especially recently, Dwarf bananas were unmarketable, resulted in huge economic losses, seriously affected the farmers enthusiasm. It is urgent to study and develop new technology, extend the dwarf banana storage period. Researchers were studied variety drying methods, in order to prolong the shelf life of dwarf banana. Microwave low temperature carbon dioxide flash puffing (MLTCDFP) is a new non-fried puffed dehydration technology. Superheated vapour were driven by heating self-contained free water and bound water, carbon dioxide gas injected into the puffing chamber, instantaneous released at the greater than standard atmospheric pressure, produced the sponge structure. The products with excellent quality, such as maintain beauty and keep young, porous texture, saving energy, protection environmental, easy storage, convenient transportation [3-8].

In this work, deep processing of rich banana varieties, improve value-added products for the farmers' income and promote sound and rapid development of the banana industry; puffed dried banana determine the optimal process parameters, to expand the MLTCDFP technology applications to develop low-power, green nutrition processing methods, extend product shelf life.

The domestic and foreign scholars have conducted a lot of research on expanded dehydration technology. Zhu [8] were used quadratic rotation orthogonal experimental design, analysis influence factors affect the airflow puffing banana crisp process, obtained the regression equation, the optimum technological conditions were determined. Puffing temperature and puffing time on puffing degree and vitamin C (Vc) loss were significantly affected in the swelling degree increases, the volume expansion at the same time, the loss rate of Vc also increased. Li [9-12] studied banana at variable temperature and pressure difference factors of explosion puffing drying, physical property analysis on the product quality of explosion puffing drying for banana chips, the method optimizes the bananas at variable temperature and pressure difference puffing drying technology, puffing drying banana crisp aroma components were analyzed and compared by solid phase micro-extraction (SPME) coupled with gas chromatography-mass spectrometry (GC-MS). Bi [13-17] were studied the explosion puffing drying and its mechanism of fruits and vegetables, such as potato, apple, pitaya fruit, peach, the drying process parameters have been optimized. Cheng [18] were studied explosion puffing drying and isothermal adsorption characteristics of winter jujube and apple slices at

[a] Corresponding author: lbysdyx@126.com

© The Authors, published by EDP Sciences. This is an open access article distributed under the terms of the Creative Commons Attribution License 4.0 (http://creativecommons.org/licenses/by/4.0/).

variable temperatures and pressure differences, established data model, optimization of process parameters. He [19] focused on quality properties of crispy winter jujube dried by explosion puffing drying, showed the more attractive crispness than those dried by freeze-drying and convectional hot air drying. Zou [20] were discussed effect of osmotic pretreatment on quality of mango chips by explosion puffing drying, osmotic pretreated decrease glass transition temperature, increase expansion ratio and rehydration ratio, improve hardness and crispness values than the non-pretreated samples.

The previous puffed dehydration technique study on more foreign research institutions in the Eastern United States Department of Agriculture Regional Research Center. Sullivan [21-23], who has to apples, carrots and puffed blueberry dehydration three products were studied in detail, and an important factor affecting puffing temperature, pressure, pre-drying moisture content were optimized to obtain the corresponding regression equation. Saca [24] compared with many other ordinary hot air drying of banana slices from color, moisture content, porosity, rehydration ratio and cell structure, results showed that puffed dried banana chips due to produce a more porous structure, thus has a shorter drying time and good rehydration ratio, and the rehydration process less filtered solids, the product can maintain its color, taste and texture. Banana puffing slices were manufactured by the high temperature and short time (HTST) combination with hot air dry,showed that the product quality and structure parameters (hardness, brittleness)were correlation with short -time high temperature process, but the nutrients and flavor substance loss more [25]. Chwastek [26] were discussed methods to increase the rate of mass transfer during osmotic dehydration of foods. Zotarelli [27] was to evaluate the application of a convective multi-flash drying process (CMFD) to producing dehydrated and crisp fruits. Banana processed by CMFD had a moisture content of 0.293 g/g (dry basis) and aw=0.272 after 3 h of processing. Mango had a moisture content of 0.09 g/g and aw=0.359 after 4h of processing. For CMFD much smaller global shrinkage was observed. Meda [28] were studied on drying kinetics and quality characteristics of microwave-vacuum dried saskatoon berries, the results shown total color difference was dependent on microwave power and vacuum pressure levels.

Dwarf banana as raw material by MLTCDFP. This paper analyzed the effect of puffing temperature, puffing pressure, vacuum drying temperature and the interaction of the three factors on color, crispness, and the effect of hardness and water content. Based on the experimental data, the quadratic regression model of four indexes was deduced, then variables were analyzed with response surface methodology (RSM). The weights of four evaluation indexes was determined by factor analysis. Aim to explore the impact of product quality, process parameters optimal level. To identify the impact of product quality dwarf banana chips significant factor, there was an interaction is significant and optimum parameters of each factor.

2 Materials and Methods

2.1 Materials and instruments

2.1.1 Materials

The dwarf banana fruit with the similar size and maturity were collected from Hainan province of China. The initial moisture content of the dwarf banana at 72.3% ± 0.86% wet basis (w.b.).

2.1.2 Instruments

Texture analyzer: TMS-PRO, food technology corporation USA; Pluripotent slicer: XL-75, Guangzhou Xu Long Machinery Co., Ltd.; Explosion puffing equipment system: Lab-made; Vacuum microwave drying ov en: BDMDMV30, Boda microwave equipment Co., Ltd. (Henan, China); Colorimeter: NR145 Shenzhen 3nh technology CO., LTD..

2.2 Experiment methods

2.2.1 Process of MLTCDFP

Detailed process: Selected Plumpness index was greater than 8.3 (dwarf banana weight g/ dwarf banana length cm) ripe dwarf bananas as a raw material from Hainan province, water content 72.3% ± 0.86%, peeled, cut into uniform size 10mm wafer with a slicer, then placed in a microwave oven, 60℃drying 30min. The sample was removed, placed in the puffing chamber, then puffing in different temperature, different puffing pressure, different evacuation temperature, respectively, residence 6min, evacuation time 70min. Study on the products quality were effect by the value of L^* (lightnes, white=100 to black=0 ,Y_1), Crispness (Y_2), hardness (Y_3) and water content (Y_4).To reduce the experimental error, all operations were carried out under the same conditions in the same puffing tank , minimize interference from other factors. All the samples were stored in dryer until used.

2.2.2 Determination of index analysis

1) Moisture content: The moisture content of dwarf banana chips was determined by weight loss after drying 5 g samples in a forced air oven at105 °C for 24 h [29].

2) Texture: Texture profile analysis was used to evaluate hardness and crispness of dwarf banana chips by using a texture analyser (TMS-PRO type). The maximum compression force and the number of peaks in the force-deformation curve of each sample were considered as an indication of hardness and crispness of sample, respectively. The analysis parameters were as following: pre-test speed: 1mm /s; test speed: 0.5mm /s; after the test speed: 5mm /s; test distance: 10mm; data acquisition rate: 200 times/s; threshold: 10g; Probe: HDP/BSK.

3) Color: The surface color of dwarf banana chips was analyzed by a colorimeter (NR145 type).The colorimeter was calibrated against a standard white plate before each actual color measurement. The measurement of brightness index $L^*$of dwarf banana chips.

4) Product Quality Evaluation Method: Banana chips product lightness index L^*value is the higher the better, the Crispness values greater, hardness is low, the moisture content less than 5% [30], the better quality of dwarf banana chips.

2.2.3 Experimental design

Based on the single factor experiment, selected puffing temperature (X_1), puffing pressure (X_2), vacuum temperature (X_3) three main factors as input variables, selected the product L^*value (Y_1), Crispness (Y_2), hardness (Y_3), moisture (Y_4) as output indicators, three factors quadratic regression orthogonal rotation design. The optimal level of the three factors in the single factor test was chosen as the zero level of the two regression orthogonal rotation combination design. Coding of factors and levels in Table 1.

Table 1. Coding of factors and levels

Coded values	X_1-puffing temperature/℃	X_2.puffing pressure/MPa	X_3-vacuum temperature /℃
-1.68	60.0	0.07	58.0
-1	70.0	0.10	65.0
0	85.0	0.15	75.0
1	100.0	0.20	85.0
1.68	110.0	0.33	92.0

Table 2. Experimental designs and results

No.	X_1 /℃	X_2 /MPa	X_3 /℃	Y_1-L^*	Y_2-Crispness/one	Y_3-Hardness/N	Y_4-Water content/%
1	1	1	1	87.61±0.04	25.22±0.05	29.28±0.11	3.7±0.05
2	1	1	-1	86.72±0.13	19.42±0.15	35.32±0.15	6.29±0.08
3	1	-1	1	82.46±0.08	21.42±0.06	35.24±0.11	3.56±0.12
4	1	-1	-1	86.27±0.11	18.75±0.06	29.3±0.16	5.92±0.06
5	-1	1	1	86.48±0.05	22.62±0.04	32.4±0.11	3.92±0.05
6	-1	1	-1	79.86±0.07	25.28±0.14	90.29±0.09	6.79±0.12
7	-1	-1	1	79.65±0.03	16.29±0.13	28.29±0.21	4.78±0.10
8	-1	-1	-1	85.43±0.02	18.19±0.05	115.53±0.17	8.51±0.11
9	-1.68	0	0	86.56±0.06	0.99±0.04	169.14±0.2	7.98±0.05
10	1.68	0	0	87.49±0.03	14.59±0.05	31.73±0.17	5.9±0.06
11	0	-1.68	0	92.19±0.08	19.44±0.48	28.52±0.12	5.02±0.11
12	0	1.68	0	86.17±0.04	18.36±0.17	35.39±0.08	4.83±0.07
13	0	0	-1.68	60.13±0.09	0.25±0.05	62.31±0.12	18.2±0.10
14	0	0	1.68	81.32±0.11	20.5±0.16	29.33±0.17	3.88±0.09
15	0	0	0	91.28±0.08	29.13±0.17	35.26±0.10	4.7±0.15
16	0	0	0	90.63±0.07	29.03±0.11	34.29±0.12	4.76±0.02
17	0	0	0	88.52±0.05	29.64±0.10	32.01±0.13	4.28±0.07
18	0	0	0	90.26±0.13	29.63±0.20	37.77±0.11	4.81±0.10
19	0	0	0	91.29±0.03	29.93±0.06	33.72±0.11	4.8±0.08
20	0	0	0	90.26±0.02	29.27±0.16	43.34±0.12	4.48±0.09
21	0	0	0	89.39±0.04	30.18±0.07	34.35±0.17	4.18±0.12
22	0	0	0	90.58±0.12	31.37±0.22	36.28±0.12	3.73±0.12
23	0	0	0	91.17±0.13	29.29±0.13	32.32±0.14	3.59±0.09

Note: X_1-puffing temperature; X_2-puffing pressure; X_3-evacuation temperature; Y_1-lightness index L^*; Y_2-product crispness; Y_3-product hardness; Y_4-product moisture content.

2.2.4 Statistical analysis

All experiments were performed in triplicated and the results were reported as mean values with standard deviations.All data were processed and analyzed on Data Processing System (DPS)V14.5 and STATISTICA 10 .

3 Results and Analysis

3.1 Test results

According to the Process of MLTCDFP methods (2.2.1), the results were shown in Table 2.

3.2 Regression and parametric analysis

The test data for regression coefficient and variance analysis(Table 3), Multiple quadratic regression equation between puffing temperature (X_1), puffing pressure (X_2), vacuum temperature (X_3) and product lightness index L^*value, crispness, hardness and moisture content as follows:

$$Y_1=90.42+0.97X_1-0.24X_2+2.45X_3-0.66X_1^2-6.43X_3^2+0.53X_1X_2-0.48X_1X_3+2.14X_2X_3 \quad (R^2= 0.80) \quad (1)$$

$$Y_2=29.42+0.27X_1-0.39X_2+3.55X_3-6.30X_1^2-2.43X_22-5.38X_3^2-3.84X_1X_2+2.92X_1X_3+1.61X_2X_3(R^2=0.76) \quad (2)$$

$$Y_3=35.69-26.98X_1-0.68X_2-14.71X_3+20.61X_1^2-3.62X_2+1.27X_3^2+2.63X_1X_2+18.14X_1X_3+2.17X_2X_3(R^2=0.89) \quad (3)$$

$$Y_4=4.39-0.59X_1-0.18X_2-2.61X_3+0.41X_1^2-0.30X_2^2+1.87X_3^2+0.38X_1X_2 \quad (R^2=0.88) \quad (4)$$

Through the regression coefficient test shows that affect product L^*value, crispness and moisture content, Primary and secondary sequence are all vacuum temperature (X_3)>puffing temperature (X_1)>puffing pressure (X_2), which vacuum temperature (X_3) affected very significantly on L^*value, effect on the moisture content significantly; The primary and secondary order of factors affecting the hardness of products is puffing temperature (X_1) > vacuum temperature (X_3) > puffing pressure (X_2), which is the most significant effect on the hardness of puffing temperature (X_1)and vacuum temperature (X_3).

3.3 Interaction analysis

From Table 3, we can see that the interaction of puffing temperature (X_1) and puffing temperature (X_3) has a very significant effect on the hardness of products, and other interaction effects are not significant. The following is only a discussion of the impact of the interaction on the product index. In order to observe the influence of interaction, the fixed one factor is 0 level, and the interaction effect equa tion (Equation 5) and response surface diagram (Figure 1) were obtained.

$$Y_3(1,3)= 1808.46-26.63 X_1-13.65 X_3+0.093 X_1 2+0.1^2 X_1 X_3 +0.013 X_3^2 \quad (5)$$

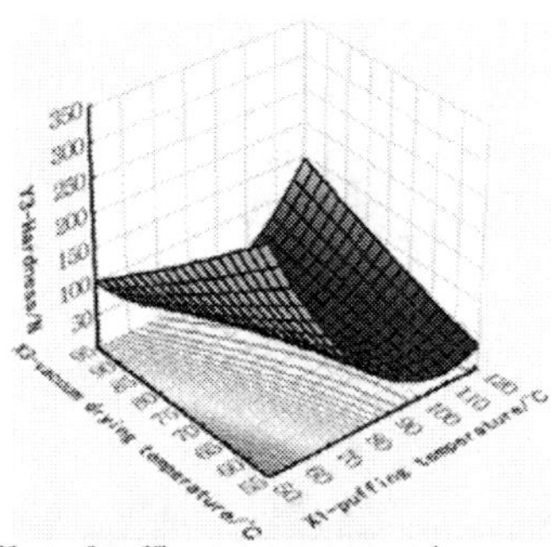

Figure 1. Effect of puffing temperature and vacuum temperature on product's hardness

Table 3. Regression coefficient and variable analysis

index	Y_1-Color	Y_2-Crisp-ness/one	Y_3-Hard-ness/N	Y_4-Water content/%
Intercept	90.423	29.424	35.687	4.399
X_1	0.966	1.866	-26.958**	-0.591
X_2	-0.238	1.209	-0.679	-0.175
X_3	2.448**	2.784	-14.693 **	-2.609**
X_1*X_1	-0.731	-5.737**	20.5044**	0.411
X_2*X_2	0.021	-1.886	-3.627	-0.302
X_3*X_3	-6.471**	-4.8161**	1.247	1.857**
X_1*X_2	0.530	-1.105	2.634	0.383
X_1*X_3	-0.478	1.613	18.144**	0.215
X_2*X_3	2.143	0.303	2.169	0.063
R^2	0.795	0.761	0.887	0.879

Note: ** indicates significance level p <0.05, * represents significance level p <0.1.

3.4 Optimization of MLTCDFP

Through the above analysis we can see that different products have different evaluation of the regression equation, the sequence of each input variable effects on product evaluation, and the presence of significant interactions are also different, which gives the resulting optimization of process parameters some difficulties related to the process parameters of multi-objective optimization problem.

3.4.1 Correlation between product evaluation

Table 4. Correlation coefficient among evaluation index

correlation coefficient	Y_1-Co lor	Y_2-Crisp-ness/one	Y_3-Hard - ness/N	Y_4-Water con-tent/%
Y_1-Color	1			
Y_2-Crispness /one	0.67**	1		
Y_3-Hardness /N	-0.2	-0.54**	1	
Y_4-Water content/%	-0.81**	-0.72**	0.44*	1

NOTE: correlation coefficient threshold value, when a=0.05, r=0.4132; when a=0.01, r=0.5256.

As can be seen from Table 4, there is a correlation between the product evaluation, product lightness index L^*value and crispness significant positive correlation, and the moisture content was significantly negatively correlated. Product friability and hardness significantly negatively correlated with the moisture content was significantly negatively correlated. Product hardness and moisture content was significantly negatively correlated. That product lightness index L^*value, the higher the product crispness, the lower the moisture content; the higher the crispness of the product, lower product hardness, the lower the moisture content. Products of high hardness, which mean high moisture content of the product.

3.4.2 Factor Analysis (FA)

FA is from the internal study variables related dependencies starting some variables with complex relationships attributed to a few factors in a comprehensive multivariate statistical analysis. Its basic idea is to classify the observed variables, the correlation is high, that is more closely linked in the same sub-class, but not the same correlation between variables is low, then the variable is actually representative of each category a basic structure, that common factors. For the problem under study is a linear function of trying to use the minimum number of unpredictable factors in the so-called public and special factors and to describe each component of the original observations.

According to the principle of FA to construct puffed products target parameter factor loading matrix, calculating the correlation coefficient matrix of eigenvalues, characteristic features of the cumulative contribution rate and the contribution rate, the results shown in Table 5.

Table 5. Eigenvalues and Cumulative contribution rate of correlated matrix

No.	Eigenvalues	Contribution rate/%	Cumulative contribution rate/%
1	3.4408	86.0191	86.0191
2	0.4325	10.8135	96.8326
3	0.0884	2.2099	99.0425
4	0.0383	0.9575	100

According eigenvalues shows the cumulative contribution rate, when taken three common factors, the cumulative variance contribution rate has reached 99.0425%. Take three common factors to build a matrix of factor loadings and variance orthogonal rotation method was a great factor loading matrix (Table 6).

By Table 6 shows that the common factor f1 product lightness index L^*, moisture dominant role of these indicators is the color of the main reaction product, it can be referred to as the color factor f1; f2 common factor plays a dominant role on the hardness of the product, So can be said for the hardness factor f2; f3 common factor from the crispness of the dominant role of the product, it can be named as crispness factor f3.

Table 6. Rotated factor loading matrix

rotated factor matrix	Factor No. f1	Factor No. f2	Factor No. f3
Color/ L^*	0.9070	0.0065	0.3295
Crispness/one	0.4562	-0.3220	0.8280
Hardness/N	-0.1245	0.9640	-0.2119
Water content/%	-0.8790	0.3100	-0.2540
Variance contribution	1.8190	1.1290	0.9037
Cumulative contribution	0.4547	0.7370	0.9629
Weight coefficient	0.45	0.29	0.22

Table 7 Compoment Score Coefficient Matrix and comprehensive evaluation value of samples

No.	Puffing temperature (X_1)	Puffing pressure(X_2)	Vacuum temperature (X_3)	Y(i,1)	Y(i,2)	Y(i,3)	Comprehensive evaluation value	Standardization
N(1)	1	1	1	0.3936	-0.5426	-0.0081	0.348889	0.9544
N(2)	1	1	-1	0.0001	-0.3558	-0.4544	0.009321	0.8213
N(3)	1	-1	1	0.1274	-0.6546	-0.5403	0.14261	0.8735
N(4)	1	-1	-1	0.0835	-0.6108	-0.7217	0.067063	0.8439
N(5)	-1	1	1	0.3765	-0.5406	-0.3771	0.255684	0.9179
N(6)	-1	1	-1	-0.9889	1.3314	1.3905	-0.551215	0.6014
N(7)	-1	-1	1	-0.1675	-1.0357	-1.2530	-0.032141	0.8050
N(8)	-1	-1	-1	-0.4159	2.2441	0.5556	-0.777456	0.5127
N(9)	-1.68	0	0	0.8935	3.4286	-2.2125	-1.203907	0.3454
N(10)	1.68	0	0	0.4335	-0.6132	-1.4840	0.049979	0.8372
N(11)	0	-1.68	0	0.8538	-0.4817	-0.9710	0.315024	0.9412
N(12)	0	1.68	0	0.3634	-0.5162	-0.9541	0.109271	0.8605
N(13)	0	0	-1.68	-4.3443	-0.1361	-0.7532	-2.084619	0.0000
N(14)	0	0	1.68	-0.0185	-0.8857	-0.6647	0.12225	0.8655
N(15)	0	0	0	0.3181	-0.0394	0.8620	0.354035	0.9565
N(16)	0	0	0	0.2228	-0.0847	0.8841	0.330761	0.9473
N(17)	0	0	0	0.1057	-0.2512	0.9218	0.339995	0.9509
N(18)	0	0	0	0.1635	0.0257	1.0263	0.3014	0.9358
N(19)	0	0	0	0.2340	-0.0512	1.0333	0.359343	0.9585
N(20)	0	0	0	0.2708	0.1672	0.9568	0.288452	0.9307
N(21)	0	0	0	0.1878	-0.1442	1.0028	0.361298	0.9593
N(22)	0	0	0	0.3602	-0.0405	1.1041	0.428993	0.9859
N(23)	0	0	0	0.5466	-0.2126	0.6568	0.465066	1.0000

Evaluation puffed food evaluation index should give priority to product color. Secondly, we should consider crisp property. Finally, considering the moisture content of the product, also hope to get the product color lightness index as much as possible higher and crispness as far as possible big, hardness as far as possible small, in view of the above analysis and factor loading matrix of the cumulative contribution rate, were given the common factor F1, common factor F2 and common factor F3, 0.45, 0.29, 0.22 heavy, and the common factor f2 gives negative weight.

After varimax orthogonal rotation sample obtained scores, the scores of each sample above right conferred by a weighted sum of the weight of each sample to obtain a comprehensive score in Table 7.

3.4.3 Regression analysis

Application of response surface methodology (RSM) to composite score as the target parameter to obtain the operating parameters of the product composite score of response surface (Figure 2) and $\alpha = 0.10$ significance level regression equation as follows:

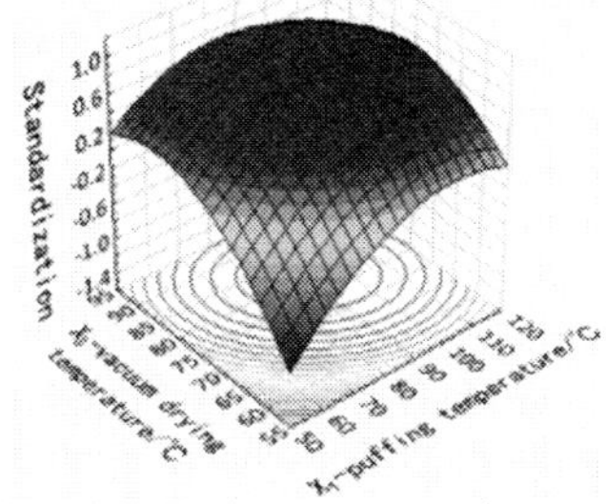

Figure 2. Effect of puffing temperature and vacuum temperature on product's comprehensive evaluation value

$Y=0.955+0.109X_1+0.16X_3-0.09X_1^2+0.015X_2^2-0.15X_3^2-0.018X_1X_2-0.056X_1X_3+0.016X_2X_3$ $(R^2=0.837)$ (6)

As can be seen from the regression equation 6, puffing temperature (X_1) and vacuum temperature (X_3) is the most important factor affecting the product's comprehensive score, and can be seen in its first term are all positive effects, which are all quadratic term negative effects, this shows puffing temperature (X_1) and vacuum temperature(X_3) on the product value comprehensive score due to their different level.

3.4.4 Optimization of process parameters based on comprehensive scoring

Combined with sensory evaluation, when the composite score ≥0.80, the product combines quality (texture, color, taste, etc.) is more desirable. Therefore, taking the composite score ≥0.80. According to the regression equation, the application of frequency analysis, computer simulation, to find the optimal combination of program operation parameters. The results showed (Table 8), a total of 40 programs to meet the requirements. From the last row of the table, while puffing temperature is 91～95℃, puffing pressure is 0.11～0.19MPa, evacuated temperature 81～85℃, the composite score is higher than 95% probability 0.80.

3.4.5 Verification test

The test verified under this condition banana chips produced by the Comprehensive score is greater than 0.80, product quality excellent to satisfactory results (Figure 3). Through the optimization of comprehensive evaluation, to multi-objective optimization purposes.

4 Conclusions

Three-factor quadratic regression orthogonal rotation design, the establishment of four indicators quadratic polynomial model by DPSV14.5 and STATISTICA 10.0

Table 8. Optimized result based on frequency analysis

Factor		Puffing temperature		Puffing pressure		Vacuum temperature	
		X_1	Frequence	X_2	Frequence	X_3	Frequence
Level	-1.6818	0	0.0000	8	0.2000	0	0.0000
	-1	0	0.0000	8	0.2000	0	0.0000
	0	15	0.3750	8	0.2000	15	0.3750
	1	15	0.3750	8	0.2000	15	0.3750
	1.6818	10	0.2500	8	0.2000	10	0.2500
Weighted average		0.7950		0		0.7950	
Standard error		0.1060		0.196		0.106	
95%Confidence interval		0.588～1.003		-0.38～0.384		0.588～1.003	
Parameter range		91～95℃		0.11～0.19MPa		81～85℃	

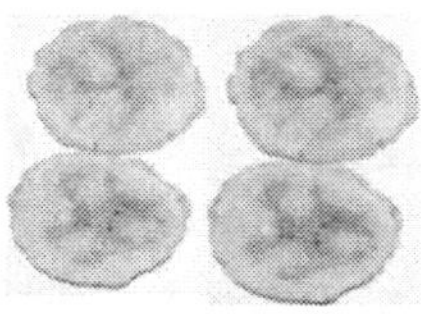

Figure 3. Photo of dwarf banana chips software.

According to the relationship between the various indicators by FA, statistical weight were given by contribution of the product quality,ultimately get the composite score and establish regression equation.

$$Y=0.955+0.109X_1+0.16X_3-0.09X_1^2+0.015X_2^2-0.15X_3^2-0.018X_1X_2-0.056X_1X_3+0.016X_2X_3 \quad (R^2=0.837).$$

Comprehensive evaluation of the results of the experiment showed that the puffing temperature and vacuum temperature, which were affected significantly on the L*value, hardness and water content. No significant interaction between puffing temperature, puffing pressure and vacuum temperature.

MLTCDFP dehydrated dwarf banana optimum parameters were: puffing temperature 91～95℃, puffing pressure 0.11～0.19MPa, vacuum temperature 81～85℃.

References

1. Y. A. Zhou, Health functions of bananas, Food and Nutrition in China, **3**: 51-52 (2006).
2. J.Cai, Talking about the health care function of banana, Food and Drugs, **7** (3A): 65-67 (2005).
3. C.M.Zmamer, Gun-puffed vegetable snacks: A new way to eat your veggies, Food Technol., **49**(10): 64-65(1995).
4. J. F. Sullivan, J. C. Craig, The development of explosion puffing, Food Technol., **38**(2): 52-55 (1984).
5. J. F. Sullivan, M.J. Egoville, R.P. Konstance, Storage stability of continuous explosion puffed potatoes,Food Sci.Technol., **16**(2):76(1983).
6. A. I. Varnalis,J. G. Brennan, D.B.MacDougall, A proposed mechanism of high-temperature puffing of puffing. Part Ⅰ.The influence of blanching and drying condition on the volume of puffed cubes , J.F.E., **48**:361-367(2001).
7. J. f. Bi, Y. M. Wei,Fruit and vegetable variable temperature differential puffing drying technology research progress , Transactions of the Chin. Soci. of Agric. Eng., **24** (6): 308-312(2008).

8. L.L.Zhu, P. Z. Zhang, K.Li, Non fried puffing banana chips technology research, Food. R.&D., **26** (2): 75-77(2006).
9. B.Y.Li, J.f.Bi,F. Fang, Research of the Impact Factors Explosion Puffing Drying by Modified Temperature and Pressure for Banana chips,Food R.&D., **31**(5):29-32(2010).
10. B.Y.Li, Physical Property Analysis on the Product Quality of Explosion Puffing Drying for Banana Chips, Farm Products Processing, **387**(7): 8-10 (2015).
11. B.Y.Li, J.Yang, K.D.Yin, Analysis of Aroma Components of Banana Slices before and after Explosion Puffing Drying at Modified Temperature and Pressure by SPME- GC-MS, Food Sci., **35**(1):184-188 (2014).
12. B.Y.Li, J.Yang, K.D.Yin, An Analysis of Flavour Compounds of Bananas under Different Drying Methods on the Basis of GC-MS Computer Technology,A.M.M., ICMTTME6,5136-5340(2014).
13. J.f.Bi, The response surface method optimization potato variable temperature differential puffing drying technology research, Food Sci., **28** (11): 236 -240 (2007).
14. J.f.Bi, Y.M.Wei, D.Wang, Cantaloupe variable temperature differential pressure extruding technology optimization study, J. Agric. Eng., **24** (3): 232 -237 (2008).
15. J.f.Bi, A.J.Y, X.Liu, Effects of pretreatments on explosion puffing drying kinetics of apple chips, LWT-food sci. technol., **60**(2):1136-1142(2015).
16. J.Y. Yi, L.Y. Zhou, J.f.Bi, Influences of microwave pre-drying and explosion puffing drying induced cell wall polysaccharide modification on physicochemical properties, texture, microstructure and rehydration of pitaya fruit chips, LWT-food sci. technol.,**70**:271-279(2016).
17. J.L.Yu, L.Y. Zhou, J.f.Bi,X.Liu, X.Y.Wu. Quality evaluation of yellow peach chips prepared by explosion puffing drying, J. Food Sci. Technol., **52**(12): 8204-8211(2015).
18. L.L.Cheng.Explosion Puffing Drying and Isothermal Adsorption Characteristics of Winter Jujube and Apple Slices at variable temperatures and Pressure differences, Hefei,: Anhui Agric. Univ.,(2012).
19. X.Y.He, J.F. Liu, L.L.Cheng, Quality Properties of Crispy Winter jujube Dried by Explosion Puffing Drying, INT. J. Food Eng., **9**(1):99 - 106(2013).
20. K.Zou, J.Teng, L.Huang, X.Dai, & B.Wei, Effect of osmotic pretreatment on quality of mango chips by explosion puffing drying. LWT-food sci. technol., **51**(1), 253- 259 (2013).
21. J.F.Sullivan, J.C.Craig, J. R, R.P.Konstance, Egoville M J. Continous Explosion-Puffing of apples ,J. Food Sci., **46**: 1550-1555(1980).
22. J.F.Sullivan, R.P.Konstance, E.S,Dellamonica,Carrot dehydration-optimization process studies on the explosion-puffing process ,J. Food. Sci., **46**:1537-1542 (1981).
23. J.F.Sullivan, J.C. Craig, E.D.Dekazos, S.M. Leiby, R.P. Konstanc, Dehydrated blueberries by the continuous explosion-puffing process, Food Sci., 1982, **47**(2): 445-448.
24. S.A.Saca, J. E. Lozano, Explosion puffing of bananas, INT. J .Food Sci. Technol., **27**: 419-426(1992).
25. K.Hofsetz, C.C.Lopes. Crispy banana obtained by the combination of HTST and a drying process, **14th IDS**: 1699-1706 (2004).
26. A. Chwastek. Methods to increase the rate of mass transfer during osmotic dehydration of foods, Acta Sci. Pol., Technol. Aliment. **13**(4):341-350(2014).
27. M. F. Zotarelli, B.D.A. Porciuncula, J.B.Laurindo. A convective multi-flash drying process for producing dehydrated crispy fruits. J. Food Eng., **108**: 523–531 (2012).
28. V.Meda,M.Gupta,A.Opoku,drying kinetics and quality characteristics of microwave-vacuum dried saskatoon berries, J. Microwave Power & Electromagnetic Energy ONLINE, **42**, 4(2008).
29. AOAC (1986) Method 986.21 Moisture in spices distillation methods.USA: offcial methods of analysis of AOAC international.
30. National health and family planning committee of the PRC, the national food safety standards puffed food GB 17401-2014, Beijing, (2014).

Purification of Food-grade Magnesium Chloride

Lianmin Ji[1], Zhiqi Liu[1, a], Lijuan Li[1], Xuexue Song[1], Zhongmin Zeng[1], Feng Nie[1]

[1]*Key Laboratory of Comprehensive and Highly Efficient Utilization of Salt Lake Resources, Chinese Academy of Sciences Qinghai Institute of Salt Lakes, Chinese Academy of Sciences, 810008 Xining Qinghai, China*

Abstract: The application of the varying weights of bischofite dissolved in the distilled water was investigated. The effects of the temperature on the rate of evaporation and the thermal precipitation time on the purity of the crystal products were fully investigated. Two validation tests including magnifying tests and recycling residue were also studied. Our results demonstrate that the contents of NaCl、KCl and $CaSO_4$ in the filtrate reached a minimum value after the pretreatment of 350 g bischofite dissolved in 100 ml distilled water. In the crystal products from the second evaporating stage of the validation tests, the contents of $MgCl_2 \cdot 6H_2O$, SO_4 and NaCl+KCl are 99%, ≤0.1±0.01% and ≤0.8±0.04%, respectively. The content of magnesium chloride in the solution was increased to a greater extent, and the impurities reduced correspondingly through the dissolution pretreatments of bischofite. This could decrease energy consumption for the impurity removing stage, evaporation and crystallization process, and thus reduce costs for the industrial production of food-grade magnesium chloride

1 Introduction

The positive correlation between diet and health has led health agencies around the world to control the intake of certain food components that are thought to promote some disorders [1].According to the national standards for food safety (GB2760－1996－2005), the additives used in food production must be food grade additives. Edible magnesium chloride ($MgCl_2 \cdot 6H_2O$) as additive has more widely application in food, salt, mineral water, medicine and other industries [1-4].

The Qinghai salt lakes in the qaidam basin are well known for their huge reserves of potassium chloride (KCl) and magnesium chloride ($MgCl_2$) in China [5]. In recent years, about 4.5 $Mt \cdot a^{-1}$ potassium fertilizer has been produced with 24-30 $Mt \cdot a^{-1}$ magnesium chloride as by-product, and a large amount of magnesium chloride has been left as the by-product or even waste in salt lakes of Qinghai [6,7]. This has caused not only the waste of magnesium resources, but also the environmental pollution [8]. The production of food grade magnesium chloride used bischofite as raw material is another effective utilization way of magnesium resources of salt lake, in which it is playing an active role.

This study investigated the performance of the varying weights of bischofite dissolved in the distilled water, determined the optimal ratio relationship between bischofite and distilled water under the condition of the least amount of impurity in the filtrate, and confirmed the reliability and repeatability of the optimal dissolution ratio.

[a] Corresponding author: zqliu@isl.ac.cn

2 Experimental

2.1 Reagents and instrumentation

The raw ore used in this study included bischofite (Golmud zhenghai magnesium science and technology development Co., Ltd.)

The instruments used for the analyses included stirrer (IKA EUROSTAR 60; Shanghai Chuangyi Science and education equipment Co., Ltd.), oil bath pan (DF-101S; Gongyi Yuhua Instrument Co., Ltd.), an inductively coupled plasma–atomic emission spectrometer (ICAP6500 DUO; America Thermo Scientific, USA).

2.2 Experiments

2.2.1 Solution analysis

The solution with a certain amount of bischofite dissolved in the distilled water was mechanically stirred at a frequency of 300 rpm for 30 min at room temperature (20 ± 2 °C). Subsequently, the residue and filtrate were separated and weighed. The filtrate was analyzed using EDTA volumetric analysis to determine the Mg^{2+}, Cl^- concentrations, and inductively coupled plasma–atomic emission spectrometry (ICP) to determine the Na^+, K^+, Ca^{2+}, As, Pb, SO_4^{2-} and B_2O_3.

2.2.2 Evaporative crystallization

© The Authors, published by EDP Sciences. This is an open access article distributed under the terms of the Creative Commons Attribution License 4.0 (http://creativecommons.org/licenses/by/4.0/).

(1) First evaporating stage

The filtrate was heated up to110±5 °C for some water loss, and then thermally precipitated for 1h, finally, conducted with thermal filtration.

(2) Second evaporating stage

In order to obtain crystals $MgCl_2{\cdot}6H_2O$, this filtrate from the first evaporating stage was heated up to140 °C again for some water loss, and then the thermal refrigerated to 40 °C for filtration. Finally, the product of crystals $MgCl_2{\cdot}6H_2O$ was obtained and analyzed using EDTA volumetric analysis to determine the Mg^{2+} concentrations, and ICP to determine the Na^+, K^+, Ca^{2+} and SO_4^{2-}concentrations.

The process flowsheet for purification of food-grade magnesium chloride was shown in Figure 1.

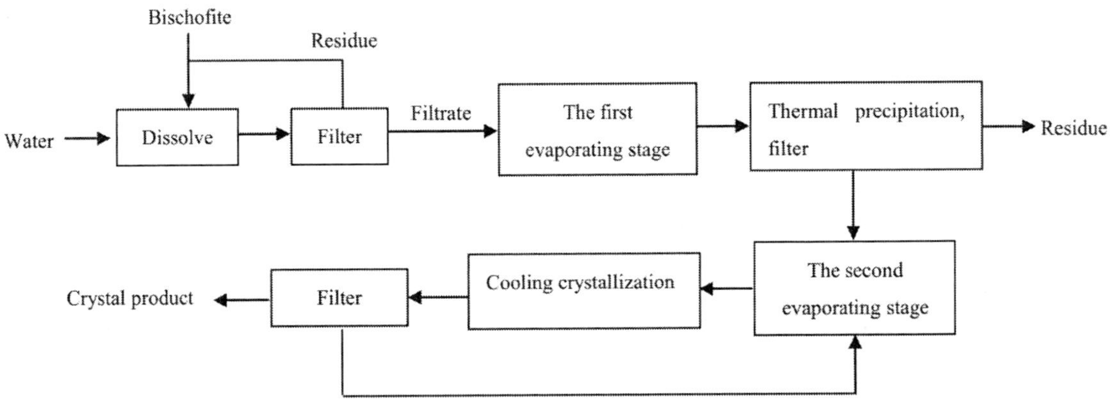

Figure 1. Process flowsheet for purification of food-grade magnesium chloride

3 Results and Discussion

3.1 Materials analysis

In order to obtain the composition of the bischofite, the solution with 50.0390 g bischofite dissolved in the distilled water was transferred into a 500 ml volumetric flask, and then analyzed using EDTA volumetric analysis and ICP. The result of such analysis is presented in table 1. The original magnesium chloride content (wt. %) is 44.77 %. The contents of impurities As, B_2O_3 and Pb in bischofite are unavailable due to extremely low values. The total contents of insoluble matter and soluble impurities ($NaCl$+KCl+$CaSO_4$) (wt. %) are < 0.10 % and 4.6 %, respectively.

Table 1. Result of materials analysis

Constitutes	Content (wt. %)
As	0.00006595
B_2O_3	0.0003732
Ca^{2+}	0.2314
K^+	0.8693
Na^+	0.8469
Pb	0.00002598
SO_4^{2-}	0.5372
$MgCl_2$	44.77
NaCl	2.15
KCl	1.66
$CaSO_4$	0.79
Insoluble matter	0.095

3.2 Varying solubility experiment

The experiments were conducted to investigate the effect of the solubility of bischofite on the contents of soluble impurities in the filtrate. The bischofites with weights varying from 100 g to 500 g were dissolved in the distilled water of 100 ml. After filtering, the density of filtrate, the concentrations of $MgCl_2$ and the components of soluble impurities were analyzed respectively.

As is shown in Figure 2, the contents of residue gradually increased with increasing the weights of bischofite. The contents increased rapidly with bischofite weights higher than 350 g. Notably, a large amounts of mineral $MgCl_2{\cdot}6H_2O$ dissolved and remained in the residue. As observed in the Figures 2 and 3, the weight contents of $MgCl_2$ steadily increased with increasing bischofite weight. However, the contents of NaCl and KCl initially increased and then rapidly decreased with increasing bischofite weights. While the $CaSO_4$ contents declined straightly. When the bischofite weight was higher than 350 g, all the NaCl, KCl and $CaSO_4$ contents reached minimum levels. Therefore, the relationship of optimal ratio between bischofite and distilled water was determined as 350 g/100ml.

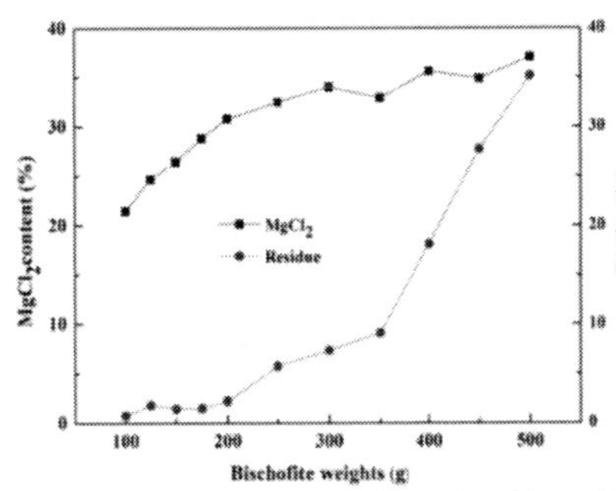

Figure 2. Relationship between $MgCl_2$ and residue with bischofite dissolved in the distilled water of 100 ml

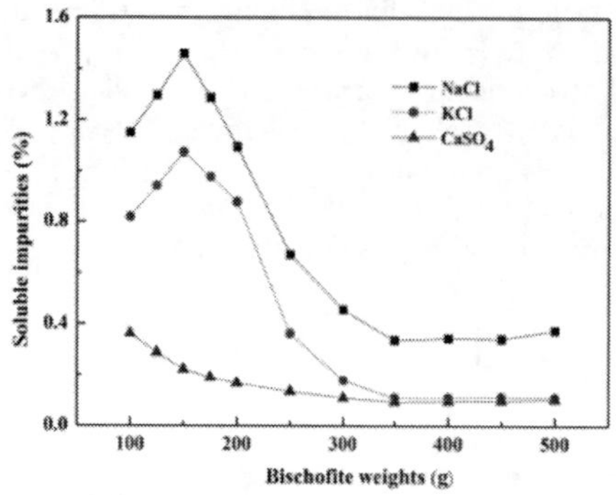

Figure 3. Relationship between soluble impurities with bischofite dissolved in the distilled water of 100 ml

3.3 Effect of temperature

To confirm the temperature for the second evaporating stage, the studies were carried out through varying temperature from 140 to 170 °C for 240 min. The results are shown in Figure 4 a and b. The evaporative loss of water initially decreased sharply and then gently declined with lengthening time.

The average speeds of evaporation during the time of 0 ~ 75 min were 0.46, 0.49, 0.53 and 0.55 g/min, respectively. While the average speeds during the time of 75 ~ 240 min were 0.051, 0.054, 0.063 and 0.053 g/min respectively. As observed, all the evaporation rates became slow after 75 min, and had nothing to do with the temperature. Therefore, the optimal evaporation temperature was 140 °C for the purpose of reducing energy consumption.

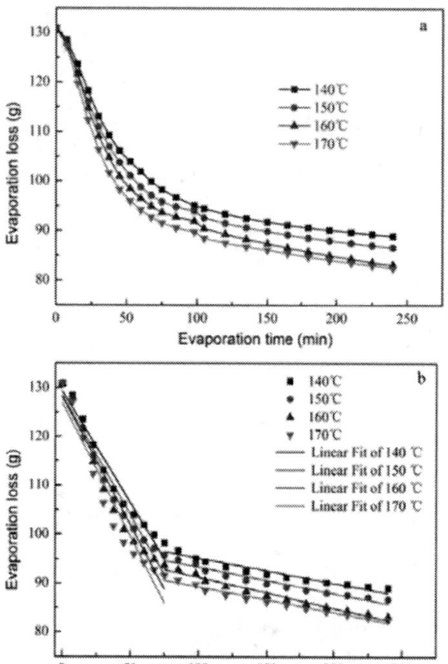

Figure 4. Effect of the temperature on the evaporation rate (a and b)

3.4 Effect of the thermal precipitation time

According to the optimal ratio of 350 g bischofite to 100 ml distilled water, the filtrate was obtained and heated up to 110±5 °C for some water loss, and then thermally precipitated for 1 ~ 5 h. Finally, it was conducted with filtration. The result for the effect of the thermal precipitation time on the purity of crystal product is listed in the Table 2. Table 2 shows that the purities of crystal product reached maximum levels when the thermal precipitation time was 1-3h. However, the purity value reduced as the thermal precipitation time was 5 h. Based on the energy consumption consideration, the optimum time is 1 h.

Table 2. Effect of the thermal precipitation time

Precipitation time (h)	**$MgCl_2$ (wt. %)**
1	46.30
2	46.18
3	46.39
5	45.37
Precipitation time (h)	**$MgCl_2$ $6H_2O$ (wt. %)**
1	98.86
2	98.61
3	99.06
5	96.88

3.5 Validation

3.5.1 Magnifying tests

The solutions were obtained with 1750 g bischofite dissolved in 500 ml distilled water, 5 times the original weights based on the optimal ratio in the section 3.2, and conducted to the first and second evaporating stages after filtration. The analyzed results of the products are shown in the tables 3 and 4. In those two products, the varying evaporation losses (17.32 g and 16.25 g) for the first stage gave rise to the different weight levels of crystals (12.68 g and 3.63 g, respectively) (Table 3). However, the total contents of $MgCl_2·6H_2O$, SO_4 and NaCl+KCl (wt. %) were > 99%, < 0.1% and < 0.6%, respectively (Table 4). This reveals that the varying evaporation losses for the first stage have no influence on the contents of the crystal products for the second stage.

Table 3. Weights of evaporation loss and crystal

Test 1 stages	**g**
Evaporation loss for the first stage	17.32
Crystal for the first stage	12.68
Evaporation loss for the second stage	45.65
Crystal for the second stage	194.79
Test 2 stages	**g**
Evaporation loss for the first stage	16.25
Crystal for the first stage	3.63
Evaporation loss for the second stage	45.34
Crystal for the second stage	203.84

DOI: 10.1051/matecconf/20166202004

Table 4. Contents of crystal products for the second stage

Product 1	**wt. %**
$MgCl_2$	46.41
$MgCl_2 \cdot 6H_2O$	99.10
NaCl	0.47
KCl	0.12
$CaSO_4$	0.022
SO_4	0.015
Product 2	**wt. %**
$MgCl_2$	46.47
$MgCl_2 \cdot 6H_2O$	99.23
NaCl	0.46
KCl	0.12
$CaSO_4$	0.031
SO_4	0.022

3.5.2 Recycling residue

(1) The solution with 350 g bischofite dissolved in the distilled water was mechanically stirred for 30 min, and then filtered.

(2) The residue from the section (1) was weighted and mixed with bischofite into mixture of total 350 g. Then, repeat the operation (1).

(3) Repeat the above (1) and (2) for 2 times. Finally, the obtained filtrate was conducted to the first and second evaporating stages after filtration. The analyzed results of the products are shown in the tables 5 and 6. In the two products, the varying evaporation losses (22.87 g and 29.62 g) for the first stage resulted in the crystals (27.78 g and 41.92 g, respectively) (Table 5). However, the total contents of $MgCl_2 \cdot 6H_2O$, SO_4 and NaCl+KCl (wt. %) are 99% approximately, ≤ 0.1±0.01% and ≤ 0.8±0.04%, respectively (Table 6). This also indicates that the varying evaporation losses for the first stage have no influence on the contents of the crystal products for the second stage.

Table 5. Weights of evaporation loss and crystal

Test 1 stages	**g**
Evaporation loss for the first stage	22.87
Crystal for the first stage	27.78
Evaporation loss for the second stage	33.16
Crystal for the second stage	261.34
Test 2 stages	**g**
Evaporation loss for the first stage	29.62
Crystal for the first stage	41.92
Evaporation loss for the second stage	30.15
Crystal for the second stage	240.57

As shown in the validation results for magnifying tests and recycling residue, the products obtained meet the product qualification (GB 25584-2010), according to the optimal ratio relationship between bischofite and distilled water (350 g: 100 ml).

Table 6. Contents of crystal products for the second stage

Product 1	wt. %
$MgCl_2$	46.20
$MgCl_2 \cdot 6H_2O$	98.65
NaCl	0.61
KCl	0.17
$CaSO_4$	0.14
SO_4	0.099
Product 2	wt. %
$MgCl_2$	46.36
$MgCl_2 \cdot 6H_2O$	98.99
NaCl	0.64
KCl	0.20
$CaSO_4$	0.16
SO_4	0.11

4 Conclusions

(1) The optimal ratio relationship between bischofite and distilled water was 350 g : 100 ml.

(2) As shown in the validation results for magnifying tests and recycling residue, the products obtained according to the optimal ratio meet the product qualification. In the products, the total contents of $MgCl_2 \cdot 6H_2O$, SO_4 and NaCl+KCl (wt. %) are 99% approximately, ≤ 0.1±0.01% and ≤ 0.8±0.04%, respectively.

Acknowledgments

The authors appreciate the financial support of Key Project supported by QingHai Science Technology Department (2014-GX-C10) and (2015-HZ-812).

References

1. C. Horita, M. Morgano, R. Celeghini, M. Pollonio, Meat Sci. **89**, 426 (2011).
2. W. Q. Yang, Chem. Technol. Market **32**, 33 (2009).
3. S. Barbut, C. Findlay, J. Food Sci. **56**, 180 (1991).
4. K. Toda, K. Yagasaki, K. Takahashi, Biosci., Biotechnol., Biochem. **72**, 2824 (2008).
5. P. Zhang, B. Zhang, Y. Tang, C. Yang, S. Huang, J. Wu, Sci. Press **99**, 107 (1999).
6. Y. Zhou, L. J. Li, Z. J. Wu, X. Li, Chem. Res. **25**, 1613 (2013).
7. W. T. Cheng, Z. B. Li, G.P. Demopoulos, Chin. J. Chem. Eng. **17**, 661 (2009).
8. P. Ma, Adv. Earth. Sci. **15**, 365 (2000).

MATEC Web of Conferences 62, 02005 (2016)
ICCFE 2016
DOI: 10.1051/matecconf/20166202005

Optimization of Processing Technology of Compound Dandelion Wine

Jixuan Wu[1], Guangren Sun[1,a], Xiuli Cao[1], Yuting Han[1], Xuesong Sun[1], Huan Zhang[1], Lei Zhang[1], and Ataer Dang[1]

[1]Department of Food Science & Engineering, Beihua University, 3999 Binjiang East RD, 132013, Jilin, China

Abstract. Exploring dandelion food has been the concern in fields of the food processing and pharmaceutical industry for playing exact curative effect on high-fat-diet induced hepatic steatosis and diuretic activity. Few dandelion foods including drinks and microencapsulation were explored and unilateral dandelion wine were less carried out for its bitter flavour. In tis paper, to optimize the processing technologies of fermented compound wine from dandelion root, the orthogonal experiment design method was used to composite dandelion root powder with glutinous rice and schisandra fruit and optimize the fermenting parameters. Four factors with dandelion content, schisandra content, acidity and sugar content were discussed. The acidity factor was firstly confirmed as 7.0 g/L. The other three factors were confirmed by a series experiments as dandelion 0.55%, schisandra 0.5%, sugar 22%. With nine step processing of mixing substrate, stirring with water, cooking rice, amylase saccharification, pectinase hydrolysis, adjusting juice, fermenting with yeast, fitering, aging, sterilization, a light yellow wine with the special taste with flavour of dandelion, schisandra and rice and less bitter, few index were determined as 14.7% alcohol , 6.85 g/L acidity. A dandelion fermented compound wine with suitable flavour and sanitarian function was developed for enriching the dandelion food.

1 Introduction

Dandelion (*Taraxacum officinale* Weber ex Wigg.) is a perennial weed widely distributed in the warmer temperate zones of the Northern Hemisphere [1]. This herb has been known since ancient times for its curative properties and has been utilized for the treatment of various ailments as a traditional medicine in China. Dandelion have long been used as medicinal herbs and young leaves are consumed fresh in salad, whereas roots are roasted and utilized as a coffee substitute and the plant extracts are used as flavor components in various food products. In China, this herbal plant used as both herbal medicine and edible vegetable [2].

Many active components were identified in dandelion including sesquiterpene lactones, taraxacoside, triterpenes, phytosterols, phenolic acids, flavonoids, polysaccharide, proline-hydroxyproline glycopeptides, vitamins, minerals and volatile components [3-7]. Dandelion has some activities and can be used medicine and food [8-11]. Its activities include protect effect on high-fat-diet induced hepatic steatosis, estrogenic activity, diuretic activity, choleretic activity, anti-inflammatory activity, anti-oxidative activity, anti-carcinogenic activity, analgesic activity, anti-allergic activity, anti-hyperglycemic activity, prebiotic activity, anti-coagulatory, anti-thrombotic activity, Empiric traditional application in humans of dandelion, in particular to treat digestive disorders.

For these activities, this herb is also used as food, including emulsion templated microencapsulation of dandelion, special drinks such as coffee from dandelion flower, green vegetable from tender dandelion leaves, functional products from dandelion root. But there is less research on wine from dandelion root especially less on fermented compound wine from dandelion root. In this paper, the technologies of processing on fermented compound wine from dandelion root were discussed for further filling lists of the food products from dandelion.

[a] Corresponding author: sungr8189@163.com

2 Materials and Methods

2.1 Materials

Dandelion (*T. officinale*) root were collected from Jiaohe city of Jilin, China in September to October, 2014. These root were washed by running water and dried in shade, then smashed by screen mesh with 250 μm mesh. Schisandra (*Schisandra chinensis*) fruit were collected from Jian city of Jlin, China in September, 2014. The fresh fruit of schisandra were dried and stored in a cool, ventilated place. The glutinous rice, saccharifying enzyme and active dry yeast was purchased from Jilin market, China, in January, 2015.

2.2 Methods

Fermentation processes included four steps. Firstly, making the composite powder as fermentation substrate mixed by dandelion powder with dried schisandra fruit and glutinous rice. Secondly, cooking the composite

© The Authors, published by EDP Sciences. This is an open access article distributed under the terms of the Creative Commons Attribution License 4.0 (http://creativecommons.org/licenses/by/4.0/).

powder after added certain amount of water. Thirdly, adjusting fermentation substrate contained 5 times water, acidity with 7.0 g/L by citric acid, sugar content with 24% by white granulated sugar. Lastly, incubating yeast with 0.01% and fermented for five to seven days under the condition of temperature between 21℃ and 28℃.

The methods of determination of sugar, acidity and alcohol. Sugar determined by the method of REDOX titration, acidity (calculated by citric acid as g/L) by acid-alkali neutralization titration, alcohol by wine viscometer method.

The methods of composite substrate by experiment designs including single factor experiments design and orthogonal experiment design. Factors included the ratio of content of dandelion powder, content of acidity, content of sugar and content of schisandra. Content of dandelion changed as 0.5%, 0.6%, 0.7%, 0.8%, 0.9%, 1.0% schisandra as 0.5%, 0.6%, 0.7%, 0.8%, 0.9%, 1.0%, sugar as 16%, 18%, 20%, 22%, 24%, 26%, acidity as 5.5 g/L, 6.0 g/L, 6.5 g/L, 7.0 g/L, 7.5 g/L, 8.0 g/L. The single experiments were evaluated by alcohol yields, residue sugar and sensory evaluation and appropriate range of these factors were determined. Based on these test, three main factors with appropriate range were selected. Three factors and three levels were used in orthogonal test design as $L_9(3^4)$. Finally, validation experiments were designed for further composite level and gained optimal conditions.

The processing of compound wine included mixing substrate, stirring with water, cooking rice, enzymatic saccharification with amylase, pectinase hydrolysis, adjusting acidity, adding sugar, fermenting with wine active dry yeast, filtering, wine aging, sterilization.

3 Experiment Results and Discussion

3.1. Results of single factor experiment

The results of single factor experiments described the effect of dandelion and schisandra was showed in the Figure 1.

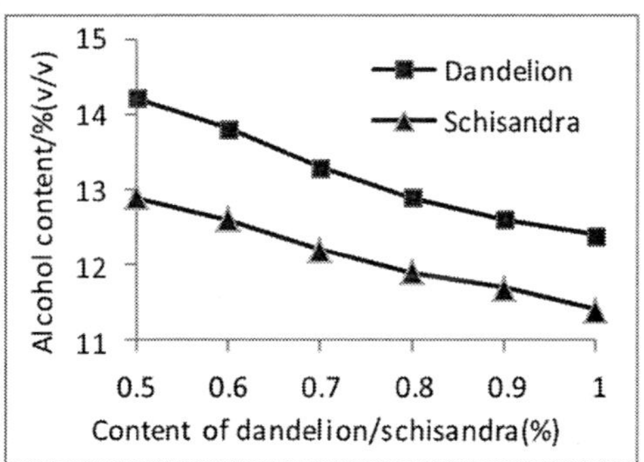

Figure 1. Effects of dandelion and schisandra on alcohol.

From Figure 1, the rules showed that the fermented alcohol gradually reduced with the increase of the substrate concentration both in schisandra on alcohol. This showed the dandelion and schisandrae had a negative effect on alcoholic fermentation and more strong effect of schisandra than dandelion. For further orthogonal test design the three levels of 0.6%, 0.7% and 0.8% were selected in with considering the sensory evaluation.

The results of single factor experiment of acidity and sugar content were showed in Table 1 and Table 2.

From Table 1, the alcohol content increased with the increase of sugar content when the sugar content changed between 16% and 22%, but over 22%, the alcohol increased in not obvious level. For further orthogonal test design the three levels of 20%, 22% and 24% were selected.

From Table 2, the alcohol content increased with the increase of acidity when the acidity more than 7.0 g/L, but acidity changed between 5.5 g/L and 7.0 g/L, the alcohol changed in not obvious level. So, select the acidity level between 5.0 g/L and 7.0 g/L. For the further test the acidity of 7.0 g/L was determined to keep high acidity for preventing bacterial contamination.

Table 1. The effects of sugar content on alcohol

No.	Sugar Content % (m/m)	Alcohol % (v/v)
1	16	12.8
2	18	13.2
3	20	14.5
4	22	14.8
5	24	14.7
6	26	14.6

Table 2. The effects of acidity on alcohol

No.	Acidity g/L	Alcohol % (v/v)
1	5.5	14.6
2	6.0	14.5
3	6.5	14.2
4	7.0	14.5
5	7.5	13.8
6	8.0	12.6

3.2 Results of orthogonal test

To set up orthogonal test design, the factor-level table was determined in Table 3. The results of this design was showed in Table 4.

Table 3. The factor-level table of substrate composite

Levelss	Factors		
	A Dandelion %	B Sugar %	C Schisandra %
1	0.5	20	0.5
2	0.6	22	0.6
3	0.7	24	0.7

Table 4 showed that the factor A had the biggest range and the factor lower than A, that to say the factor A (dandelion content) is the main factor to fermentation.

Table 5 showed the variance analysis of this test, the results showed that the factor A had a significant difference. This conclusion was the same as Table 4. But the factor C (schisandra content) had low level to improve the taste.

Table 4. The analysis results of orthogonal test

No	A	B	C	D	Scores
1	1	1	1	1	75.7
2	1	2	2	2	77.4
3	1	3	3	3	70.2
4	2	1	2	3	87
5	2	2	3	1	85.6
6	2	3	1	2	90.6
7	3	1	3	2	75.6
8	3	2	1	3	80.8
9	3	3	2	1	78.2
K1	223.3	238.3	247.1	239.5	
K2	263.2	243.8	242.6	243.6	
K3	234.6	239.0	231.4	238.0	
k1	74.43	79.43	82.37	79.83	
k2	87.73	81.27	80.87	81.20	
k3	78.20	79.67	77.13	79.33	
R	13.30	1.83	5.23	1.87	

Table 5. Analysis table of variance on orthogonal test

Source	Sum	DF	Variance	F value
A	281.96	2	140.98	50.33*
B	5.98	2	2.99	1.07
C	43.58	2	21.79	7.78
Error	5.60	2	2.801	

Note: * $F_{0.05}$=19, significant

From the above analysis, factor A was the most major in the three factors and this factor should be further determined by single factor test.

In above analysis the most combination of these factors is $A_2B_2C_1$, that is, dandelion content as 0.6%, sugar content as 22% and schisandra content as 0.5%.

3.3 Results of further single factor test

To determine the best range of dandelion content, a further single factor test was used in Table 6 with sugar content as 22% and schisandra content as 0.5% the results were evaluated by taste evaluation.

Table 6. The effects of dandelion content on taste and alcohol

No.	Dandelion Content % (m/m)	Alcohol % (v/v)	Taste Scores
1	0.45	14.8	89.7
2	0.50	14.2	90.5
3	0.55	14.5	92.1
4	0.60	14.8	90.1
5	0.65	14.7	89.7
6	0.70	14.6	84.3

From Table 6, the alcohol content had nor obviously significant difference, but the taste scores had obviously significant difference. The biggest dandelion content was about 0.55%.

3.4 Results of verification experiment

To determine the best fermentation condition and best range of substrate content, a verification experiment was worked out in Table 7, the results were evaluated by taste evaluation and the analysis table of variance on verification experiment were showed in Table 8.

Table 7. The results of verification experiment

Method	A(%)	B(%)	C(%)	Scores*
1	0.55	22	0.5	92.2±0.3[a]
2	0.6	22	0.5	90.8±0.7[a]
3	0.6	24	0.5	91.1±0.7[a]
4	0.5	20	0.5	76.2±1.1[c]
5	0.7	24	0.7	82.4±0.7[b]

Note: * different letter showed significant difference

Table 8. Analysis table of variance on verification experiment

Source	Sum	DF	Variance	F value
Method	783.36	4	195.8	353.5**
Error	8.31	15	0.554	
Total	791.67	19		

Note: Repeat n=3, $F_{0.01}(4,15)$=4.89

In verification test, choose the best in further single factor test as method 1, best in orthogonal test as method 2, results of orthogonal test as method 3, the lowest level as method 4 and the highest level as method 5.

From Table 8, these methods have obviously significant difference, and the multiple comparison method can be used in Table 7. By the analysis of multiple comparison method, method 1, method 2 and method 3 were not significant but significant to method 4 and 5.

3.5 Optimization of composite substrate

From above discussion, four factors with dandelion content, schisandra content, acidity and sugar content were discussed. The acidity factor was firstly confirmed as 7.0 g/L. The other three factors were confirmed by a series experiments as dandelion 0.55%, schisandra 0.5%, sugar 22%.

In order to facilitate application of optimized formula, reorganize the making process of the fermentation substrate are listed in Table 9.

Table 9. The list of substrate formula

Material	State	Content	Amount kg/t
Dandelion	Dried root powder	0.55%	5.5
Schisandra	Dried fruit powder	0.50%	5.0
Glutinous rice	powder	10%	100
Sugar	powder	22%	220
SO_2	6.0% liquid	120 mg/L	2.0
Citric acid	powder	7.0 g/L	7.0
Water	Liquid		662.5
Total			1000

3.6 Optimization of processing fermented wine

Based above tests, the substrate formula of compound wine was determined, and the processing technologies were determined according to these experiments. The compound wine processing included four steps.

Step 1: Mixing substrate – Powder of dried dandelion root, dried schisandra fruit and rice powder were mixed as composite substrate.

Step 2: Stirring with water – Above composite substrate was stirred with water at 3 times of powder as composite juice.

Step 3: Cooking rice – Above composite juice was cooked at 121℃ for 30 min as cooked juice.

Step 4: Enzymatic saccharification with amylase – Above cooked juice cooled about 60℃ and keep this temperature for 2 hr., adding amylase to make the starch from dandelion and rice change to sugar as saccharification juice.

Step 5: Pectinase hydrolysis – Keep saccharification juice at 50℃, adding pectinase for 5 hr. as purification juice.

Step 6: Adjusting juice – adjusting 7.0 g/L acidity with citric acid, 22% sugar with white granulated sugar, 120mg/L SO_2 with 6.0% sulfurous acid solution as adjusted juice

Step 7: Fermenting with wine active dry yeast – Adding wine active dry yeast after activated to adjusted juice and fermented for five to seven days with stirring every day. When there is less gas in the fermentation, the main fermentation had a stop and the fermentation would go into after fermentation. First fermented juice was obtained by the main fermentation.

Step 8: Filtering, wine aging – In this transition process of first fermented juice to after fermentation, the filtering operation is necessary to prevent bacteria pollution. After filtering, the fermenting process change into wine aging for three to five months as fermented juice.

Step 9, Sterilization – Above fermented juice should be operated with sterilization by high temperature instantaneous fungicide in 135℃ for 3-5S.

Some components were determined as 14.7% alcohol, 6.85 g/L acidity (calculated by citric acid), 0.67% residue sugar.

3.7 Further exploration on dandelion wine

Processing of fermented wine processing traditionally contains material treating, fermenting by yeast, and wine aging. And in this paper, the processing of fermented dandelion wine included material treating, material composing, enzyme hydrolysis, fermenting by yeast and wine aging [12]. The difference between above processes is filling material composing and enzyme hydrolysis as special two steps from dandelion wine. This innovation improved the flavour of fermented dandelion wine.

Fermentation using microorganismreactor synthesis natural on dandelion can improve product chlorogenic acid possibility and flavour [13]. In this paper, the fermentation technologies included enzyme and yeast, the flavour was improved, if there is any improving components in the fermenting process should be further studied.

4 Conclusion

Processing technologies of fermented dandelion wine were optimized by single factor experiment and orthogonal test design. Among four factors affecting compound wine, the acidity factor was firstly confirmed as 7.0 g/L. The other three factors were confirmed by a series experiments as dandelion 0.55%, schisandra 0.5%, sugar 22%. Through nine-step processing of mixing substrate, stirring with water, cooking rice, amylase saccharification, pectinase hydrolysis, adjusting juice, fermenting with yeast, fitering, aging and sterilization, a light yellow wine with the special taste with flavour of dandelion, schisandra and rice and less bitter. Some index were determined as 14.7% alcohol, 6.85 g/L acidity (calculated by citric acid), 0.67% residue sugar.

Bitter taste from the fermented dandelion wine was solved by composite technology. Unilateral dandelion wine were less carried out for its bitter flavour. Although fermented dandelion root was less bitter than un fermented, but the bitter taste was still unpleasant. To solve this problem, schisandra fruit with high level acidity content was selected to improve acid content. Based on compositing dandelion with schisandra, sugar was another high sugar level seasoning. To develop the compound fermented wine, glutinous rice was used as a sugar substrate through the fermented method by enzyme. By above measures, the bitter taste of fermented wine from dandelion root was developed.

Dandelion food was compositing with schisandra to strengthen the effect on protecting liver function. Schisandra has strong antioxidant activities and significant protective effect against acute hepatotoxicity induced by CCl_4. The effect on protecting liver function of dandelion wine had been on protecting liver functioned by compositing with schisandra.

Operation steps of processing systems were disposed. A nine-step operation system was arranged for exploring the fermented compound dandelion wine. The series operation techniques included crushing, compositing, cooking, amylase saccharification, pectinase hydrolysis, inoculating, fermenting, filtering, blending, and other operations. In these techniques, cooking and enzymolysis were used to ferment dandelion wine at first time.

Fermented compound wine has enriched dandelion food and composited with schisandra to strengthen the function of this dandelion food.

Both dandelion and schisandra have the function of the protective effect on fatty liver [14, 15]. This fermented wine composed dandelion and schisandra. Whether this compound wine have improved protective effect on fatty liver should be further tested by additional experiments.

References

1. Z. Kabir and R.T. Koide, Agriculture, Ecosystems and Environment, 78 (2000)

2. L. H. Tahtamouni, N. M. Alqurna, M. Y. Al-Hudhud, H. A. Al-HajjSchütz, R. Carle and A. Schieber, Journal of Ethnopharmacology, 135 (2011)
3. B. Simandi, Sz.T. Kristo , A´. Kery, L.K. Selmeczi, I. Kmecz and S. Kemeny, Journal of Supercritical Fluids, 23 (2002)
4. E. Racz-Kotilla, G. Racz and A. Solomon, Planta Med. 26 (1974)
5. T. Kuusi, H. Pyysalo, K. Autio and Lebensm.-Wiss. Technol. 18 (1985)
6. H. B. Wang, International Journal of Biological Macromolecules, 68 (2014)
7. A. A. Astafieva, A. A. Enyenihi, E. A. Rogozhin, S. A. Kozlov, E. V. Grishin, T. I. Odintsova, R. A. Zubarev and T. A. Egorov, Plant Science, 238 (2015)
8. A. Belscak-Cvitanovic, A. Busic, L. Barisic, D. Vrsaljko, S. Karlovic, I. Spoljaric, A. Vojvodi, G. Mrsic and D. Komes, Food Hydrocolloids, 57 (2016)
9. K. Schütz, R. Carle, A. Schieber, Journal of Ethnopharmacology, 107 (2006)
10. S. M. Oh, H. R. Kim, Y. J. Park, Y. H. Lee and K. H. Chung, Chinese Journal of Natural Medicines, 13 (2015)
11. M. Davaatseren, H. J. Hur, H. J. Yang, J. T. Hwang, J. H. Park, H. J. Kim, M. J. Kim, D. Y. Kwon and M. J. Sun, Food and Chemical Toxicology, 58 (2013)
12. Y. S. Kumar, R. S. Prakasam and O. V. S. Reddy, International Journal of Food Science & Technology, 1(2009)
13. G. Liu, K. Zong, L. Zhang, Y. Jin, W. Pan and X. Yang, Innovational Edition of Farm Products Processing, 12(2009)
14. M. Davaatseren, H. J. Hur, H. J. Yang, Jin-Taek Hwang, J. H. Park,Hyun-Jin Kim, M. J. Kim, D. Y. Kwon and M. J. Sung, Food and Chemical Toxicology, 58(2013)
15. H. J. Park, S. J. Lee, Y. Song, S. J. Jang, Y. G. Ko, S. M. Kang, B. Y. Chung, H. D. Kim, G. S. Kim, and J. H. Cho, Journal of Medicinal Food, (1)2014

MATEC Web of Conferences **62**, 02006 (2016) DOI: 10.1051/matecconf/20166202006
ICCFE 2016

The Storage Stability of Anthocyanins in Mao (*Antidesma thwaitesianum* Müll. Arg.) Juice and Concentrate

Arunya Prommakool[1,a], Kriangkrai Phattayakorn[1]

[1]*Kasetsart University, Chalermphrakiat Sakon Nakhon Province Campus, Thailand*

Abstract. Mao or Makmao (*Antidesma thwaitesianum* Müll. Arg.) is a wild plant found in the northeast of Thailand. Mao is one of fruits which are source of anthocyanins. Mao fruits are used for juice and concentrate which are consumed for healthy drinks. Determination of the kinetic parameters is essential to predict the quality changes and stability of anthocyanins in Mao juice and concentrate that occur during storage. The purpose of this research was to study the degradation of anthocyanins in Mao juice and concentrate during storage at 5, 30 and 37°C for 15 days. The storage stability of Mao anthocyanins was studied in 15°Brix juice and 45°Brix concentrate. The degradation kinetic (k), half-life ($t_{1/2}$), activation energy (E_a) and Q_{10} values for Mao anthocyanins degradation were determined. The results indicated that analysis of kinetic data for the degradation of anthocyanins followed a first-order reaction. An increase storage temperature from 5 to 30 and 37°C increased k value of anthocyanins in Mao juice and concentrate. Increasing storage temperature decreased $t_{1/2}$ value of anthocyanins in both concentrations. At 5, 30 and 37°C, the $t_{1/2}$ of anthocyanins decreased from 35 to 13 and 5 days for Mao juice and 32 to 25 and 21 days for Mao concentrate. The E_a value of the anthocyanins degradation in Mao juice and concentrate were 38.03 and 8.42 kJ/mol, respectively. Q_{10} values of both Mao juice and concentrate at 30-37°C were higher than those were storaged at 5-30°C. Thus higher stability of anthocyanins was achieved by using concentration and storage at lower temperature.

1 Introduction

Mao or Makmao (*Antidesma thwaitesianum* Müll. Arg.) is a wild plant found in the northeast of Thailand. However, Mao plant also grows well in Africa, Australia, tropical Asia and islands in the Pacific Oceans. Mao is one of fruits which are source of anthocyanins and antioxidants [1-3]. Mao fruits are used for juice and concentrate which are consumed for healthy drinks and theirs have become more popular in Thailand.

Anthocyanins are a good source of natural antioxidant, in addition, anthocyanins have been reported to be highly effective in reducing cholesterol levels in the human body including preventing cardiovascular diseases and anti-carcinogenic activity [4, 5]. Anthocyanins are natural pigments which are responsible for the attractive red, purple and blue colors in Mao fruits. However, anthocyanin are rather unstable during processing and storage, which readily degrade and form colorless or undesirable brown-colored compounds [6, 7]. Temperature, time, solid content and storage conditions are considered to be important factors influencing its stability [8].

In previous studies, thermal and storage degradation kinetics of fruits and their products have been reported. Kirca and Cemeroğlu [9] studied stability of anthocyanins in blood orange juice and concentrate. They found that anthocyanin degradation followed a first-order reaction with the activation energies (E_a) ranged from 73-89 kJ mol^{-1}. Wang and Xu [10] indicated that thermal degradation of blackberry anthocyanins at 60-90 °C followed first-order reaction kinetics with E_a value of anthocyanin degradation was 59 kJ mol^{-1} for blackberry juice. In addition, Kirca et al. [8] studied stability of black carrot anthocyanins during storage at 4-37 °C. The studied resulted the degradation fitted a first-order reaction with E_a values ranged from 62-86 kJ mol^{-1}. The anthocyanins decreased with increasing solid content during storage. Similarly, Kara and Erçelebi [11] reported the degradation of anthocyanins in Urmu mulberry showed a first-order reaction as well as the degtadation of anthocyanins in blood orange juice [12]. Hellström et al. [13] studied the stability of anthocyanins in Chokeberry, Blackcurrant, Crowberry and Mixed Berry juice stored at 4, 9 and 21 °C. The results suggested that anthocyanins stability affected by type of anthocyanin, the origin of the juice and especially the storage temperature. The $t_{1/2}$ value of anthocyanins was much shorter at room temperature than at cold storage.

Determination of the kinetic parameters is essential to predict the quality changes that occur during storage. However, a study of storage stability of anthocyanains in Mao juice and concentrate has never been reported. Therefore, the objective of our study was to determine the stability of anthocyanins in Mao juice and concentrate during storage at various temperatures.

[a] Corresponding author: csnayp@ku.ac.th

© The Authors, published by EDP Sciences. This is an open access article distributed under the terms of the Creative Commons Attribution License 4.0 (http://creativecommons.org/licenses/by/4.0/).

2 Materials and Methods

2.1 Preparation of Mao juice and concentrate

Fruits of Mao (*Antidesma thwaitesianum* Mull. Arg) at a commercially mature stage were purchased from a local market in SakonNakhon, Thailands. The fruits were homogeneously with a blender and filtered through cheesecloth to remove pulp. 60°Brix of sucrose solution were added to the juice gave the finally concentration to 15°Brix. The juice was concentrated to 45°Brix by rotary evaporator (BUCHI Rotavapor R-114 model, Switzerland) at 60°C. The pasteurization of the juice and concentrate were carried out at 85°C for 15 min transferred to sterile glass bottles and followed by rapid cooling in tap water.

2.2 The study of storage stability and degradation kinetic of anthocyanins in Mao juice and concentrate

The storage stability of Mao anthocyanins was studied in 15°Brix juice and 45°Brix concentrate. The Aqueous of 10 mL Mao juice and concentrate were put into screw-cap test tubes and then incubated at 5, 30 and 37°C for 15 days. Degradation kinetic of anthocyanins was analyzed. The zero and first order reaction rate constant (k) was calculated by the following equations [14]:

$$Q_t = Q_0 - kt_s \quad (1)$$

$$ln\,(Q_t/Q_0) = -kt_s \quad (2)$$

Where k is the rate constant (day^{-1}), Q_t is the anthocyanins content after t day of storage at a given temperature, Q_0 is the initial anthocyanin content, t_s is time (day).

k value was calculated from the slope of the straight lines given by equation (1 or 2), using a linear regression procedure of the sigmaplot.

Half life ($t_{1/2}$) which is the times needed for 50% degradation of anthocyanins was calculated by the following equations:

$$t_{1/2} = Q_0/2k \quad (3)$$

$$t_{1/2} = -ln0.5/k \quad (4)$$

The dependence of the degradation of Mao anthocyanins on temperature was determined by calculating the activation energy (E_a) and Temperature coefficient (Q_{10}) value from the following equation:

$$k = k_0\,x\,e^{-Ea/RT} \quad (5)$$

$$Q_{10} = (k_2/k_1)^{10/(T_2 - T_1)} \quad (6)$$

Where k_0 is the frequency factor (day^{-1}), E_a is activation energy (J/mol), R is the universal gas contant (8.314 J/mol K) and T is the absolute temperature (Kelvin, K).

The E_a value was calculated from the slope of the straight lines given by equation (5), using a linear regression procedure of the sigmaplot.

2.3 Determination of anthocyanins

The total anthocyanin content was determined by the pH-differential method [10] using two buffer systems; 0.025 M potassium chloride (KCl) buffer at pH 1.0 and 0.4 M sodium acetate ($NaC_2H_3O_2$) buffer at pH 4.5. A 0.1 ml of sample was transferred to 10 ml volumetric flash and made up to 10 ml with corresponding buffer and the absorbance was masured at 510 and 700 nm. The total anthocyanins were calculated as cyanindin-3-gluciside according to the following equation:

$$\text{Total anthocyanins (mg/L)} = \frac{AxMWxDFx1000}{\varepsilon\,x\,l} \quad (7)$$

Where $A = (A_{510} - A_{700})_{pH1.0} - (A_{510} - A_{700})_{pH4.5}$; MW = 449.2 g/mol for cyaniding-3-glucoside; DF = dilution factor; l=pathlength in cm; ε=26,900 molar extinction coefficient in L/mol/cm for cyaniding-3-glucoside; 1000=conversion from g to mg. All analyses were done in triplicate (n=3).

2.4 Determination of antioxidant activity

The antioxidant activity of juice was evaluated using the stable radical DPPH, according to the method of Deng et al. [15]. Briefly, 0.1 ul of the juice was transferred into a test tube and then mixed thoroughly with 5 ml of DPPH solution (1.0 mM in methanol). After incubation for 40 min, the absorbance of the mixtures at 517 nm was determined using spectrophotometer. The percentage of DPPH radical scavenging activity of juice was calculated as:

$$\text{DPPH scavenging effect (\%)} = \left(\frac{Abs_0 - Abs_1}{Abs_0}\right) X100 \quad (8)$$

Where Abs_0 = absorbance value of control sample; Abs_1 = absorbance value of Mao juice

2.5 Other analyses

Total soluble solid was used a refractometer (Atogo, Tokyo, Japan). The pH was measured by pH meter (Methohm IKA-MF10basic, Germany)

2.6 Statistical analysis

Data was analyzed by one-way analysis of variance (ANOVA) with the application of SPSS software version 10.0.1 (SPSS Inc., Chicago, USA) at 5% significance level.

3 Results and Discussion

The total soluble solid, pH, anthocyanin content and antioxidant activity of Mao juice and concentrate showed in Table 1. It was found that there was the loss of anthocyanins during concentration process ($p \leq 0.05$). The loss of anthocyanins due to retention time of concentration more than 3 hours. Wang and Xu [10] reported that there was less than 5% loss of anthocyanins from the concentration step. The time was needed for juice concentration was less than 100 min at 60 °C. The oxidation activity of Mao juice and concentrate were not significantly affect (p>0.05).

Table 1: Chemical composition of Mao juice and concentrate

Sample	Total soluble solid (°Brix)	pH	Anthocyanin content (mg/L)	Antioxicant activity (%)
Juice	15^{b}±0.41	2.94^{b}±0.28	33.48^{a}±2.01	87.98^{a}±2.25
Concentrate	45^{a}±0.37	3.52^{a}±0.69	23.21^{b}±0.94	89.69^{a}±0.80

Mean values within a column superscripted by the same letter are not significantly different at P<0.05.

3.1 Stability of anthocyanins during storage

The degradation of anthocyanins from Mao juice and concentrate was studied during storage at 5, 30 and 37 °C. The anthocyanins content from Mao juice and concentrate were plotted as a function of time (Figure 1 and 2). It showed that storage temperature affected on degradation of anthocyanins. The results clearly indicated that anthocyanin degradation increased with increasing storage temperature and time. An increase in temperature might be attributed to the degradation of anthocyanin from juice and concentrate. Anthocynains are a bioactive compound which can be reduced due to heated degradation. Mao juice anthocyanins were faster degradation as compared to concentrate.

Anthocyanin contents during storage decreased with increasing temperature and time. Linear regression confirmed that the degradation of anthocyanins in Mao juice and concentrate followed by a first-order reaction. Previous studies showed that degradation of anthocyanins in many juice and their products followed a first-order reaction [7, 10-13, 16].

The kinetic parameters of anthocyanins degradation in Mao juice and concentrate are shown in Table 2. The k value varies between (2-13.2)x10^{-2} and (2.2-3.3)x10^{-2} day^{-1} for juice and concentrate, respectively. The $t_{1/2}$ values at 5, 30 and 37°C were calculated as 35, 13 and 5 days for juice samples and 32, 25 and 21 days for concentrate samples, respectively. The results revealed that anthocyanins in Mao concentrate are more stable than juice during storage at 30 and 37 °C. The storage at 30 and 37°C resulted in a much faster anthocyanin degradation as compared to storage at 5°C. Thus, an increase of storage temperature trends to increased degradation rate while as $t_{1/2}$ values decreased. These results were somewhat similar to the results of Wang and Xu [10] who studied anthocyanin degradation in blackberry juice and concentrate over the tempature ranged within 5-37°C. The results found that increasing anthocyanin degradation incresed with increasing storage temperature but $t_{1/2}$ values decreased. The $t_{1/2}$ value of anthocyanins was much shorter at room temperature than at cold storage. Silmilarly, the stability of black carrot anthocyanins at various solid contents between 11-64°Brix during storage at 4-37°C [8] and the stability of anthocyanins in berry juices stored at 4, 9 and 21 °C [13].

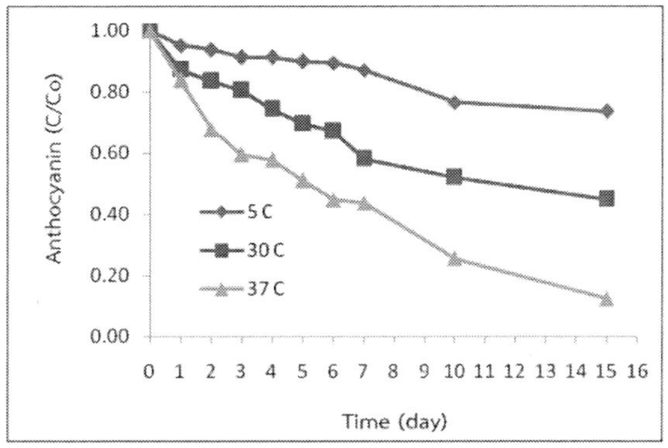

Figure 1. Relation between anthocyanin content (C/Co) and time (day) of Mao juice at 5, 30 and 37 °C ;where C/Co as ratio of anthocyanins and initial content.

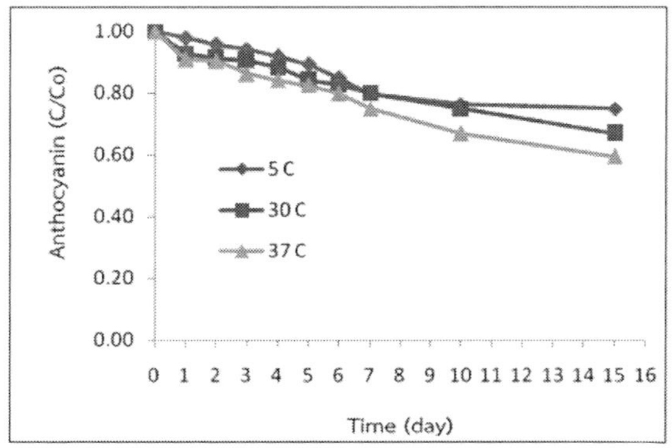

Figure 2. Relation between anthocyanin content (C/Co) and time (day) of Mao concentrate at 5, 30 and 37 °C ;where C/Co as ratio of anthocyanins and initial content.

In this study, increasing solid content resulted in a slower degradation of anthocyanins during storage. Similar trends were observed for black carrot anthocyanin [8] and blood orange anthocyanins [9]. However, Wang and Xu [10] found that anthocyanins degraded much faster in blackberry concentrate (65 °Brix) than in blackberry juice (8.9 °Brix) during storage at 5-37°C.

The stability of anthocyanins depends on type and structure of anthocyanins, origin of juice including polysacharides in fruit. These factors might be attributed to the degradation rate and $t_{1/2}$ value of anthocyanins in various juice and concentrate.

Table 2. Effect of storage temperature on the rate constant (k), half life ($t_{1/2}$) of anthocyanins degradation in Mao juice and concentrate

Sample	Temp (°C)	k x 10^{2} (day^{-1})	$t_{1/2}$ (days)
Juice	5	2.00(0.94)*	35
	30	5.30(0.96)	13
	37	13.20(0.99)	5
Concentrate	5	2.20(0.91)	32
	30	2.80(0.98)	25
	37	3.30(0.98)	21

* Numbers in parentheses are the determination coefficients

Kirca et al. [8] studied stability of black carrot anthocyanins during storage at 4-37 °C. The degradation fitted a first-order reaction while as anthocyanins decreased with increasing solid content during storage. Higher E_a values were obtained during storage with increasing solid content.

3.2 Temperature dependence

To determine the effect of temperature on the parameter studies, the constants obtained from Eq. (5) were fitted to an Arrhenius-type equation. The dependence of the degradation of Mao juice and concentrate anthocyanins on temperature was shown in Figure 3. At 5, 30 and 37°C, the calculated activation energy (E_a) (Table 3). The result showed that Mao juice has higher E_a value than concentrate. High E_a value implied that the degradation of anthocyanins in Mao juice was more susceptible to temperature elevation than those in concentrate. Similar to results were observed for black carrot anthocyanins [8] and blackberry anthocyanins [10].

Temperature coefficient (Q_{10}) was calculated from Eq. (6). The higher Q_{10} value also obtained for anthocyanins in juice samples. The results clearly indicated the concentration and low temperature storage are able to prevent the degradation of anthocyanins in Mao juice and concentrate.

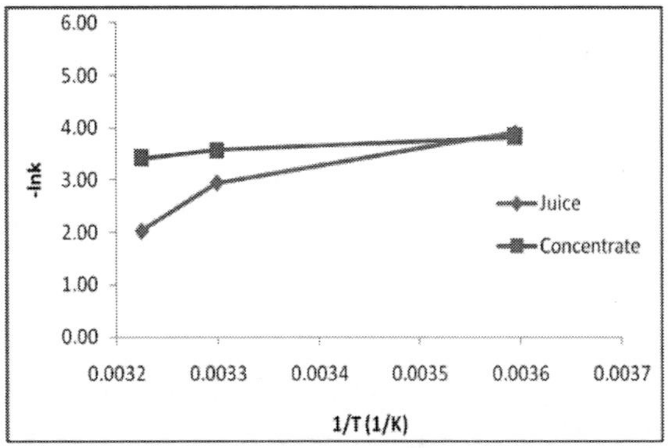

Figure 3. The Arrhenius plots for degradation of anthocyanins in Mao juice and concentrate during storage at 5, 30 and 37 °C

Table 3. Effect of storage temperature on activation energy (E_a) and Temperature coefficient (Q_{10}) of anthocyanins degradation in Mao juice and concentrate

Sample	E_a (kJ/mol)	Q_{10}	
		5-30 °C	30-37 °C
Juice	38.03	1.48	3.69
Concentrate	8.42	1.10	1.26

4 Conclusions

The kinetic degradation of anthocyanins in Mao juice and concentrate followed a first-order reaction during storage at 5-37°C. The Stability of Mao anthocyanins depended on storage temperature and solid content. Increasing anthocyanin degradation increased with increasing storage temperature but $t_{1/2}$ values decreased. The $t_{1/2}$ value of anthocyanins was much shorter at room temperature than at cold storage. Increasing solid content resulted in a slower degradation of anthocyanins during storage. Mao juice has higher E_a and Q_{10} value than concentrate. Thus higher stability of anthocyanins was achieved by using concentration and storage at lower temperature. In this study, the kinetic stability could be used to predict the anthocyanins degradation in Mao juice and concentrate during storage.

Acknowledgments

The authors would like to express their sincere appreciation to Kasetsart University Research and Development Institute (KURDI) and Faculty of Natural Resource and Agro-Industry, Kasetsart University Chalermphrakiat SakonNakhon Provinces Campus for supporting financially.

References

1. D. Puangpronpitag, P. Areejitranusorn, P. Boonsiri, M. Suttajit, P. Yongvanit, J Food Sci **73**, 6 (2008)
2. N. Nuengchamnong, K. Ingkaninan, Food Chem **118**, 6 (2010)
3. D. Puangpronpitag, P. Yongvanit, P. Boonsiri, M. Suttajit , P. Areejitranusorn , H.K. Na, Y.J. Surh, Food Chem **127**, 9 (2011)
4. Q. Du, J. Zheng, Y. Xu, J Food Comp Anal **21**,6 [2008]
5. P. Aramwit, N. Bang , T. Srichana, Food Res Int 43, 5 [2010]
6. P.J. Tsai, L. Delva, T.Y. Yu, Y.T. Huang, L. Dufossé, Food Res Int. **38**, 7 [2005]
7. C. Shao-qian,L. Liang, P. Si-yi, Agri Sci China **10**, 6 [2011]
8. A. Kirca, M. Ozkan, B. Cemeroğlu, Food Chem **101**, 7 (2007)
9. A. Kirca, B. Cemeroğlu, Food Chem **81**, 5 (2003)
10. W.D. Wang, S.Y. Xu, J Food Eng **82**, 5 (2007)
11. S. Kara, E.A. Erçelebi, J Food Eng **116**, 7 (2013)
12. C. Shao-qian, L. Liang, P. Si-yi, Agri Sciences in China **10**, 12 (2011)
13. J. Hellström, P. Mattila , R. Karjalainen, J Food Comp Anal **31**, (2013)
14. D. Kilcast , P. Subramaniam, New York;CRC press (2000)
15. J. Deng , W. Cheng, G. Yang, Food Chem **125**, 4 (2011)
16. A. Kırca, M. Özkan, B. Cemeroğlu, Food Chem **97**, 8 (2006)

Empirical Modeling on Hot Air Drying of Fresh and Pre-treated Pineapples

Yardfon Tanongkankit [1,a], Kanjana Narkprasom[1], Nukrob Narkprasom[1]

[1]Maejo University in Chiang Mai, Thailand

Abstract. This research was aimed to study drying kinetics and determine empirical model of fresh pineapple and pre-treated pineapple with sucrose solution at different concentrations during drying. 3 mm thick samples were immersed into 30, 40 and 50 Brix of sucrose solution before hot air drying at temperatures of 60, 70 and 80°C. The empirical models to predict the drying kinetics were investigated. The results showed that the moisture content decreased when increasing the drying temperatures and times. Increase in sucrose concentration led to longer drying time. According to the statistical values of the highest coefficients (R^2), the lowest least of chi-square (χ^2) and root mean square error (RMSE), Logarithmic model was the best models for describing the drying behavior of soaked samples into 30, 40 and 50 Brix of sucrose solution.

1 Introduction

Pineapple (*Anana comosus*) is a tropical fruit belonging to the *Bromeliaceae* family, and originated from South America [1]. This fruit has been reported to be low in calories and a source of nutritional components such as fiber, mineral and antioxidants [1-2]. Pineapple is commonly consumed in a form of fresh product. Nevertheless, it has a short storage life approximately 5 days at 4°C and 1 day at 15°C [3]. Therefore, to extend its shelf life pineapple is usually dried.

Drying is a process to remove moisture content in products to a level at which deterioration from chemical reaction and microorganism are minimized. However, this method involving in high temperatures and times can cause changes in final quality of products when compared to the fresh fruit [4]. To avoid this problem, pre-drying treatments such as addition of sugars are needed for fruit drying in order to the improvement in texture and the stability of the color pigment during storage [5].

Pre-treatment with sugar solution has been claimed to be very effective for producing high quality dried fruits and vegetable. Since sugar especially non-reducing sugar does not react with amino acids or proteins to cause Maillard reaction, browning and loss of nutrition in products can be eliminated [5]. Giovanna [6] reported that immersing pineapple in 50 Brix of sucrose solution for 30 min protected the color of pineapple during drying at 70, 75 and 80°C.

The objective of this study was to determine kinetic dryings of fresh pineapple and pre-treated pineapple with sucrose solution. Empirical modeling for describing drying behavior of pineapples was also investigated.

2 Materials and Methods

[a] Corresponding author: yardfon@hotmail.com

2.1 Experimental material

Fresh pineapple cv. Nanglae were obtained from local market in Chiang Mai. An initial moisture content of pineapple was 80 ± 2% (d.b.), which was determined by drying the fresh pineapple in the electric thermal dryer at 105°C for 24 h until the weight was constant. Each run in the experiment was performed in triplicate.

2.2 Sample preparation

The fruits were hand peeled and sliced into a rectangular shape of 5×10 cm and 5 mm thickness. A batch each of 250 g of slices were pre-treated accordingly with sucrose osmosis method by immersing fresh slices into sucrose solution (Food grade of 98% purity) with concentration of 30, 40 and 50 Brix for 30 min. The samples were drained on a screen and reweighed.

2.3 Drying experiment

The prepared samples were dried in hot air dryer (Memmert, 500/108I, Germany). Approximately 250 g of the sample was placed as single layer on a perforated tray. The experiments were performed at temperature of 60, 70 and 80 °C. The samples were dried until a final moisture content of less than 18% (d.b.) was reached. During each experiment 3–5 g of the sample was taken out at various intervals to determine its moisture content. The moisture content of the sample was determined using a gravimetric method at 105 °C [7].

2.4 Mathematical modelling

© The Authors, published by EDP Sciences. This is an open access article distributed under the terms of the Creative Commons Attribution License 4.0 (http://creativecommons.org/licenses/by/4.0/).

The moisture ratio (MR) of the pineapple was defined as follows [8]:

$$MR = \frac{M_t - M_e}{M_0 - M_e} \quad (1)$$

Where M_t, M_0 and M_e are the moisture content at any time of drying (kg water/kg dry mass), initial moisture content (kg water/ kg dry mass) and equilibrium moisture content, respectively. The experimental data at different thicknesses of paddy were fitted into 6 thin-layer drying models that commonly used in most food and biological materials (Table 1).

Table 1. Mathematical models applied the moisture ratio

Model name	Model equation	Reference
1. Lewis	$MR = \exp(-kt)$	[8]
2. Page	$MR = \exp(-kt^n)$	[9]
3. Modified Page	$MR = \exp(-kt)^n$	[9]
4.Wang and Singh	$MR = a\exp(-kt^n) + bt$	[10]
5.Handerson and Pabis	$MR = a.\exp(-kt)$	[11]
6.Logarithmic	$MR = a\exp(-kt) + c$	[12]

2.5 Correlation coefficients and error analyses

The determination of coefficients (R^2), reduced chi-square (χ^2) and root mean square error (RMSE) were used to evaluate the goodness fit. These parameters were calculated as follows:

$$\chi^2 = \frac{\sum_{i=1}^{N} \left(MR_{\exp,i} - MR_{pre,i}\right)^2}{N - z} \quad (2)$$

$$RMSE = \sqrt{\left[\frac{1}{N}\sum_{i=1}^{N}\left(MR_{pre,i} - MR_{\exp,i}\right)^2\right]} \quad (3)$$

Where N is the number of observations, z is the number of constants, MR_{exp} are the experimental and predicted moisture ratios, respectively.

2.6 Statistical analysis

The experiments were designed in complete random. The data were subjected to an analysis of variance (ANOVA) and are presented as mean values with standard deviations. Differences between mean values were established using Duncan's multiple range tests. Values were considered at a confidence level of 95%. All statistical analyses were performed using SPSS® software (version 17) (SPSS Inc., Chicago, IL). All experiments were performed in duplicate unless specified otherwise.

3 Results and Discussion

3.1. Drying kinetics of pineapples

Fig. 1 shows the drying curves of unpretreated and pre-treated with different sucrose concentration samples. The results showed that the drying rates at higher temperatures were higher than those at lower temperatures. This is due to drying at higher temperature led to higher moisture diffusivity values and larger driving force for heat/mass transfer than at lower temperatures. The equilibrium moisture content (EMC) and time needed to dry a sample to the desired moisture content of less than 0.18 g/g dry basis are listed in Table 2. EMC of the samples were in the range of 0.023-0.141 g/g dry basis. It was also observed that higher concentration of sucrose resulted in longer drying time.

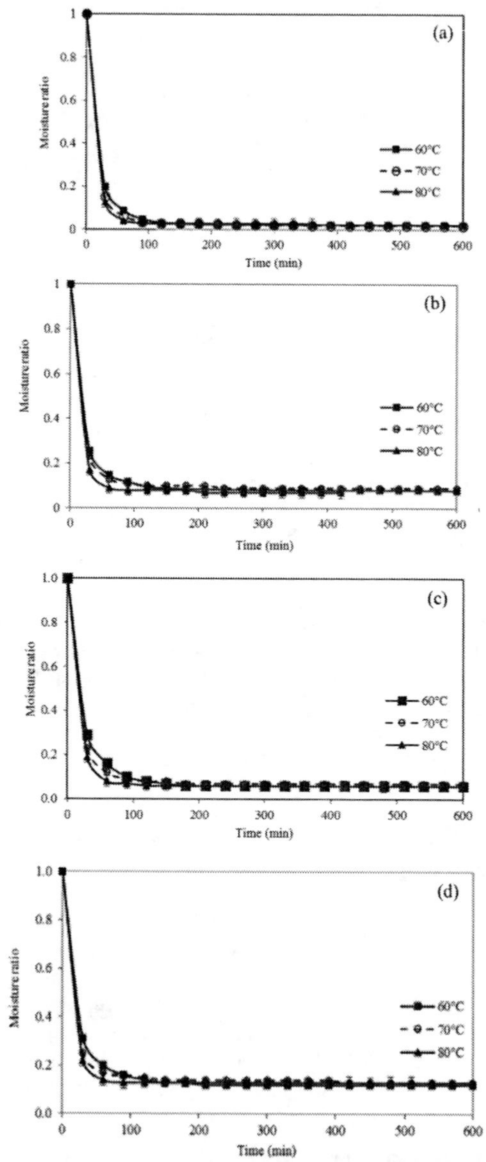

Figure 1. Drying curves of (a) unpretreated (b) pre-treated with 30 Brix of sucrose (c) pre-treated with 40 Brix of sucrose and (d) pre-treated with 50 Brix of sucrose samples during drying

Table 2. Time to dry samples to the final moisture content of less than 18% dry basis.

Drying temperature (°C)	Sucrose concentration (Brix)	Drying time (min)	Equilibrium moisture content (g/g d.b.)
60	0	300	0.023
	30	720	0.080
	40	690	0.089
	50	900	0.075
70	0	240	0.023
	30	540	0.093
	40	600	0.077
	50	810	0.122
80	0	210	0.028
	30	300	0.073
	40	390	0.066
	50	480	0.141

3.2. Empirical modeling of drying curves

Drying curves of the unpretreated and pre-treated pineapples under hot air at 60, 70 and 80°C were fitted with six different moisture ratio models shown in Table 1. The statistical results of the different models, including the drying model coefficients and the comparison criteria used to evaluate goodness of fit, namely, R^2, χ^2 and RSME, are summarized in Table 3. The R^2 values for the models were in the rage of 0.7507-0.9997. χ^2 values were varied between 0.0001 and 0.1081 and RMSE values between 0.122 and 0.981. By comparing the criteria values among four drying models, it can be seen that Logarithmic equation was the best descriptive model for all treatments, since it exhibited the highest average value of R^2, the lowest average values of χ^2 and RMSE. Thus, it was selected to represent the drying characteristics of pineapples.

Table 3. Statistical results obtained from different drying models.

Model	Sucrose concentration (Brix)	Temperature (°C)	Parameter	R^2	χ^2	RMSE
Newton	0	60	k = 0.0172	0.9691	0.0490	0.0668
		70	k = 0.0246	0.9536	0.0078	0.0834
		80	k = 0.0263	0.9908	0.0012	0.0325
	30	60	k = 0.0057	0.7507	0.0370	0.1884
		70	k = 0.0061	0.8451	0.0336	0.1775
		80	k = 0.0125	0.8573	0.0274	0.1578
	40	60	k = 0.0053	0.7770	0.0359	0.1855
		70	k = 0.0066	0.8565	0.0226	0.1468
		80	k = 0.0159	0.8029	0.0275	0.1597
	50	60	k = 0.0046	0.8731	0.0184	0.1334
		70	k = 0.0030	0.8528	0.0323	0.1764
		80	k = 0.0042	0.9028	0.0267	0.1587
Page	0	60	k = 0.0527, n = 0.7182	0.9888	0.0020	0.0403
		70	k = 0.1230, n = 0.5710	0.9870	0.0025	0.0442
		80	k = 0.0649, n = 0.7554	0.9967	0.0005	0.0196
	30	60	k = 0.3208, n = 0.2360	0.9880	0.0017	0.0391
		70	k = 0.3170, n = 0.2366	0.9919	0.0016	0.0379
		80	k = 0.3977, n = 0.2319	0.9932	0.0015	0.0349
	40	60	k = 0.2779, n = 0.2578	0.9879	0.0018	0.0405
		70	k = 0.0933, n = 0.4681	0.9730	0.0043	0.0627
		80	k = 0.3741, n = 0.2589	0.9870	0.0020	0.0414
	50	60	k = 0.0353, n = 0.6089	0.8958	0.0163	0.1234
		70	k = 0.1257, n = 0.3572	0.9739	0.0050	0.0679
		80	k = 0.1095, n = 0.3945	0.9800	0.0051	0.0672
Wang and Singh	0	60	k = 0.0108, n = 2.84×10^{-5}	0.9328	0.0188	0.1093
		70	k = 0.0137, n = 4.49×10^{-5}	0.9058	0.0316	0.1325
		80	k = 0.0209, n = 1.18×10^{-4}	0.9925	0.0014	0.0298
	30	60	k = 0.0039 n = 4.53×10^{-6}	0.8688	0.0331	0.1667
		70	k = 0.0065, n = 1.15×10^{-5}	0.8618	0.0318	0.1558
		80	k = 0.0093 n = 2.47×10^{-5}	0.9132	0.0317	0.1420
	40	60	k = 0.0040, n = 4.82×10^{-6}	0.8816	0.0306	0.1597
		70	k = 0.0048, n = 6.45×10^{-6}	0.9195	0.0170	0.1273
		80	k = 0.0077, n = 1.60×10^{-5}	0.8611	0.0184	0.1663
	50	60	k = 0.0032, n = 2.94×10^{-6}	0.9185	0.0170	0.1216
		70	k = 0.0033, n = 2.95×10^{-6}	0.9230	0.0213	0.1351
		80	k = 0.0077, n = 1.60×10^{-5}	0.9072	0.0334	0.1598
Logarithmic	0	60	k = 0.9172, a = 121.16, c = 0.2513	0.8735	0.0320	0.1527
		70	k = 1.0168, a = 158.43, c = 0.1637	0.9249	0.0207	0.1175
		80	k = 1.0356, a = 443.81, c = 0.1621	0.9271	0.0146	0.1032
	30	60	k = 0.7307, a = 0.0268, c = 0.2677	0.9998	0.0001	0.0053
		70	k = 0.6968, a = 0.0328, c = 0.3032	0.9997	0.0001	0.0106
		80	k = 0.7395, a = 0.0438, c = 0.1621	0.9991	0.0003	0.0151
	40	60	k = 0.7228, a = 0.0249, c = 0.2750	0.9998	0.0000	0.0053
		70	k = 0.7806, a = 0.0164, c = 0.2326	0.9981	0.0003	0.0167

		80	k = 0.7395, a = 0.0438, c = 0.1621	0.9991	0.0003	0.0151
	50	60	k = 0.8355, a = 0.0092, c = 0.2015	0.9922	0.0012	0.0330
		70	k = 0.7032, a = 0.0152, c = 0.3194	0.9968	0.0006	0.0237
		80	k = 0.6720, a = 0.0190, c = 0.3479	0.9962	0.0010	0.0291
Henderson and pabis	0	60	k = 0.9915, a = 0.0169	0.9696	0.0054	0.0662
		70	k = 0.9950, a = 0.0244	0.9538	0.0089	0.0832
		80	k = 0.9974, a = 0.0262	0.9027	0.0231	0.1373
	30	60	k = 0.7369, a = 0.0029	0.8584	0.0208	0.1385
		70	k = 0.8057, a = 0.0042	0.8854	0.0231	0.1423
		80	k = 0.9115, a = 0.0101	0.8742	0.0260	0.1459
	40	60	k = 0.7479, a = 0.0029	0.8707	0.0203	0.1363
		70	k = 0.8793, a = 0.0050	0.8870	0.0183	0.1288
		80	k = 0.9438, a = 0.0135	0.8149	0.0278	0.1543
	50	60	k = 0.9248, a = 0.0039	0.8875	0.0167	0.1250
		70	k = 0.7610, a = 0.0019	0.9082	0.0182	0.1301
		80	k = 0.8322, a = 0.0030	0.9298	0.0185	0.1277
Modified Page	0	60	k = 0.0166, n = 0.7182	0.9643	0.0092	0.0866
		70	k = 0.0254, n = 0.5710	0.9375	0.0170	0.1150
		80	k = 0.0267, n = 0.7554	0.9849	0.0025	0.0452
	30	60	k = 0.0081, n = 0.2360	0.8836	0.0871	0.2831
		70	k = 0.0077, n = 0.2366	0.8886	0.1402	0.3502
		80	k = 0.0187, n = 0.2319	0.8671	0.0846	0.2656
	40	60	k = 0.0069, n = 0.2578	0.8951	0.0925	0.2912
		70	k = 0.0063, n = 0.4681	0.9039	0.0487	0.2098
		80	k = 0.0224, n = 0.2589	0.8461	0.0537	0.0537
	50	60	k = 0.0041, n = 0.6089	0.9012	0.0314	0.1714
		70	k = 0.0030, n = 0.3572	0.9169	0.1074	0.3158
		80	k = 0.0036, n = 0.3945	0.9267	0.1081	0.3089

4 Conclusion

The experiments were performed to determine drying characteristics of unpretreated pineapple and pre-treated pineapple with sucrose solution at 30, 40 and 50 Brix. It was evidenced that higher drying temperature resulted in shorter drying time since moisture diffusivity values and larger driving force for heat and mass transfer were larger at higher drying temperatures Increase in sucrose concentration resulted in longer drying time. Different mathematical models were also determined with the drying behavior of pineapples. The results indicated that the Logarithmic model could present better predictions for the moisture transfer than others. Studies on physical and nutritional properties of the pineapple after pre-treatment and drying are suggested as a future work.

Acknowledgement

The authors express their sincere appreciation to Upper Northen Research Administration network for supporting the study financially and Faculty of Engineering and Agro-Industry, Maejo University for laboratory equipment.

References

1. S. W. Wijeratnam, Ref. Modul. Food Sci. Ency. Food Health **380-384** (2015)
2. D. I. S. da Silva, G. D. R. Nogueira, A. G. Duzzioni, M. A. S. Barrozo, *Indus. Crop. Product.* **50** (2013)
3. S. Chen, B. Mulgrew, and P. M. Grant, IEEE Trans. on Neural Network. **4** (1993)
4. M. K. Krokida, V. T. Karathanos, Z. B. Maroulis, D. Marinos-Kouris, J. Food Eng. **59** (2002).
5. G. E. Lombard, J. C. Oliveira, P. Fito, A. Andre´s, J. Food Eng. **85** (2008)
6. G. Cortellino, P. Pani, D. Torreggiani, *Procedia Food Sci.*, **1** (2011)
7. Association of Official Analytical Chemists (AOAC), *Official methods of analysis*, 15th ed., The Association of Official Analytical Chemists: Washington D.C (1995).
8. I. Ceylan, M. Aktas, H. Dogan, *Appl. Themal. Eng.* **27** (2007)
9. M. Rasouli, S. Seiiedlou, H. R. Ghasemzadeh, H. Nalbandi,. Austra. J. Agri. Eng. **2** (2011)
10. C. Y. Wang, R. P. Singh. ASAE, (1978)
11. M. A. Basunia, T. Abe, J. Food Eng. **47** (2001)
12. Z. Wang, J. Sun, X. Liao, F. Chen, G. Zhao, Food Res. Inter. **40** (2007)

Extraction of Coconut Oil from Coconut Milk Foulants Using Enzyme

Phanida Saikhwan[1,a], Chanokchat Nuchnet[1], Wannakarn Wanakayont[1], Angkana Suksa-nga[1]

[1]*Thammasat University, Thailad*

Abstract. Coconut milk manufacturing process encounters problems with foulants formed during pasteurization process. For example, fouling layers reduce heat transfer efficiency of a heat exchanger. As the fouling layers are considered as waste, this research aimed at extracting coconut oil from the foulants to produce a product from the waste. A model coconut milk foulant was used to simulate foulants formed during batch pasteurization process and coconut oil was extracted from the foulant using celloulase enzyme. The extracted oil then was evaluated in terms of fatty acid composition and antioxidant properties (total phenolic and flavonoid contents). The antioxidant activities were evaluated using DPPH (1,1-diphenyl-2-picrylhydrazyl) radical scavenging and FRAP (Ferric reducing antioxidant power) methods. Results showed that the oil extracted from the foulants appeared similar to virgin coconut oil (VCO); the extracted oil appeared as clear viscous liquid with aroma associated with roasted coconut. The oil extracted using enzyme contained all fatty acids found in VCO in lower proportions but large extent of linoleic acid was found. Antioxidant capacity was similar to that of VCO. The foulants after the extraction of fat using enzyme were easier to clean suggesting the possibility to couple cleaning of coconut milk foulants and oil extraction in the same process.

1 Introduction

Nowadays, most coconut milk manufactures encounter problems with foulants formed during pasteurization process. Fouling is generally defined as the accumulation and formation of unwanted materials on the surfaces of processing equipment, which leads to reduction in heat transfer efficiency. Other problems associated with foulants are loss of energy, cost of maintenance, cost of cleaning, and waste management.

Coconut milk foulants contain fat in large proportion and the extracted fat has physical properties (colour and aroma) similar to virgin coconut oil (VCO) [1]. VCO is distinguished from another type of coconut oil (Refined, Bleached and Deodorized: RBD) by its colour (colourless) and a coconut aroma. In addition, VCO contains a large amount of medium-chain fatty acids such as lauric, capric, caproic and caprylic acids which were reported to have antimicrobial and antiviral effects [2]. It has been claimed that VCO has several beneficial health effects such as antioxidant capacity and high level of vitamin E [3-6]. Extracting coconut oil from coconut milk foulants using petroleum ether yielded oil with fatty acid profile lying in the standard accepted for VCO [7]. Although the extracted oil had comparable antioxidant capacity to virgin olive oil, the antioxidant capacity was on the lower bound reported for VCO [7]. The lower antioxidant capacity observed with the coconut oil extracted from coconut milk foulants using petroleum ether could be due to the exposure to the chemical and high temperature used during the extraction.

Recently, enzymes have been employed in oil extraction, e.g. extraction of sesame oil [8], extraction of peanut oil [9], and extraction of oil from coconut meat [10]. Major benefit of using enzymes in oil extraction that have been reported is that quality of oil is not altered by the mild condition used during the extraction [11]. Enzymes have also been used in cleaning of milk foulants [12, 13]. The efficiency was reported to be comparable to that of chemical cleaning and using enzyme is more environmental friendly. Hence, this work aimed to investigate properties of coconut oil extracted from coconut milk foulants using enzyme. As extracting oil during cleaning would be more practical for use with a continuous pasteurization process, this paper also reported results of preliminary study of possibility of combining the oil extraction using enzyme with cleaning of the foulants.

2 Materials and Methods

2.1 Preparations of coconut milk foulants

Fresh grated coconuts were purchased from local markets and coconut milk was extracted by mixing 500 g grated coconut meat with 200 ml distilled water. The coconut milk was used to prepare a model coconut milk foulant using the method reported by [14] where pasteurization was conducted using lab-scaled apparatus simulating batch and continuous pasteurization processes (Figure 1).

[a] Corresponding author: psaikhwan@engr.tu.ac.th

© The Authors, published by EDP Sciences. This is an open access article distributed under the terms of the Creative Commons Attribution License 4.0 (http://creativecommons.org/licenses/by/4.0/).

To produce foulants from a batch pasteurization, coconut milk was heated in a container (Figure 1(a)) placing on a hot plate. Three sample plates (SS304) were placed at the bottom of the container where foulants were formed. An overhead stirrer (100 rpm) was used to improve heat transfer and prevent film forming at a surface of coconut milk. The coconut milk was heated from a room temperature and once the milk reached 70°C, temperature was controlled at this temperature and heating was continued for 20 minutes. After that the coconut milk was poured away and sample plates with coconut milk foulants formed on the plates were removed.

The apparatus used to simulate a continuous pasteurization (Figure 1(b)) is a heat exchanger with sample plates separating coconut milk flow (0.048 m/s) and hot water (95°C). To produce the foulants, coconut milk was preheated (in a beaker immersed in hot water at 50°C) and was fed into the heat exchanger. Thermocouples were used to monitor temperatures of the system. At steady state, coconut milk temperature was at 70°C and the process was continued for 20 minutes after that. At the end of pasteurization, coconut milk was drained out and sample plates with coconut milk deposits were collected.

(a)

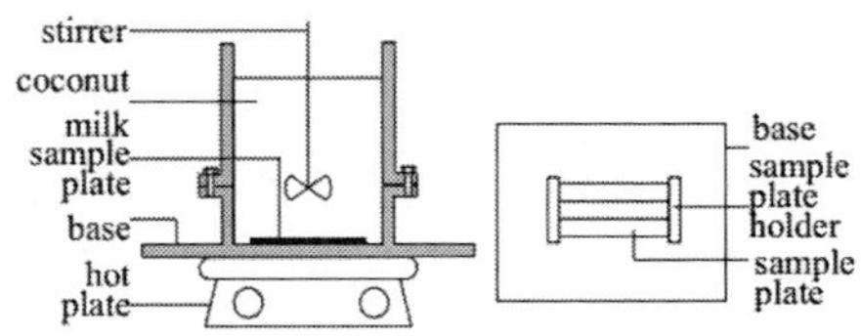

(b)

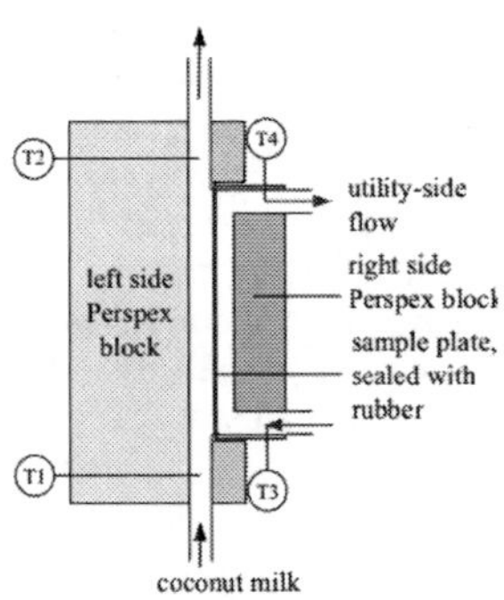

Figure 1. Schematic of apparatus used in formation of coconut milk foulants: (a) batch process; (b) continuous process (T1-T4 are thermocouples used for temperature monitoring).

2.2 Oil extraction

Since coconut milk foulants contain carbohydrate which is hard to remove during cleaning [14], celloulase (Brentag, Thailand) was chosen as the enzyme used to extract coconut oil in this study. The extraction protocol was modified from the protocol used by [7] where celloulase was employed in oil extraction from sesame seeds.

First, foulants obtained from a model batch pasteurizer were scraped from sample plates. Distilled water was added to the foulants at a ratio of 3 ml of water per 1 g of foulant. Then this mixture was boiled for 10 minutes and left to cool down to room temperature. Celloulase then was added to the mixture. Varying ratios of celloulase mass to foulant mass (2-20%) were used in order to find the optimum enzyme-to-foulant mass ratio. After adding the enzyme, if pH of the mixture was not in a range of pH 5.0-5.5 (optimum pH of celloulase activity), pH of the mixture would be adjusted using 0.2N HCl. The mixture containing water, foulant and enzyme was then placed in a hot water bath (45-50°C) for 2 hours and left overnight. Finally, the mixture was centrifuged and oil layer was found. The oil was then obtained and kept for further testing.

As amounts of oil extracted using different enzyme-to-foulant mass ratios showed that the ratio of 6% was the optimum ratio (See 3.1), only this ratio was used in oil extraction from coconut milk foulants from a continuous pasteurization. The extraction steps were the same as those used in the extraction from the foulants from a batch pasteurization.

2.3 Characterization of extracted oils

2.3.1 Determination of fatty acid profiles

Determination of fatty acids was supplied by The Institute of Biotechnology and Genetic Engineering (IBGE), Chulalongkorn University, Thailand. Preparation of fatty acid methyl ester (FAME) was carried out according to IUPAC standard method 2.301 (IUPAC, 1979) and FAME standards used in this analysis were obtained from an AccuStandard. FAME was determined on an Agilent Technologies 6890 N gas chromatography equipped with INNOWAX capillary column (30 m × 0.3 mm i.d., 0.25 μm film thickness) and a Flame Ionization Detector (FID) system. The injector and detector temperatures were set at 150°C and 250°C, respectively. Helium was used as the carrier gas at a flow rate of 2.3 mL/min. The Gas Chromatographic temperature was set as follows: initial temperature of 150°C, increasing to 180°C at 10°C/min, then to 200°C at 5°C/min and then held for 5 min (total runtime of 33 min). Individual FAMEs were identified by comparison with reference standards [15].

2.3.2 Antioxidant capacity measurements

Antioxidant capacity of the extracted oil was measured using DPPH (2, 2-diphenyl-1-picrylhydrazyl) radical scavenging assay, FRAP (Ferric reducing antioxidant power) assay and determination of total phenolic and determination of flavonoid content. The antioxidant capacity measurements were supplied by Excellent Center for Cosmetic and Wellness (ECCW), Mae Fah Luang University, Thailand.

2.4 Preliminary study of extracting oil and cleaning coconut milk foulants using enzyme

In practice, it is not possible to remove foulants from a continuous pasteurizer (or a heat exchanger) manually

DOI: 10.1051/matecconf/20166202008

before extracting oil from the foulants. So celloulase was mixed with water at 6% by weight and was fed into a continuous pasteurizer (with foulants inside) to simulate Cleaning-In-Place (CIP), a cleaning process generally used in cleaning of a heat exchanger. This experiment, hence, studied the possibility of cleaning the foulants and extracting oil during the cleaning.

The CIP protocol used in this study involves: (i) soaking the mixture of water and enzyme in the heat exchanger fouled with coconut milk deposits at 45-50°C for 2 hours and (ii) circulating the mixture in the heat exchanger and observe any removal of foulants. CIP studies were also conducted using water as circulating fluid.

In addition, swelling studies of the deposits soaked in the enzyme solution were conducted to elucidate effects of enzyme on cleaning of the foulants. The swelling studies were done using a gravimetric method. A deposit was immersed in a beaker containing the enzyme solution and it mass was measured at every 5 minutes for 2 hours. Before weighing the deposit, excess liquid was removed from the deposit using filter paper.

3 Results

3.1 Optimum enzyme-to-foulant mass ratio

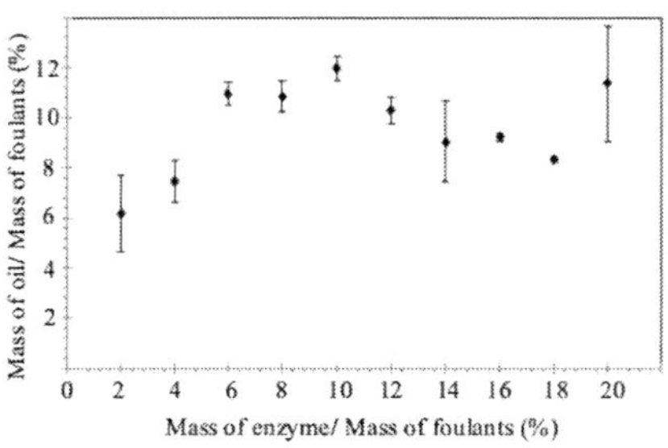

Figure 2. Amounts of oil extracted from coconut milk foulants using different enzyme-to-foulants mass ratios.

Figure 2 depicts oil yields from using various enzyme-to-foulants mass ratios (2-20%). According to the figure, there is a peak at enzyme-to-foulants mass ratio of 10% and when enzyme-to-foulants mass ratio increases beyond 10%, there is a drop in oil yield. The yield then rises again at enzyme-to-foulants mass ratio of 20%. The initial increase in oil yield at enzyme-to-foulants mass ratios below 10% could be because there was unused substrate in the foulants. Hence, when more enzyme was added, more oil was obtained. It was also observed that when more enzyme was used, foulants were digested to smaller size. As foulants acted as a barrier between oil layer and aqueous layer (Figure 3), a layer of foulants that were digested to small sizes was quite unstable. Consequently, it was more difficult to remove all oil without interfering the foulants and aqueous layers. Some oil was also in the digested foulants layer. This explains the drop in oil yield and varying results when enzyme-to-foulants mass ratio was higher than 10%.

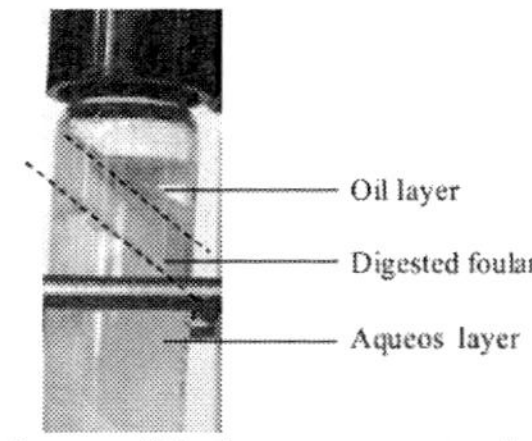

Figure 3. A mixture of foulants, enzyme and water after centrifugation (Dotted line shows a foulant layer).

Consider the results in Figure 2, enzyme cost and the ease of removing an oil layer, it was decided that the optimum enzyme-to-foulants mass ratio was 6%. The extraction yield was approximately 11%. This is quite low compared to the yield obtained when extracting using petroleum ether (50% [6]). The low yield is typical in extracting oil using enzymes [7-9].

3.2 Properties of extracted oils

3.2.1 Fatty acid profiles

Table 1 shows fatty acid profiles of various samples of coconut oil. As fatty acid profiles of oil extracted using enzyme from foulants found in batch and continuous pasteurizer were comparable, their profiles are reported as a profile of oil extracted using enzyme in the table.

Table 1. Fatty acid profiles of coconut oil samples extracted from coconut milk foulants using enzyme and petroleum ether compared with fatty acid profile of APPC standards for VCO.

Fatty acids	Oil extracted using enzyme	Oil extracted using petroleum ether [6]	APPC standards for VCO
C6:0	-	-	0.4-0.6
C8:0	5.01±1.32	7.10±0.03	5.0-10.0
C10:0	4.32±0.86	5.88±0.08	4.5-8.0
C12:0	32.26±2.44	45.84±1.36	45.1-53.2
C14:0	12.25±0.41	17.91±0.23	16.8-21.0
C14:1	1.25±0.36	-	NA
C15:0	-	0.38	NA
C16:0	5.64±0.59	8.24±0.36	7.5-10.0
C16:1	-	0.35±0.01	NA
C18:0	1.76±0.36	2.24±0.38	2.0-4.0
C18:1	3.53±0.44	5.48±0.51	5.0-10.0
C18:2	19.13±2.71	3.06±0.01	1.0-2.5
C22:2	2.53±0.30	-	<0.5 (C18:3-C24:1)
Other	4.33	0.1	NA

While the fatty acid profile of the oil extracted from foulants using petroleum ether fits the APPC standards quite well, the fatty acid profile of the oil extracted using enzyme had lower contents of the fatty acids specified in the standards. This could be because other fatty acids not specified in the standards were found in the oil extracted using enzyme. These other fatty acids have been reported for health benefits. For instance, myristoleic acid (C14:1) has been reported to have activity against prostate-cancer cells [16]. Docosadienoic acid (C22:2) is an essential polyunsaturated fatty acid in a group of omega 6 fatty acids which has a high nutritive value [17]. It should also be noted that the oil extracted using enzyme contained strikingly large content of linoleic acid (C18:2). In a recent study, intake of linoleic acid can reduce total and LDL cholesterol, improve insulin sensitivity and blood pressure and lower risk of coronary disease [18, 19].

3.2.2 Antioxidant capacity

Table 2. Antioxidant capacities of coconut oil obtained from coconut milk foulants using enzyme (EB-from batch pasteurization; EC-from continuous pasteurization) compared with that of coconut oil obtained from an extraction from foulants from a batch pasteurization using petroleum ether (PEB) and results from the literature.

Oil sample	Antioxidant activity		Total phenolic mgGAE/g	Flavonoid content mgQE/g
	DPPH mgTEAC/g	FRAP mgTEAC/g		
EB	0.934 ±0.066	1.699 ±0.00416	0.196 ±0.009	6.491 ±0.882
EC	0.126 ±0.016	0.571 ±0.034	0.169 ±0.005	0.132 ±0.009
PEB [6]	0.079 ±0.0296	0.935 ±0.071	0.061 ±0.006	0.546 ±0.015
VCO [20, 21]	0.154	NA	0.078-0.29	NA

From Table 2, consider antioxidant capacities of EB and PEB, it can be concluded that the extraction using petroleum ether significantly reduced antioxidant capacity of the oil. It can also be seen that coconut oil extracted from foulants from a batch pasteurization generally had higher antioxidant capacity compared to the oil from foulants from a continuous process. Particularly, flavonoid contents found in both EB and BPE (both samples were extracted from foulants found in a batch pasteurization) were larger than that associated with EC. As a wide spectrum polyphenolics from the class of flavonoids can be found in coconut oil [22, 23], the higher flavonoid content of EB compared to that of EC explains the slightly higher total phenolic content in the EB. Moreover, as it has been shown a positive relationship between total phenolic content and antioxidant activity measured by FRAP assay [24], the higher activity of EB measured by FRAP assay could be due to the higher total phenolic content of EB.

Nevertheless, it should be noted that antioxidant activity of EC obtained from DPPH assay was slightly higher than that of EB. Comparing antioxidant capacities of EB, EC and VCO reported by the literature, the oils extracted from foulants formed in batch and continuous pasteurizers had antioxidant capacities comparable to that of VCO.

3.3 Preliminary results of cleaning coconut milk foulants using enzyme

Two sets of CIP studies were done: (i) use the enzyme solution as a cleaning liquid and (ii) use water. At the same flow rate, an extent of removal of deposits observed with the use of enzyme was approximately 3 times of an extent of removal associated with CIP using water. It was also noticed that an onset of removal was observed at a flow rate of 163 g/s whereas an onset of removal was observed at a flow rate of 435 g/s and 372 g/s when water and sodium hydroxide (pH 12) were used in cleaning respectively [14]. The results suggest that using enzyme could improve cleaning efficiency of foulants. In the cleaning effluent, oil film was observed. Hence, oil extraction also occurred during the cleaning.

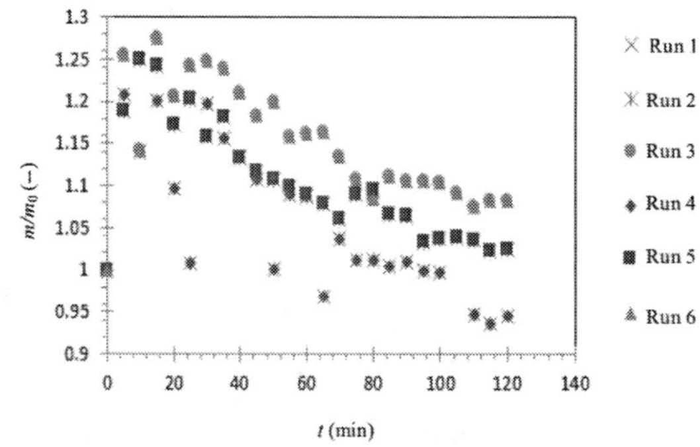

Figure 4. Swelling of coconut milk deposits (from a continuous pasteurization) in a mixture of celloulase and water: m – mass of deposits; m_0 – initial mass of deposits; t – time.

Swelling studies of coconut milk deposits in the mixture of enzyme and water are illustrated by Figure 4. An initial increase in mass was observed and then mass of the deposits decrease with soaking time. The decrease is due to removal of the deposits as some debris were observed during the swelling studies. The removal like this was not observed in swelling of coconut milk deposits in sodium hydroxide [14]. This is probably why cleaning efficiency seemed to be better when using enzyme in CIP studies.

4 Conclusions

Celloulase can be used to extract oil from coconut milk foulants formed during pasteurization of coconut milk. The oil contains all fatty acids found in VCO but at lower proportion. This is because the oil contains significantly larger content of linoleic acid and other fatty acids which have health benefits.

Although the yield obtained when extracting oil using enzyme was lower than that obtained when using extraction with petroleum ether, antioxidant capacity of the oil was preserved when extracting using enzyme. The type of pasteurization also affects antioxidant capacity of the oil extracted. The oil extracted from foulants from a batch pasteurizer had higher antioxidant activity than the oil from foulants from a continuous pasteurizer. This could be because of more flavonoid content found in the oil from foulants formed in a batch pasteurizer.

Preliminary studies of cleaning of the foulants using enzyme suggest a better cleaning efficiency compared to cleaning by NaOH, the typical chemical used in cleaning. The better efficiency was explained by digestion of the foulants by enzyme which helped removing the foulants during CIP. As oil film was observed in the cleaning effluents, oil was also extracted from the foulants during cleaning. Further work should be done to optimize the CIP using enzyme so that oil extraction and cleaning can be done in the same stage. This in turn will generate a useful product from waste, reduce cleaning cost/time and lower a load for wastewater treatment.

References

1. P. Saikhwan, S. Thongchan, Proceedings of World Academy of Science, Engineering and Technology (2013)
2. B.J. Villarino, D.L. Marsha, M. Concepc'ion, C. Lizada, LWT-Food Sci. Technol. **40**, 193-199 (2007)
3. R. Arlee, R., S. Suanphairoch, P. Pakdeechanuan, Int. Food Res. J. **20**, 2103-2109 (2013)
4. D.D. Bawalan, K.R. Chapman, Virgin Coconut Oil production manual for micro- and village-scale processing. FAO Regional Office for Asia and the Pacific (2006)
5. A.S. Librado, M.V. Von Luigi, Int. Food Res. J. **20**, 1683-1689 (2013)
6. K.N. Seneviratne, C.D. Hapuarachchl, S. Ekanayake, Food Chem. **114**, 1444-1449 (2009)
7. P. Saikhwan, K. Jarungsorasak, V. Opaspong, C. Nuchnet, W. Wanakayont, A. Suksa-nga, Proceedings of ICEAS (2015)
8. S. Latif, F. Anwar, Food Chem. **125**, 679-684 (2010)
9. D.H.L. Jiang, Z. Wang, S. Xu, Food Bioprod. Process. **88**, 233-238 (2010)
10. Y.O.K. Tano-Debrah, J. Sci. Food Agric. **74**, 497-502 (1997)
11. H.B. Petrus, H.L.V. Chen, N. Norazman, J. Membr. Sci. **325**, 783-792 (2008)
12. A. Grasshoff Food Bioprod. Process. **88**, 233–238 (2002)
13. A. Boyce, A.V. Piterina, G. Walsh, Biofouling **26(7)**, 837-850 (2010)
14. P. Saikhwan, S. Thongchan, N. Jumwan, P. Thungsiabyuan, J. Sakdanuphap, S. Boonsom, P. Kraitong, P. Danwanichakul, Food and Bioprod. Process. **93**, 166-175 (2015)
15. M. Limsuwatthanathamrong, S. Sooksai, S. Chunhabundit, S. Noitung, N. Ngamrojanavanich, A. Petsom, Asian J. Anim. Sci. **6**, 65-72 (2012)
16. K. Iguchi, N. Okumura, S. Usui, H. Sajiki, K. Hirota, K. Hirano, Prostate **47(1)**, 59-65 (2001)
17. F. Jabeen, A.S. Chaudhry, Food Chem. **125(1)**, 991-996 (2011)
18. M.S. Farvid, M. Ding, A. Pan, Q. Sun, S.E. Chiuve, L.M. Steffen, W.C. Willett, F.B. Hu, Circulation **130(18)**, 1568-1578 (2014)
19. E. Phares, The nutrition source: Harvard T.H. Chan School of public health **November 5** (2014)
20. A. M. Marina, Y. B. Che Man, I. Amin, Trends Food Sci. Tech. 20, 481-487 (2009)
21. K. Ratanapradit, S. Palakas, Determination of anti-radicals and anti-oxidatioin reaction properties of virgin coconut oil and virgin coconut oil supplemented functional foods products, Burapha University, Thailand (2008)
22. K.G. Nevin, T. Rajamohan, Food Chem. **99(2)**, 260-266 (2006)
23. K.G. Nevin, T. Rajamohan, e-SPEN **3(1)**, e1-e8 (2008)
24. K. Thaipong, Boonprakob, U., Crosby, K., L. Cisneros-Zevallos, D.H. Byrne, J. Food Compost. Anal. **19**, 669-675 (2006)

Effect of Pre-treatment Methods on the Color Changes during Drying of Red Chilli (*Capsicum frutescens* L.)

Rattapon Saengrayap [1,a], Natthida Boonlap [1], and Uthumporn Boonsorn [1]

[1]*Department of Agro-Industry Technology and Management, Faculty of Agro-Industry, King Mongkut's University of Technology North Bangkok (KMUTNB) Prachinburi Campus, Prachinburi, Thailand*

Abstract. Chilli was dried using conventional tray dryer with 3 different drying conditions, i.e., 60°C, 80°C and 2-stage drying (80°C and 60°C). The effects of pre-treatment methods were evaluated by soaking chilli in pre-treatment solutions prior to drying. In terms of drying characteristics, the drying time found to be shortest at 80°C. It was also found that the drying rate of pre-treated chilli increased compared with non-treated one. The modified Page's model was suitable for describing drying characteristics at all conditions (R^2=0.97-0.99). Furthermore, the color change during drying was investigated using image analysis methods. It was found that the pre-treatment methods prevented the color change. The combined kinetic model was the best model for predicting the change of color during drying (R^2=0.98-0.99). A pre-treated chilli soaking in a mixed solution of 0.3% (w/w) $Na_2S_2O_5$ and 1% (w/w) $CaCl_2$ showed the better quality. In addition, the combined kinetics model could describe the change of L^* a^* and b^* values in all conditions.

1 Introduction

Chilli is one of spice products that used as the condiments for flavoring and coloring in cuisines. Chilli is a good source of flavonoids, i.e., carotenoids and vitamin C, which their antioxidant activities confer the protection against carcinogenic compounds and delay the aging process [1, 2].

Generally, the conventional drying method for chilli is sun drying, and remains the most practical method throughout chilli producing countries including Thailand, that takes several days to obtain the required moisture content (4–11% d.b.) [3, 4]. However, sun drying method causes a major loss of qualities of the final product, due to the long drying time and direct exposure to the surroundings. Undesirable changes in the color attributes of the chilli may lead to a decrease in its quality and marketing value [5-7]. Moreover, hot air drying is another common method for drying chilli.

Browning reaction is the important phenomenon that tends to decrease the quality of dried products [8-10]. Many methods were applied to improve the quality of dried products. The pre-treatment step was achieved to prevent the change in qualities of many processed food. Blanching with hot water and soaking in chemical solutions were extensively used for solving the change of dried product [11-14].

Color is considered to be a major criterion which affects the consumer's perception. Normally, the colorimeter is successfully achieved to measurement the color but the limitation of colorimeter can be found when determine the non-homogeneous color of solid samples. Moreover, low repeatability unrepresentative results are found when use colorimeter on small surface area. Therefore, an alternative method is proposed which is the use of digital image analysis. The digital image analysis has been successfully applied in many food industrial applications such as classification, grading and estimating qualities [15-18].

Therefore, the objectives of this study were to study the effect of pretreatment methods on the change in drying characteristics and to determine the change in color during drying using image analysis method.

2 Materials and Methods

2.1 Materials

Fresh red chilli (*Capsicum frutescens* L.) was procured from a local market then washed and graded prior to storage in a refrigerator at 5°C. Initial moisture content of chilli was determined using AOAC method 920.151A [19]. The initial moisture content of chilli was 85.53±0.04% wet basis.

2.2 Pre-treatment methods

Chilli was blanched using hot water at 90°C for 3 min [10] and then cooled in cold water at 5°C for 5 min and drained on a perforated tray prior to soak in pretreatment solutions. Blanched chilli was soaked in the controlled

[a] Corresponding author: rattapon.s@agro.kmutnb.ac.th

© The Authors, published by EDP Sciences. This is an open access article distributed under the terms of the Creative Commons Attribution License 4.0 (http://creativecommons.org/licenses/by/4.0/).

temperature pre-treatment solutions of 25±1°C for 10 min The operating ratio was 1:2 (w/w) for chilli and solution, respectively. The mixed pre-treated solutions were consists of (1) 0.1% (w/w) $Na_2S_2O_5$ and 1% (w/w) $CaCl_2$; (2) 0.2% (w/w) $Na_2S_2O_5$ and 1% (w/w) $CaCl_2$; (3) 0.3% (w/w) $Na_2S_2O_5$ and 1% (w/w) $CaCl_2$; and (4) blanched chilli without soaking were used as a control group. However, in this study, according to the food safety and to avoid an allergic reaction in those who are sensitive to the residues of $Na_2S_2O_5$ and $CaCl_2$; the concentration of $Na_2S_2O_5$ and $CaCl_2$ was then controlled to be lower than the limitation of food additive usage of $Na_2S_2O_5$ and $CaCl_2$ by Thailand FDA.

2.3 Drying procedures

A laboratory hot-air dryer (Memmert, UF750, Germany) was used to dry chilli. The temperature was control at 60°C and 80°C with air velocity 2.0 m/s. Another 2-stage drying was performed using initial temperature of 80°C for 5 hr following by 60°C for a final stage. Temperature was monitored using a type-T thermocouple which connected with a data logger (Fluke, 714B, USA). The drying process continues until the sample mass was constant. After drying process finished, the sample was placed in a desiccator for cooling down temperature prior to pack in an aluminum bag, and then kept in a desiccator for further quality evaluations.

2.4 Drying characteristics

The effects of drying methods and drying conditions on drying characteristics were determined. The average moisture ratio (*MR*) was calculated using Eq. (1):

$$MR=M-M_e/M_0-M_e \quad (1)$$

Where M is sample moisture content (d.b.), M_0 is initial moisture content (d.b.), and M_e is the equilibrium moisture content.

2.5 Color measurement

Surface colors of chilli were measured using a colorimeter (HunterLab, ColorQuest XE Plus, Reston, VA, USA). Color was expressed in CIELAB, i.e., $L*$ (whiteness or brightness), $a*$ (redness/greenness) and $b*$ (yellowness/blueness) coordinates, standard illuminant D65 and observer 10°. The calorimeter was calibrated against a standard white plate ($L*$=91.7, $a*$=-1.16, $b*$=1.06) before color measurement. The sample was measured three times with three different positions and then reported the data as an average value of the measurements. Total color changes (ΔE), Chroma (C), hue angle (°h) and browning index (BI) were calculated using following by Eqs. (2-5):

$$\Delta E=\sqrt{\left(L^*-L_0^*\right)^2+\left(a^*-a_0^*\right)^2+\left(b^*-b_0^*\right)^2} \quad (2)$$

$$C=\sqrt{a^{*2}+b^{*2}} \quad (3)$$

$$h^\circ=\tan^{-1}\left(b^*/a^*\right) \quad (4)$$

$$BI=\left[100\left(x-0.31\right)\right]/0.71 \quad (5)$$

Where L^*_0, a^*_0 and b^*_0 are initial color parameters of the chilli; $x=(a^*+1.75L^*)/(5.645L^*+a^*-0.3012b^*)$

2.6 Image analysis

2.6.1 Image acquisition system

The image acquisition system (Figure 1.) consists of an illumination system, a digital camera and a computer to acquire the RGB images. A light box (Medalight Model DIB-0806, Hong Kong) was used as a light source. A digital camera (NIKON, D60, Japan) was mounted with the built-in camera holder on the top of the light box and connected with a computer (Fujitsu, Lightbook B531, Japan) via a USB port for directly capturing the images by a remote control utility software (NIKON, Camera Control Pro 2 - Trial version, Japan).

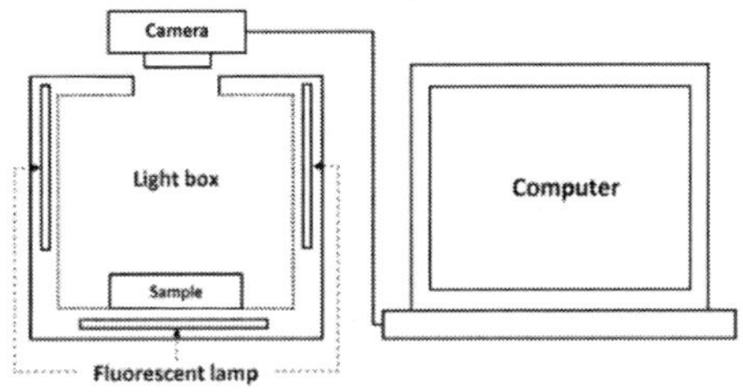

Figure1. Image acquisition system

2.6.2 Calibration and color conversion

Calibration of the image acquisition system was conducted to compare the results between the image analysis method with those of conventional one. The standard color sheets were measured using both a colorimeter (HunterLab, ColorQuest XE Plus, Reston, VA, USA) and image analysis method. The images from image acquisition system were then analyzed using image processing algorithm which developed in MATLAB (V2008a, MathWork, Inc., MA, USA). The accuracy for the use of machine vision could be determined using the following equations:

$$\Delta E=\sqrt{\left(L_c^*-L^*\right)^2+\left(a_c^*-a^*\right)^2+\left(b_c^*-b^*\right)^2} \quad (6)$$

Where $L*$, $a*$ and $b*$ were the actual lightness, redness and yellowness values from colorimeter, respectively; L_c*, $a_c*$and b_c* were the converted lightness, redness and yellowness values, respectively.

2.7 Kinetics model

2.7.1 Drying kinetics

The obtained drying curves undergoing different methods were fitted with five different models given in Table 1. The models were tested for their validity in describing drying characteristics of chilli. The coefficients in each model were determined using STATISTICA® (Version 6.0, StatSoft, Tulsa, Okla., U.S.A.). The suitable model was selected using the standard error (*SE*) and the coefficient of detrmination (R^2). The best fit model would provide a lowest *SE* with highest R^2 [20].

Table 1. Mathematical models applied to the drying curves

Model	Model equation
Newton	$MR = exp(-kt)$
Modified Page	$MR = exp[-(kt)^n]$
Henderson and Pabis	$MR = a\ exp\ (-kt)$
Logarithmic	$MR = a\ exp\ (-kt) + c$
Wang and Singh	$MR = 1+at+bt^2$

Note: *k*, Drying coefficient in min^{-1}; *n*, exponent (dimensionless); *t*, time in min; *a*, *b* and *c*, coefficient (dimensionless), specific to individual equations.

2.7.2 Color kinetics

In order to determine the change of color during drying, chilli was taken from the dryer 30 min interval for each drying condition. Chilli was taken an image using the developed image acquisition system. The RGB values were then determined and converted to CIELAB using MATLAB. The converted CIELAB values in terms of the normalized color differences were plotted as a function of drying time. Several equations were reported to describe the color kinetics [21-23]. Generally, the rate of change of color can be expressed by the following equation:

$$\frac{dC}{dt} = \pm kC^n \tag{7}$$

Where k is the kinetic rate constant; C is the concentration of quality factor at time t and n is the order of reaction.

The time dependent relationships for most foods appear to be described by zero-order or first order kinetic model. By integrating Eq. (7), the zero order equation, the first order equation and the combined kinetic models equation could be derived as [24]:

The zero order equation: $C = C_0 \pm k_0 t$ (8)

The first order equation: $C = C_0 \exp(\pm k_1 t)$ (9)

The combined kinetic models equation:

$$C = \frac{k_0}{k_1} - \left[\left(\frac{k_0}{k_1} - C_0\right)\exp(\pm k_1 t)\right] \tag{10}$$

Where C_0 is the initial value of color and C is the color value at a specified time. Moreover, in the equation, a positive (+) and a negative (-) values indicated the formation and degradation of any quality parameter, respectively [24].

3 Results and Discussion

3.1 Drying characteristics of chilli

The results showed that the temperature affected the drying characteristics of chilli. Drying at a higher temperature of 80°C presented lower drying time for all drying conditions. Drying time of chilli is shown in Table 2. It was observed that the pre-treated chilli showed the lower rate of moisture removal than those of non-treated one. This might due to the fact that the blanching and soaking processes increased the moisture content of chilli. A significant amount of moisture was absorbed into chilli and then resulted in a higher time for drying chilli to reach the desired moisture content (9-11% w.b.). In addition, the concentration levels of the mixed pre-treatment solutions on the drying time had no significant difference in each drying temperature.

Table 2. Drying time of chilli undergoing different pre-treatment conditions

Pre-treatment	Drying time (h)		
	60°C	80°C	2-stage
Non-treated	12	4.5	8
Blanching	14	5.5	9.5
0.1% $Na_2S_2O_5$ and 1% $CaCl_2$	13.5	5	8.5
0.2% $Na_2S_2O_5$ and 1% $CaCl_2$	13.5	6	8.5
0.3% $Na_2S_2O_5$ and 1% $CaCl_2$	13	6	8

Table 3. The prediction errors of modified Page models for describing the drying characteristics of chilli undergoing 80°C

Pre-treatment	*SE*	R^2
Non-treated	0.02	0.98
Blanching	0.01	0.99
0.1% $Na_2S_2O_5$ and 1% $CaCl_2$	0.01	0.98
0.2% $Na_2S_2O_5$ and 1% $CaCl_2$	0.01	0.97
0.3% $Na_2S_2O_5$ and 1% $CaCl_2$	0.01	0.98

Among the pre-treatment methods, chilli soaking in 0.3% (w/w) $Na_2S_2O_5$ and 1% (w/w) $CaCl_2$ showed lower drying time at 60°C and 2-stage drying for 13 h and 8 h, respectively. On the other hand, according to the drying time, chilli soaking in 0.1% (w/w) $Na_2S_2O_5$ and 1% (w/w) $CaCl_2$ showed the lowest drying time at 80°C for 5 h. The chemical components in the pre-treatment solutions, especially, Na and Ca salts were reported to alter the cellular structure of chilli resulting in increased the moisture removal [3, 11].

Table 4. The coefficient of modified Page models for describing the drying characteristics of chilli undergoing 80°C

Pre-treatment	*k*	*n*
Non-treated	0.13	1.19
Blanching	0.10	1.24
0.1% $Na_2S_2O_5$ and 1% $CaCl_2$	0.22	1.27
0.2% $Na_2S_2O_5$ and 1% $CaCl_2$	0.18	1.19
0.3% $Na_2S_2O_5$ and 1% $CaCl_2$	0.19	1.21

Table 3 shows the results for selecting the suitable model for describing the drying behaviour of chilli undergoing different pre-treatment conditions. It was observed that the modified Page model was the best model for predicting the change of moisture removal comparing with other models. It provided the lowest *SE* and highest R^2 in all condition of drying. Therefore, modified Page model was then selected for predicting drying characteristics of chilli. Table 4 shows the parameter and drying coefficient of the selected model.

3.2 Effect of pre-treatment on color change

Figure 2 shows the change of color parameters of chilli undergoing different pre-treatment methods at 60°C. The ΔE was found to be lower when chilli was soaked in pre-treatment solutions. The lowest change of ΔE was at 0.3% (w/w) $Na_2S_2O_5$ and 1% (w/w) $CaCl_2$ with had no significant different between levels of concentration. On the other hand, the fresh chilli and only fresh blanched chilli had higher ΔE values.

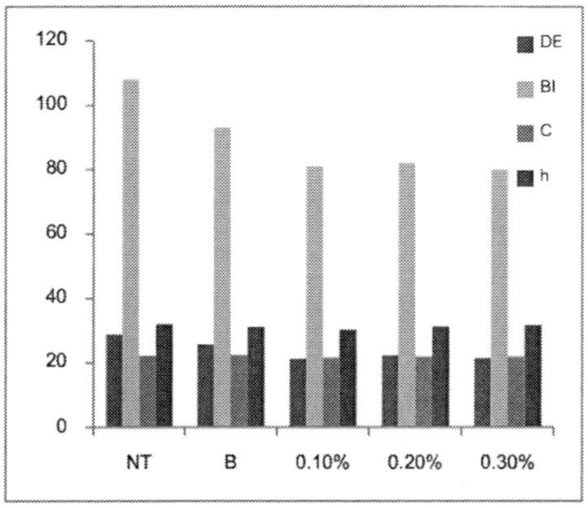

Figure 2. Color attributes of chilli undergoing different pre-treatment methods at 60°C (DE, ΔE; BI, Browning index; C, Chroma; h, hue angle; NT, Non-treated; B, Blanching; 0.10%, 0.1% $Na_2S_2O_5$ and 1% $CaCl_2$; 0.2%, 0.3% $Na_2S_2O_5$ and 1% $CaCl_2$; 0.3%, 0.3% $Na_2S_2O_5$ and 1% $CaCl_2$)

Drying at higher temperature (80°C) resulted in higher change of ΔE as expected. The surface of chilli became darker due to a larger extension of non-enzymatic browning reaction, especially at higher drying temperature [25-29]. According to the 2-stage drying, higher temperature was applied at the initial stage of drying (the first 4 h.) and a lower temperature was then applied. It was observed that a color was found to be better compared to a 1-stage drying at 80°C while the ΔE values were still found to be larger than those of 1-stage drying at 60°C. This might be due to the fact that the thermal degradation of some color pigments and non-enzymatic browning reaction were occurred during the initial stage of drying.

Furthermore, in terms of *BI*, the results revealed that non-treated sample presented a significant higher *BI* values compared to the pre-treated one. It was also found that 0.3% (w/w) $Na_2S_2O_5$ and 1% (w/w) $CaCl_2$ provided lowest BI value while the level of concentration did not have significant effect on the *BI*. The presence of browning reaction during drying resulted in the higher *BI* values of chilli drying at 80°C and 2-stage drying when compared to drying at 60°C as well. Moreover, all chroma values were in the same range (20.58-22.49) with no significant different. Higher value refers to the purity of red color relating with the present of carotenoids within the chilli [3, 11]. In terms of $h°$ values, it showed similar results with chroma values. There were no significant different between all drying condition. The values of h° were in the range of 30.99-33.36 which exhibited a reddish-orange color.

3.3 Image analysis

The results showed the compromise in the using of image analysis. The conversion of CIELAB values from RGB values showed a high accuracy and had no significantly different between the converted CIELAB values and those of conventional method.

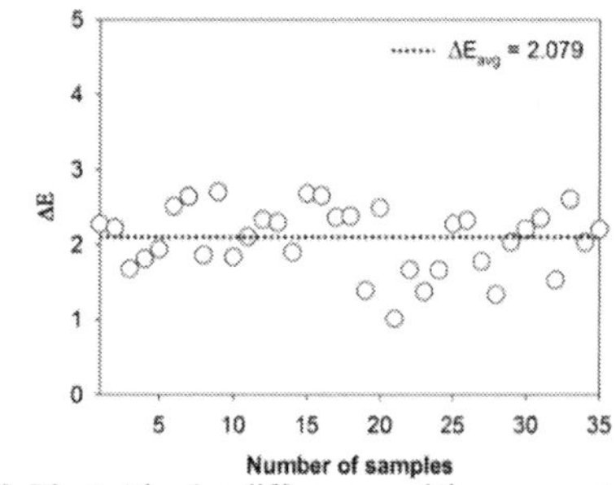

Figure 3. The total color difference and the average total color difference between converted and actual values of the dried chilli

The average ΔE was 2.079 (Figure 3) with standard deviation of 0.48. The average ΔE was below the threshold value of 2.2; which shows the effectiveness of developed image analysis process and machine vision system.

3.4 Color change kinetics

In terms of lightness changes chilli became darker than the fresh one. Figure 4 shows the kinetics of the lightness change as a function of drying time at 80°C. The lightness values decreased during drying due to the browning reactions during drying. The rate of lightness change was also greater at higher temperature.

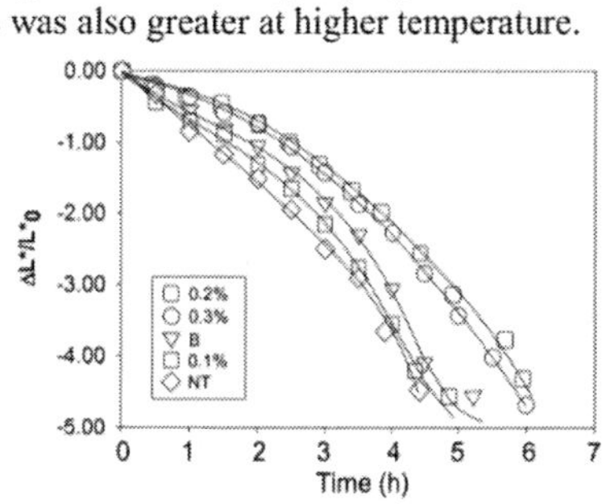

Figure 4. Kinetics of the lightness change as a function of drying time undergoing different pre-treatment methods at 80°C (DE, ΔE; BI, Browning index; C, Chroma; h, hue angle; NT, Non-treated; B, Blanching; 0.10%, 0.1% $Na_2S_2O_5$ and 1% $CaCl_2$; 0.2%, 0.3% $Na_2S_2O_5$ and 1% $CaCl_2$; 0.3%, 0.3% $Na_2S_2O_5$ and 1% $CaCl_2$)

Table 5-6 show the estimated parameters and prediction error values of the combined kinetics models for lightness values of chilli undergoing 80°C.

Table 5. The estimated parameter of the combined kinetics models for lightness values of chilli undergoing 80°C

Pre-treatment	k_0	k_1
Non-treated	-0.082	-0.078
Blanching	-0.061	-0.076
0.1% $Na_2S_2O_5$ and 1% $CaCl_2$	-0.137	-0.067
0.2% $Na_2S_2O_5$ and 1% $CaCl_2$	-0.111	-0.061
0.3% $Na_2S_2O_5$ and 1% $CaCl_2$	-0.109	-0.063

For redness and yellowness changes, the increase of both values was observed. Drying at higher temperature provided the higher change of these values. The rate of color changes seemed to be faster when above drying condition was applied. The redness values gradually increased as the enzymatic browning reaction occurred

The determination of color change during drying showed the power of image analysis process which had a capability to evaluate the change of product properties which gradually changed during drying. Therefore, this method can be used as a guideline for the on-line quality monitoring of many food processing.

Table 6. The prediction errors of the combined kinetics models for lightness values of chilli undergoing 80°C

Pre-treatment	*SE*		R^2
	k_0	k_1	
Non-treated	0.03	0.04	0.99
Blanching	0.05	0.06	0.99
0.1% $Na_2S_2O_5$ and 1% $CaCl_2$	0.03	0.04	0.98
0.2% $Na_2S_2O_5$ and 1% $CaCl_2$	0.04	0.05	0.99
0.3% $Na_2S_2O_5$ and 1% $CaCl_2$	0.02	0.02	0.98

4 Conclusions

The fastest drying time was found at 80°C. The rate moisture removal of pre-treated chilli was found to be decrease compared to non-treated samples (fresh) at all drying conditions due to the absorption of moisture and solutions during the blanching and soaking process. On the other hand, the chemicals pre-treatment would enhance the rate moisture removal since the Na and Ca salt components of mixed solution altered the chilli structures. However, the levels of concentration had no significant effect on the rate of moisture removal. In terms of modeling, the modified Page model was selected to describe the drying characteristics of chilli since it provided the lowest *SE* (0.01-0.02) and highest R^2 (0.97-0.99) for all drying conditions. The application of pre-treatment on the chilli could improve the change of dried chilli color. The *BI* was found to be lower when chilli had been blanched in hot water and then soaked in the mixed solution of 0.3% (w/w) $Na_2S_2O_5$ and 1% (w/w) $CaCl_2$ prior to drying. The higher drying temperature (80°C) resulted in the higher degradation of color change of dried product. Moreover, the combined kinetic model was the best fit model that could be used to predict the change of color during drying. It provided the lowest *SE* (0.02-0.04) and highest R^2 (0.98-0.99); the estimated kinetic parameters were in the range of -0.082 to -0.137 and -0.061 to -0.078 for k_0 and k_1, respectively.

Acknowledgement

The authors would like to thanks Agro-Industry Research Center and Faculty of Agro-Industry, King Mongkut's University of Technology North Bangkok for the financial support through the 2015 Research Funding.

References

1. M.A. Islam, S.S. Sharma, P. Sinha, M.S. Negi, B. Neog, S.B. Tripathi, Sci. Hort., **183**, 66-71(2015)
2. N. Toontom, W. Posri, S. Lertsiri, M. Meenune, Int. Food Res. J., **23**, 1, 289-299 (2016)
3. W. Phomkong, S. Soponronnarit, P. Thammarutwasik, Drying Tech, **28**, 12, 1466-1476 (2010)
4. R. Jinorose, R. Assawarachan, Int. Soc. Hort. Sci., 595-598 (2015)
5. H.S. Santos, M.A. Silva, Drying Tech., **26**, 12, 1421-1437 (2008)
6. E. Uribe, D. Marín, A. Vega-Gálvez, I. Quispe-Fuentes, A. Rodríguez, Food Chem., **190**, 559-565 (2016)
7. G.O. Sigge, C.F. Hansmann, E. Joubert, E., J. Food Qual. **22**, 4, 439-452 (1999)
8. A. Hossain, L. Woods, B.K. Bala, Intl. J. Food Sci&Tech. **42**, 11, 1367–1375 (2007)
9. C. Helou, P. Jacolot, C. Niquet-Léridon, P. Gadonna-Widehem, F.J. Tessier, Food Chem., **190**, 904-911 (2016)
10. P. Gupta, J. Ahmed, U.S. Shivhare, G.S.V. Raghavan, G.S.V. Drying Tech., **20**, 10, 1975–1987 (2002)
11. Vega-Gálvez, K. Di Scala, K. Rodríguez, R. Lemus-Mondaca, M. Miranda, J. López, M. Perez-Won, Food Chem., **117**, 4, 647-653 (2009)
12. G. Adiletta, P. Russo, W. Senadeera, M. Di Matteo, J. Food Eng., **172**, 9-18 (2016)
13. J. Carranza-Concha, M. Benlloch, M.M. Camacho, N. Martínez-Navarrete, Food Bio. Proc., **90**, 2, 243-248(2012)
14. S.M. Oliveira, I.N. Ramos, T.R.S. Brandão, C.L.M. Silva, J. Food Process. Preserv., **39**, 8, 2485-2496 (2015)
15. P. Komarnicki, R. Stopa, D. Szyjewicz, M. Młotek, Post. Bio. Tech., **114**, (36-44 (2016)
16. S. Savadkoohi, H. Hoogenkamp, K. Shamsi, A. Farahnaky, Meat Sci., **97**, 410-418 (2014)
17. Girolami, F. Napolitano, D. Faraone, G. Di Bello, A. Braghieri, Meat Sci, **96**, 610-616 (2014)
18. P.M. Szczypiński, A. Klepaczko, P. Zapotoczny, Comp. Elec. Agri., **110**, 1-8 (2015)
19. AOAC, *Official methods of analysis of AOAC International*, **17**, (2000)
20. S.D.F. Mihindukulasuriya, H.P.W. Jayasuriya, Agri. Eng. Int.: CIGR Journal, **15**, 154-166 (2013)

21. M.L.B. Ávila, C.L.M. Silva, J. Food Eng., **39**, 2, 161-166 (1999)
22. C.R. Chen, H.S. Ramaswamy, LWT - Food Sci &Tech., **35**, 5, 415-419 (2002)
23. M. Maskan, J. Food Eng., **48**, 2, 169-175 (2001)
24. G. Dadali, E. Demirhan, B. Özbek, Drying Tech., **25**, 10, 1703-171 2 (2007)
25. R. Saengrayap, A. Tansakul, G.S. Mittal, J Food Sci&Tech., **52**, 5, 2610-2621 (2015)
26. N.C. Acevedo, C. Schebor, P. Buera, Food Chem., **108**, 3, 900-906 (2008)
27. L. Manzocco, S. Calligaris, D. Mastrocola, M.C. Nicoli, C.R. Lerici, Trends in Food Sci& Tech., **11**, 9–10, 340-346 (2000)
28. S. Azeez, O. Lasekan, S. Jinap, R. Sulaiman, J. Food Sci.&Tech., **52**, 8050-8058 (2015)
29. G. Sacchetti, F. Ioannone, M. De Gregorio, C. Di Mattia, M. Serafini, D. Mastrocola, J. Food Eng., **169**, 44-52 (2016)

 DOI: 10.1051/matecconf/20166202010

Feasibility Study of Aseptic Homogenization: Affecting Homogenization Steps on Quality of Sterilized Coconut Milk

Chanthima Phungamngoen[1,a], Tippunsa Asawajinda[1], Rujira Santad[1], Wanticha Sawedboworn[1]

[1]*King Mongkut's University of Technology North Bangkok, 129 M. 21 Noenhom, Maung, Prachinburi, 25230 Thailand*

Abstract. Coconut milk is one of the most important protein-rich food sources available today. Separation of an emulsion into an aqueous phase and cream phase is commonly occurred and this leads an unacceptably physical defect of either fresh or processed coconut milk. Since homogenization steps are known to affect the stability of coconut milk. This work was aimed to study the effect of homogenization steps on quality of coconut milk. The samples were subject to high speed homogenization in the range of 5000-15000 rpm under sterilize temperatures at 120-140 °C for 15 min. The result showed that emulsion stability increase with increasing speed of homogenization. The lower fat particles were generated and easy to disperse in continuous phase lead to high stability. On the other hand, the stability of coconut milk decreased, fat globule increased, *L* value decreased and *b* value increased when the high sterilization temperature was applied. Homogenization after heating led to higher stability than homogenization before heating due to the reduced particle size of coconut milk after aggregation during sterilization process. The results implied that homogenization after sterilization process might play an important role on the quality of the sterilized coconut milk.

1 Introduction

Coconut milk is a liquid produced by soaking the white inner flesh of the seed of the coconut in water and straining. Coconut milk is commonly used in South East Asian cuisine for its aromatic property in curry and other gravy. The major components of coconut milk are water and fat which fat globules disperse throughout the water phase [1].

Sterilized coconut milk required high temperature for long processing time because it is an emulsion and low acid food (pH~7). Separation of an emulsion into an aqueous phase and cream phase is commonly occurred and this leads an unacceptably physical defect of either fresh or processed coconut milk [2]. Moreover, the protein denaturation occurred leading to curding in liquid system occured when temperature reaching to 80 °C [3] which can affect the stability of the milk. Previous research works have demonstrated that fat particle size, dispersion, homogenization and temperature had significant effects on a stability of coconut milk [2, 4, 5]. However, after heated coconut milk product with high temperature leads to much quality degradation, i.e., separation and brown color in product. In the sterilization process of coconut milk, there is no homogenization again therefore had the opportunity to make the product less stable.

Since homogenization steps are known to affect the stability of coconut milk, it is interesting to study the effects of homogenization steps on the stability, fat particle size and color of sterilized coconut milk. In this study, attempt was made to feasibility study of aseptic homogenization. The opjective of this work was to study the effect of homogenization steps on somes quality of coconut milk.

2 Materials and Methods

2.1 Sample preparation

Coconut milk without added water from a local market was used in the experiments. The initial fat content of coconut milk was determined using Rose–Gottlieb method [6]. The fresh coconut milk was diluted by distilled water to obtain the coconut milk with 20% w/v fat. Emulsifier stabilizer mixture of Montanox 60 (Adinop, Thailand) and CMC (Thai Food and Chemical, Thailand) with a ratio of 0.6:0.6 %w/v [3] were added while the samples were heated and stirred continuously until sample temperature reached 70 °C.

2.2 Homogenization steps

2.3.1 Effect of homogenization speed and heating temperature

Applied the single step of homogenization (Polytron, model PT-MR 3100, Germany) at speeds of 5,000,

[a] Corresponding author: chanthimap@kmutnb.ac.th

© The Authors, published by EDP Sciences. This is an open access article distributed under the terms of the Creative Commons Attribution License 4.0 (http://creativecommons.org/licenses/by/4.0/).

10,000 and 15,000 rev/min for 2 minutes (1st steps). Then heat sample in oil bath at a temperature of 120, 130 and 140 °C for 15 minutes.

2.3.2 Effect of homogenization steps

Applied the two steps of homogenization at speeds of 5,000, 10,000 and 15,000 rev/min for 2 minutes (1st step) and heating sample at 120, 130 and 140 °C for 15 minutes after that homogenization again at speeds of 500, 1,000 and 5,000 rev/min for 1 minutes (2nd step).

2.3 Determine quality of coconut milk

2.3.1 Emulsion stability (%ES)

To evaluate the stability, the sterilized samples were kept at room temperature (~ 30 °C) for 3 days and determined the stability of emulsion following the method described by Phungamngoen and others [4]. The emulsion stability can be calculated by following formula:

$$\%\,ES = \frac{\text{Height of cream phase}}{\text{Height of whole coconut milk}} \times 100 \quad (1)$$

2.3.2 Droplet size

A few drops of oleoresin dye were added to 10 mL coconut milk sample which was stirred for at least 1 min to disperse the dye. A few drops of the sample were transformed to the slide and a cover slip was placed over the sample. An optical standard microscope (Olympus Model CH30, Japan) was used to determine the fat structure at a magnification of 400× and photographs were taken from typical fields.

2.3.3 Color

Color of samples was measured in a Hunter colorimeter (Hunter Lab, Model Colorflex 45/0, Virginia) Reflection spectra were registered and Hunter Lab color parameters for 10° vision angle and D65 illuminant (*L* : brigthness, *a* : redness, *b* : yellowness) were calculated. Measurements were done in triplicate on each of two subsamples.

2.4 Statistical analysis

All data were analyzed using the analysis of variance (ANOVA) and were presented as mean values with standard deviations. Differences between mean values were established using Duncan's new multiple range test; values were considered at a confidence level of 95%. SPSS (version 10.0 statistics software, SPSS Inc., Chicago, IL) was used to perform all statistical calculations.

3 Results and Discussion

3.1. Effect on emulsion stability

Table 1 shows emulsion stability (% ES) of coconut milk as affect of homogenizing steps and sterilizing temperature. The result showed that the non homogenized sample had lower stability than homogenized ones. The % ES value of coconut milks increased as the homogenization speed increased. Increaseing heating temperature exhibited low stabity of product. This observation was consistent with a previous work which studied the effect of homogenizing pressure and sterilizing condition on quality of canned high fat coconut milk [2].

Homogenization after heating (2 steps homogenization) leading to higher % ES at the same hogenization speed when compare with hogenization only one step. It was observed that increasing 2nd step of homogenization speed did not have significant effect on %ES of coconut milk.

Table 1. Effect of homogenizing steps and sterilizing temperature on emulsion stability (%ES) of coconut milk.

Homogenization steps	Temperature		
	120 °C	130 °C	140 °C
Non-homogenized sample (control)	70.12^{e} (3.01)	68.33^{e} (1.97)	66.41^{e} (4.20)
5,000 rev/min	78.85^{cde} (5.77)	78.31^{cde} (6.31)	78.23^{cde} (1.31)
10,000 rev/min	82.31^{cd} (2.31)	81.92^{cd} (2.69)	79.20^{cde} (2.28)
15,000 rev/min	82.28^{cd} (4.80)	83.86^{cd} (0.14)	80.77^{bc} (3.85)
5,000 rev/min-Heat-500 rev/min	85.54^{bcd} (2.46)	82.92^{cd} (3.00)	78.85^{cd} (1.92)
5,000 rev/min-Heat-1000 rev/min	86.31^{bc} (1.69)	83.34^{cd} (2.35)	77.69^{cd} (0.77)
5,000 rev/min-Heat-5000 rev/min	86.54^{bc} (1.92)	82.57^{cd} (4.11)	79.43^{cd} (0.73)
10,000 rev/min-Heat-500 rev/min	87.98^{bc} (0.56)	89.43^{ab} (1.54)	90.38^{ab} (1.92)
10,000 rev/min-Heat-1000 rev/min	89.75^{ab} (2.25)	89.77^{ab} (3.14)	90.00^{b} (0.40)
10,000 rev/min-Heat-5000 rev/min	88.14^{b} (3.53)	89.47^{ab} (2.81)	90.79^{ab} (2.22)
15,000 rev/min-Heat-500 rev/min	92.11^{a} (1.13)	91.01^{ab} (0.10)	90.37^{ab} (4.92)
15,000 rev/min-Heat-1000 rev/min	92.25^{a} (1.48)	91.99^{ab} (1.08)	90.84^{ab} (3.55)
15,000 rev/min-Heat-5000 rev/min	92.00^{a} (0.33)	90.80^{ab} (2.00)	90.51^{ab} (2.97)

Values in the same column with different superscripts mean that the values are significantly different ($p<0.05$).

3.2 Effect on fat particle size

The effect of homogenization speed on fat particle size of coconut milk were conducted using optical standard microscope (Figure 1-4). During the homogenization, the high shear forces acted on dispersed phase to reduce droplet size [7]. Small fat globule sizes were obtained at higher homogenization speed. Reduction in the fat

 DOI: 10.1051/matecconf/20166202010

particle diameters resulted in an increase in %ES value. When the homogenized coconut milk samples were subjected to heat treatments, small fat globules formed irregular rearrangement of aggregates (Figure 2). Naturally, coconut milk composes of fat globules surrounded by the aqueous protein solution [2, 8, 9].

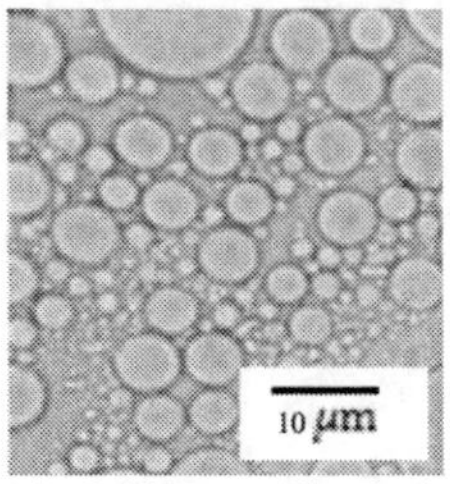

Figure 1. Micrographs (×400 magnification) of fresh coconut milk sample.

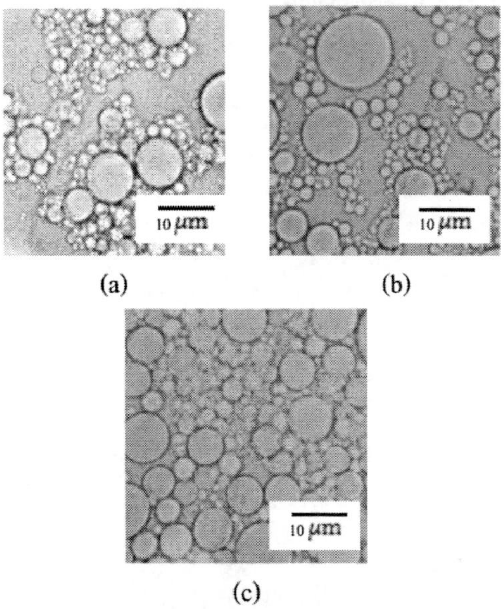

Figure 2. Micrographs (×400 magnification) of coconut milk samples passing homogenization at 5000 rev/min with different heating temperature; (a) 120 °C, (b) 130 °C and (c) 140 °C.

Figure 3 shows micrographs of coconut milk samples passing different conditions. At the same heating temperature, higher 2nd steps of homogenization speeds resulted in the same size of fat particle size. There is no benefit in increasing stability of coconut milk product when it used 2 steps homogenization. This was due to the ideal second-step homogenization should be between 10 and 15% of the total homogenization speed. It has been demonstrated with emulsion that passage through the second-stage valve, itself, does not significantly alter the condition of the emulsion. Sampling the emulsion both before and after the second-stage did not reveal any substantial change in emulsion quality [10, 11]. The temperature differences between the homogenization speed at higher heating temperatures gave larger fat particle size than that at a lower temperature (data not shown).

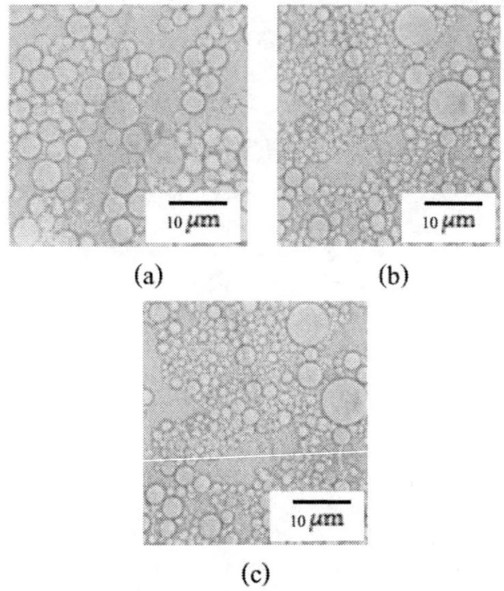

Figure 3. Micrographs (×400 magnification) of coconut milk samples passing different conditions; (a) 5,000 rpm-heat 120 °C-500 rpm, (b) 5,000 rpm-heat 120 °C-1,000 rpm and (c) 5,000 rpm-heat 120 °C-5,000 rpm.

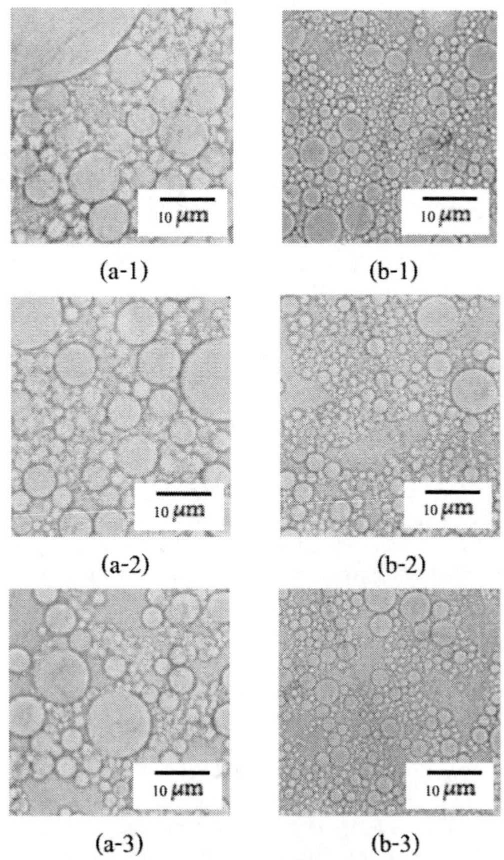

Figure 4. Micrographs (×400 magnification) of coconut milk samples passing different conditions; homogenized before heating at 120 °C (a-1) 5,000 rpm, (a-2) 10,000 rpm, (a-3) 15,000 rpm; homogenized before and after heating at 120 °C (b-1) 5,000 rpm-heat-500 rpm, (b) 10,000 rpm-heat-500 rpm and (c) 15,000 rpm-heat -500 rpm.

Figure 4 shows exemplifies comparing effect of one step (Figure 4: a-1, a-2 and a-3) and two steps of homogenization speed (Figure 4: b-1, b-2 and b-3). It was observe that two steps of homogenization speed led to smaller fat particle size than one step homogenization speed. This might be due to the fact that during heating the samples with high temperature gained fat globules tended to form aggregates [2]. At this temperature, the protein denaturation occurred leading to curding in liquid system [3]. Therefore, applying homogenization speed after heating was applicable for improving the stability of coconut milk samples. The micrographs supported the results from the emulsion stability studies that increasing in %ES of heated treated homogenized coconut milk was caused from the change in microstructure.

3.3 Effect on color

It is seen in Table 2-4 that the applied homogenization speed and heating temperature showed significant effect on the color of coconut milk. In general, lowering the speed of homogenization and lower heating temperature helped to prevent browning reaction and color degradation in coconut milk [4]. It was obvious that high homogenization speed resulted in an increase in the *L* value and a decrease in the *b* value. However, it was observed in this work that increasing 2nd steps of homogenization speed did not have a significant effect on the color of coconut milk.

Table 2. Effect of homogenizing steps on color of sterilized coconut milk at 120 °C.

Homogenization steps	Color		
	L	*a*	*b*
Non-homogenized sample	77.72^{f} (0.03)	0.87^{a} (0.03)	7.73^{b} (0.02)
5,000 rev/min	80.38^{e} (0.22)	0.46^{b} (0.03)	7.38^{a} (0.03)
10,000 rev/min	82.52^{d} (0.33)	-0.74^{g} (0.12)	5.59^{b} (0.23)
15,000 rev/min	85.51^{c} (0.10)	0.30^{def} (0.02)	4.93^{cd} (0.07)
5,000 rev/min-Heat-500 rev/min	85.83^{bc} (0.24)	0.23^{f} (0.02)	5.04^{cd} (0.10)
5,000 rev/min-Heat-1000 rev/min	86.66^{b} (0.06)	0.28^{def} (0.02)	5.14^{cd} (0.01)
5,000 rev/min-Heat-5000 rev/min	86.65^{ab} (0.42)	0.33^{cde} (0.02)	5.00^{cd} (0.17)
10,000 rev/min-Heat-500 rev/min	86.40^{b} (0.17)	0.34^{cde} (0.02)	4.81^{cde} (0.14)
10,000 rev/min-Heat-1000 rev/min	86.56^{b} (0.35)	0.39^{c} (0.02)	4.91^{bc} (0.14)
10,000 rev/min-Heat-5000 rev/min	87.87^{a} (0.13)	0.27^{def} (0.04)	4.59^{cd} (0.10)
15,000 rev/min-Heat-500 rev/min	88.07^{a} (0.04)	0.24^{f} (0.03)	4.56^{cd} (0.06)
15,000 rev/min-Heat-1000 rev/min	88.36^{a} (0.22)	0.23^{f} (0.03)	4.48^{cd} (0.08)
15,000 rev/min-Heat-5000 rev/min	88.23^{a} (0.31)	0.25^{ef} (0.03)	4.72^{c} (0.07)

Values in the same column with different superscripts mean that the values are significantly different (p<0.05).

Among the studied two steps homogenization speed had higher *L* and lower *b* values compared to the one step homogenization speed. Smaller droplets were produced when the higher homogenization speeds were applied. The reflectance increased with increasing droplet concentration and decreasing droplet size [12, 13].

4 Conclusion

The effects of homogenization steps (before sterilization process and homogenization before and after sterilization) on the stability, fat particle size, viscosity and color of sterilized coconut milk were investigated. The result showed that emulsion stability increase with increasing speed of homogenization because the higher speed of homogenization. High sterilization temperature led to *L* value decreased and *b* value increased. Homogenization after heating gave higher stability than homogenization before heating. The results implied that homogenization after sterilization process might play an important role on the quality of the sterilized coconut milk. The results could be used as a guidance for producing sterilized coconut milk with aseptic homogenizer.

Table 3. Effect of homogenizing steps on color of sterilized coconut milk at 130 °C.

Homogenization steps	Color		
	L	*a*	*b*
Non-homogenized sample	75.56^{f} (0.14)	1.01^{a} (0.03)	8.16^{a} (0.02)
5,000 rev/min	78.38^{c} (0.22)	0.46^{b} (0.03)	5.21^{b} (0.10)
10,000 rev/min	80.60^{b} (0.37)	-0.55^{d} (0.08)	5.04^{bc} (0.07)
15,000 rev/min	82.21^{ab} (0.18)	-0.45^{c} (0.04)	4.93^{cb} (0.10)
5,000 rev/min-Heat-500 rev/min	81.97^{ab} (0.36)	-0.52^{cd} (0.08)	5.14^{b} (0.01)
5,000 rev/min-Heat-1000 rev/min	81.57^{b} (0.22)	-0.58^{de} (0.05)	5.00^{bc} (0.17)
5,000 rev/min-Heat-5000 rev/min	82.90^{a} (0.07)	-0.55^{d} (0.03)	4.81^{c} (0.14)
10,000 rev/min-Heat-500 rev/min	82.23^{ab} (0.38)	-0.57^{de} (0.09)	4.91^{bc} (0.14)
10,000 rev/min-Heat-1000 rev/min	82.84^{a} (0.31)	-0.54^{d} (0.04)	4.60^{d} (0.10)
10,000 rev/min-Heat-5000 rev/min	83.16^{a} (0.06)	-0.44^{c} (0.03)	4.48^{de} (0.08)
15,000 rev/min-Heat-500 rev/min	83.12^{a} (0.80)	-0.59^{de} (0.09)	4.56^{e} (0.06)
15,000 rev/min-Heat-1000 rev/min	83.37^{a} (0.32)	-0.51^{cd} (0.10)	4.72^{cd} (0.07)
15,000 rev/min-Heat-5000 rev/min	83.87^{a} (0.89)	-0.54^{cde} (0.05)	4.67^{cd} (0.10)

Values in the same column with different superscripts mean that the values are significantly different (p<0.05).

 DOI: 10.1051/matecconf/20166202010

Table 4. Effect of homogenizing steps on color of sterilized coconut milk at 140 °C.

Homogenization steps	Color		
	L	*a*	*b*
Non-homogenized sample	72.07[d] (0.11)	1.08[a] (0.03)	8.97[a] (0.02)
5,000 rev/min	76.36[c] (0.07)	-0.50[de] (0.03)	6.45[b] (0.03)
10,000 rev/min	80.07[b] (0.30)	-0.44[d] (0.08)	5.77[c] (0.03)
15,000 rev/min	80.81[ab] (0.30)	-0.51[d] (0.05)	5.30[d] (0.11)
5,000 rev/min-Heat-500 rev/min	80.03[b] (0.49)	-0.50[de] (0.08)	5.48[d] (0.10)
5,000 rev/min-Heat-1000 rev/min	80.21[b] (0.38)	-0.46[d] (0.07)	5.37[d] (0.13)
5,000 rev/min-Heat-5000 rev/min	79.26[b] (0.18)	-0.52[de] (0.05)	5.44[d] (0.04)
10,000 rev/min-Heat-500 rev/min	81.01[ab] (0.18)	-0.47[d] (0.03)	5.62[c] (0.09)
10,000 rev/min-Heat-1000 rev/min	81.53[a] (0.34)	-0.03[c] (0.01)	5.58[c] (0.06)
10,000 rev/min-Heat-5000 rev/min	81.54[a] (0.55)	-0.02[c] (0.05)	5.23[d] (0.02)
15,000 rev/min-Heat-500 rev/min	81.35[a] (0.13)	0.05[b] (0.07)	5.25[d] (0.04)
15,000 rev/min-Heat-1000 rev/min	81.23[a] (0.17)	0.06[b] (0.04)	5.26[d] (0.03)
15,000 rev/min-Heat-5000 rev/min	81.36[a] (0.07)	-0.50[d] (0.03)	5.21[c] (0.06)

Values in the same column with different superscripts mean that the values are significantly different ($p<0.05$).

Acknowledgements

The authors express their sincere appreciation to the King Mongkut's University of Technology North Bangkok for financially supporting the study in the form of the General Researcher Grant. Contract no. KMUTNB-GEN-57-39.

References

1. S.P. Ng, O.M. Lai, F. Abas, H.K. Lim, C.P. Tan, Food. Res. Int. **64**, 919-930 (2014)
2. N. Chiewchan, C. Phungamngoen, S. Siriwattanayothin, J. Food Eng. **73(1)**, 38-44 (2006)
3. N. Tangsuphoom, J.N. Coupland, Food Hydrocoll. **23**, 1792-1800 (2009)
4. S.N. Raghavendra, K.S.M.S. Raghavarao, J Food Eng. **94**, 341-347 (2010)
5. K. Jirapeangtong, S. Siriwatanayothin, N. Chiewchan, J. Food Eng. **87(3)**, 422-427 (2008)
6. Association of official analytical chemistry (AOAC). *Official method of analysis*. 15th ed. The association of official agricultural chemists. Virginia, (1990).
7. Z.Y. Yu, S.W. Jiang, X.M. Cao, S.T. Jiang, L.J. Pan, J. Food Eng. **177**, 1-8 (2016).
8. T. Peamprasart, N. Chiewchan, J. Food Eng. **77**, 653-658 (2006)
9. P.R. Borwankar, A. Gonsalves, T.D. Wasan, A. Nikolov, W. Xu, J. Food Sci. **63**, 183-188 (1998)
10. N. Tangsuphoom, J.N. Coupland, J. Food Sci. **70**, E466-E470 (2005)
11. P. Dhankhar, Int. Org. Sci. Res. J.Eng. **54**, 2278-8719 (2014)
12. D.J. McClements, *Food emulsions: Principles, practice, and techniques*. CRC Press LLC, 161-265 (2005)
13. W. Chantrapornchai, F. Clydesdale, D.J. McClements, *Colloids and Surfaces A: Physicochemical and Engineering Aspects*, 155, 373-382 (1999)

MATEC Web of Conferences **62**, 02011 (2016) DOI: 10.1051/matecconf/20166202011
ICCFE 2016

The Organic Food Purchase Behaviour: Using ECT to Explore Customers' Satisfaction

Hui Hsin Huang[1,a]

[1]*Aletheia University, New Taipei City, Taiwan*

Abstract. This paper bases on expectation confirmation theory (ECT) to propose a stochastic model of predicting organic food purchase behaviour. In the model, the expectation of pre-purchase and outcome performance of post-purchase are considered as relative variables. We consider these two variables as two different normal distribution and their probability density function (pdf) and cumulative distribution function(cdf) will be demonstrated to display their correlation. Then the joint density can be computed to show the results of positive or negative disconfirmation. This study also conducts an empirical data for parameter estimation and model application. We use the 0.5 as threshold level of probability to test the fitness of real overall satisfaction and the calculation results of probability. The results of chi-square testing show good fitness. Finally, the conclusion is made.

1 Introduction

The growing trend in the demand for organic food or product has received increased attention by researchers [1-5]. There are many studies [2, 4-10] to investigate the characteristics of organic food producer [7, 9] and consumption [5-12] on both supply and demand [5, 11] side. This research focus on the consumption topic to explore the customer purchase behaviour of organic food product. In this topic, the customers' satisfaction is one of the curial factors to predict the customer purchase tendency [13, 14].

The customer's satisfaction results from a comparison between expectation and outcome performance [14-16]. The comparison of prior expectations with observed performance cases disconfirmation. The disconfirmation is positive, when the outcome performance of product is higher than the expectation. In the contrary, the disconfirmation is negative, if the outcome performance of product is lower than the expectation. The positive disconfirmation reflects satisfaction and negative disconfirmation demonstrates dissatisfaction [15, 17-20]. Oliver [21-22] proposed expectation confirmation theory (ECT) to describe these series process.

Bases on ETC, expectation and outcome performance are two important variables which can influence the judgment of satisfaction measure. Thus, this research proposes two stochastic models of expectation and outcome performance. And through combining these two models into a full distribution, the satisfaction can be predicted.

The remainder of this paper is organized as follows: Firstly, the literature review of ETC is demonstrated in the next section. Secondly, we introduce the expectation and outcome performance models. The probability density function (pdf) and cumulative distribution function (cdf) will be demonstrated in this section. Thirdly, the results of the parameter estimation and empirical test are report. Finally, the conclusion is made.

2 The Research of ECT

ECT is constructed by Oliver [21-22] and is used popularly to discuss the customer satisfaction process. In the process of satisfaction judgments, first buyers may experience from expectations of the specific product or service prior to purchase. Second, consumption reveals a perceive performance level of product which is influence by expectations if difference between actual performance and expectations is perceived as being small. Hence, perceived performance may increase or decrease directly with expectations as indicated by the arrow drawn from expectations to perceived performance. Third, perceived performance may either confirm or disconfirm prepurchase expectation [16-17, 19, 23-24].

Thus, in this theory, satisfaction is a function which includes the difference between observed outcome performance of post-purchase and prior expectations about the outcome's performance of pre-purchase [17,25-7]. The expectations provide a baseline or anchor level of satisfaction. Some research find that the perceived performance which falls short of expectations has greater impact on satisfaction and repurchase intentions than performance which exceeds expectations [27-28]. Thus, the expectation of pre-purchase and outcome performance of post-purchase are relative variables. We consider this characteristic of correlation ship into model development.

[a] Corresponding author: hoyasophia@gmail.com

© The Authors, published by EDP Sciences. This is an open access article distributed under the terms of the Creative Commons Attribution License 4.0 (http://creativecommons.org/licenses/by/4.0/).

3 The Model

In the customer satisfaction of organic food purchase behaviour, we base on ECT to construct the stochastic expectation model.

3.1 The expectation

We denote w is the volume of one's expectation of organic food purchase which is a random variable with its probability density function (p.d.f.) as

$$f_W(w)=\frac{1}{w\sigma_a\sqrt{2\pi}}exp\left[-\frac{(log\,w-\mu_a)^2}{2\sigma_a^{\ 2}}\right],\quad w>0 \tag{1}$$

$$F_W(w)=\Phi\left[\frac{(log\,w-\mu_a)}{\sigma_a}\right],\quad W>0 \tag{2}$$

3.2 The outcome performance

The outcome performance of organic food is considered as an random variable k with its probability density function(p.d.f.)as

$$f_K(k)=\frac{1}{k\sigma_b\sqrt{2\pi}}exp\left[-\frac{(log\,k-\mu_b)^2}{2\sigma_b^{\ 2}}\right],\quad k>0 \tag{3}$$

$$F_K(k)=\Phi\left[\frac{(log\,k-\mu_b)}{\sigma_b}\right],\quad K>0 \tag{4}$$

3.3 The full model

The models of expectation and outcome performance of organic food can be combined into a joint distribution as

$$f_{WK}(w,k)$$
$$=\frac{1}{2\pi wk\sigma_a\sigma_b\sqrt{1-\gamma^2}}\times exp\left[-\frac{1}{2(1-\gamma^2)}\right]\times$$
$$\left[\left(\frac{log\,w-\mu_w}{2\sigma_a}\right)^2-2\rho\left(\frac{log\,w-\mu_w}{2\sigma_a}\right)\left(\frac{log\,k-\mu_k}{2\sigma_b}\right)+\left(\frac{log\,k-\mu_k}{2\sigma_b}\right)^2\right] \tag{5}$$

$$\gamma=\frac{\operatorname{cov}(w,k)}{\sigma_a\sigma_b} \tag{6}$$

$$f_{k|w}(k|w)=\frac{1}{k\sigma_{b|a}\sqrt{2\pi}}exp\left[-\frac{(log\,k-\mu_{b|a})^2}{2\sigma_{b|a}^{\ 2}}\right] \tag{7}$$

where $\mu_{b|a}=\mu_b-\left[\gamma\frac{\sigma_b}{\sigma_a}(log\,w-\mu_a)\right]$

and $\sigma_{b|a}=\sigma_b\sqrt{(1-\gamma)^2}$.

3.4 The model of ECT

In ETC, if the outcome performance of product is higher than the expectation, then the disconfirmation is positive. On the other hand, if the outcome performance is lower than the expectation, then negative disconfirmation happens. The probability of these two scenarios will be demonstrated as bellow.

3.4.1 Positive disconfirmation

The customer feels satisfaction if the disconfirmation is positive. According equation (1)-(4), the positive disconfirmation can be demonstrated as

$$P(K>w_0|W=w_0)$$
$$=1-\int_0^{w_0}f_K(k|w_0)\,dk \tag{8}$$
$$=1-\int_0^{w_0}\frac{1}{k\sigma_{b|a}\sqrt{2\pi}}exp\left[-\frac{(log\,k-\mu_{b|a})^2}{2\sigma_{b|a}^{\ 2}}\right]dk$$

3.4.2 Negative disconfirmation

The customer feels dissatisfaction if the disconfirmation is negative. According equation (1)-(4), the negative disconfirmation can be demonstrated as

$$P(K<w_0|W=w_0)$$
$$=\int_0^{w_0}f_K(k|w_0)\,dk \tag{9}$$
$$=\int_0^{w_0}\frac{1}{k\sigma_{b|a}\sqrt{2\pi}}exp\left[-\frac{(log\,k-\mu_{b|a})^2}{2\sigma_{b|a}^{\ 2}}\right]dk$$

4 The Empirical Data Analysis

We use a real database of an organic food chain store. There are 2566 customers in this database.

4.1 The analysis process

The analysis process is following(see figure1):

Step1:

First, we separate these samples into two parts. One is for parameters estimation. Another is for model application. There are 1280 samples for parameters estimation and 1286 samples for model validation.

Step2:

In step2, we use MLE (maximum likelihood estimate) to estimate the parameters of full model. Then we put the results of parameters estimation into the proposed model.

Step3:

In this step, we use the full model without parameters to calculate the probability of customer satisfaction and compare these results with the real data of satisfaction.

Finally, we use chi-square test to fitness between the proposed model and the real data of customer satisfaction.

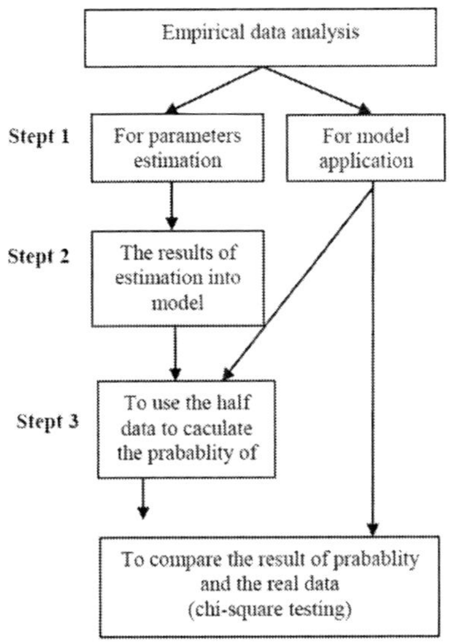

Figure 1. The Analysis Process.

The results of each step will be demonstrated in section 4.2 and 4.3. Section 4.2 shows the result in step2 and section 4.3 shows the result in step3.

4.2 The parameters estimation

The MLE (maximum likelihood estimate) is used to estimate the parameters in our model. Let L denote the total likelihood and n is the sample size.

$$L(\mu_a,\mu_b,\sigma_a^2,\sigma_b^2,\rho) = \prod_{i=1}^{n} f_{WK}(\mu_a,\mu_b,\sigma_a^2,\sigma_b^2,\rho) = \left(\frac{1}{2\pi wk\sigma_a\sigma_b\sqrt{1-\rho^2}}\right)^n \times \left\{exp\, n\left[-\frac{1}{2(1-\rho^2)}\right]\right\} \times \left[\left(\frac{\log w-\mu_a}{2\sigma_a}\right)^2 - n\rho\left(\frac{\log w-\mu_a}{\sigma_a}\right)\left(\frac{\log k-\mu_b}{\sigma_b}\right) + \left(\frac{\log k-\mu_b}{2\sigma_b}\right)^2\right] \quad (10)$$

We differentiate $L(\mu_a,\mu_b,\sigma_a^2,\sigma_b^2,\rho)$ respectively regarding μ_w, σ_w^2, μ_k, σ_k^2 and ρ set them equal to zero. That is,

$$\begin{cases} \frac{\partial\mu_a}{\partial} L(\mu_a,\mu_b,\sigma_a^2,\sigma_b^2,\rho)=0 \\ \frac{\partial\mu_b}{\partial} L(\mu_a,\mu_b,\sigma_a^2,\sigma_b^2,\rho)=0 \\ \frac{\partial\sigma_a^2}{\partial} L(\mu_a,\mu_b,\sigma_a^2,\sigma_b^2,\rho)=0 \\ \frac{\partial\sigma_b^2}{\partial} L(\mu_a,\mu_b,\sigma_a^2,\sigma_b^2,\rho)=0 \\ \frac{\partial\rho}{\partial} L(\mu_a,\mu_b,\sigma_a^2,\sigma_b^2,\rho)=0 \end{cases} \quad (11)$$

We take the solutions of (11) as MLE for μ_w, σ_w^2, μ_k, σ_k^2 and ρ. The details of left-hand side of (11) are given in the Appendix.

Finally, we use our empirical data which has been described in previous section to estimate these five parameters.The results are in Table 1.

Table 1. The Parameters Estimate.

μ_w	σ_a^2	μ_k	σ_b^2	ρ
4.001	0.857	4.897	2.345	0.521

Then, we use the result of parameters estimation to the full model. It shows in equation(12).

$$f_{WK}(w,k) = wke^{-0.686} \times \left[0.0271(\log w-\mu_w)^2 - 0.017(\log w-\mu_w)(\log k-\mu_k) + 0.872(\log k-\mu_k)^2\right] \quad (12)$$

4.3 The model application

We use the 0.5 as threshold level of probability to test the fitness of real overall satisfaction and the calculation results of probability. The results of chi-square testing are shown in table 2.

Table 2. The results of chi-square testing.

	satisfaction	**dissatisfaction**
Fitness rate	93.67%	92.26%
X^2=18589 (p<.05)	---	---

The rate of fitness is higher than 90% in all scenarios and the chi-square testing are significant. Then prediction analysis shows good fitness between the model and the real data.

5 Conclusion

This research is to propose a stochastic model based on ECT to predict customer satisfaction of organic food. According to the empirical data application, the results of chi-square testing show good fitness of this model. Thus, this model can be used to explore the purchase behaviour of organic food and its impact variables such as expectation evaluation to control the level of making balance or disconfirmation of satisfaction. The company manager or researchers of organic food can decompose the ECT model into prepurchase expectation and postpurchase evaluation to discuss each part of positive or negative disconfirmation according our mathematic model to maximum the profit. For example, managers can manipulate the expectation to be lower than postpurchase performance to achieve the positive disconformation(satisfaction) or to enhance the postpurchase performance higher than expectation to make result of positive disconformation.

The proposed model provides the view of point on mathematical concept to discuss the organic food research. In the future, other impact factors can be involved to explain the purchase behaviour. The researchers can also consider other form of distribution such as log normal or exponential to test mutli-model fitness. For the further research, the relationship of repurchase expectation and postpurchase performance can be consider no relative or nonlinear relative to find different effect of ECT model.

Acknowledgements

The author would like to thank the National Science Council of the Republic of China, Taiwan (NSC 99-2410-H-156-013) and Aletheia University(AU-AR-104-014) for financially supporting this research.

References

1. C. P. Bravoa, A. Cordtsa, B. Schulzeb, A. Spillera, Food Qual. Prefer. **28**,1 (2013)
2. R. M. Ness, M. Ness, M. Brennan, E. Oughton, C. Ritson, E. Ruto, Food Qual. Prefer. **21**, 1 (2010)
3. D. Pearsona , J. Henryks, H. Jones, Renew Agr. Food Syst. **26**, 2 (2011)
4. J. Chen, A. Lobo , N. Rajendran, International Journal of Consumer Studies **38**, 4 (2014).
5. U. Hjelmar, Appetite **56**,2 (2011)
6. P. Justin, R. Jyoti, Journal of Consumer Marketing **29**, 6 (2012)
7. L. Marian, , P. Chrysochou , A. Krystallis , J.Thøgersen, Food Quality and Preference **37**(2014)
8. A. Khanlari, Strategic Customer Relationship Management in the Age of Social Media .
9. P. Kriwy, R. A. Mecking, International Journal of Consumer Studies **36**,1 (2012)
10. M. Yazdanpanah, M. Forouzani, Journal of Cleaner Production **107**, 16(2015)
11. S. K. Pandey , A. Khare, Journal of Indian Business Research 7 ,4 (2015)
12. J. Hwang, Journal of Retailing and Consumer Services **28**(2016)
13. S. Halilovica, M. Cicica, Behaviour & Information Technology **32**,4 (2013)
14. C. L. Hsu, J.C. Lin,Electronic Commerce Research and Applications **24**,3 (2015)
15. Y. Sun, L. Liu, X. Peng, Y. Dong, Electronic Markets **24**, 1(2014)
16. C. L. Hsu, J.C. Lin,Electronic Commerce Research and Applications **24**,3 (2015)
17. P. K.Kopalle , D. R. Lehmann, J. M. R. **38**, 3 (2001)
18. Y. M. Cheng, Information Technology & People, **27**, 3 (2014)
19. S. Halilovica, M. Cicica, Behaviour & Information Technology **32**,4 (2013)
20. I. L. Wu, International Journal of Information Management **33**, 1(2013)
21. R. L. Oliver, J. Appl. Psychol. 62, 4 (1977)
22. R. L. Oliver, J. M. R. **17**,4 (1980)
23. W. E. Anderson, W. M. Sullivan, Marketing Sci. **12**, 2 (1993)
24. H. H. Huang, C. H. Liu, Taiwan J. M. S. **10**, 1 (2014)
25. W. E. Anderson, W. M. Sullivan, Marketing Sci. **12**, 2 (1993)
26. Y. Sun, L. Liu, X. Peng, Y. Dong, Electronic Markets **24**, 1(2014)
27. C. L. Hsu, J.C. Lin,Electronic Commerce Research and Applications **24**,3 (2015)
28. S. Walther, R. Eden , G. Phadke, E. Torsten, Lecture Notes in Business Information Processing **198**(2015)

Appendix

1.

$$\frac{\partial \mu_a}{\partial} L(\mu_a, \mu_b, \sigma_a{}^2, \sigma_b{}^2, \rho)$$

$$= \left(\frac{1}{2\pi wk\sigma_a\sigma_b\sqrt{1-\rho^2}} \right)^n \times$$

$$\left\{ exp\, n\left[-\frac{1}{2(1-\rho^2)} \right] \right\}$$

$$\times \left[\left(\frac{\mu_a - \log w}{\sigma_a} \right) + n\rho \left(\frac{\log k - \mu_b}{\sigma_a\sigma_b} \right) \right]$$

2.

$$\frac{\partial \mu_b}{\partial} L(\mu_a, \mu_b, \sigma_a{}^2, \sigma_b{}^2, \rho)$$

$$= \left(\frac{1}{2\pi wk\sigma_a\sigma_b\sqrt{1-\rho^2}} \right)^n \times$$

$$\left\{ exp\, n\left[-\frac{1}{2(1-\rho^2)} \right] \right\}$$

$$\times \left[n\rho \left(\frac{\log w - \mu_a}{\sigma_a\sigma_b} \right) + \left(\frac{\mu_b - \log k}{\sigma_b} \right) \right]$$

3.

$$\frac{\partial \sigma_a{}^2}{\partial} L(\mu_a, \mu_b, \sigma_a{}^2, \sigma_b{}^2, \rho)$$

$$= \left(\frac{1}{2\pi wk\sigma_b\sqrt{1-\rho^2}} \right)^n \times$$

$$\left\{ exp\, n\left[-\frac{1}{2(1-\rho^2)} \right] \right\}$$

$$\times \left[\begin{array}{l} n\rho(1+n)\left(\frac{\log w - \mu_a}{\sigma_a{}^{n+2}} \right)\left(\frac{\log k - \mu_b}{\sigma_b} \right) \\ -\frac{(\log w - \mu_a)^2(8+4n)}{\sigma_a{}^{n+3}} - \frac{n(\log k - \mu_b)^2}{4\sigma_b{}^2\sigma_a{}^{n+1}} \end{array} \right]$$

4.

$$\frac{\partial\sigma_b^{\ 2}}{\partial} L(\mu_a,\mu_b,\sigma_a^{\ 2},\sigma_b^{\ 2},\rho)$$

$$=\left(\frac{1}{2\pi wk\sigma_a\sqrt{1-\rho^2}}\right)^n \times$$

$$\left\{exp\, n\left[-\frac{1}{2(1-\rho^2)}\right]\right\}$$

$$\times\left[\begin{array}{l} n\rho(1+n)\left(\frac{log\, w-\mu_a}{\sigma_a}\right)\left(\frac{log\, k-\mu_b}{\sigma_b^{\ n+2}}\right) \\ -\frac{(log\, k-\mu_b)^2(8+4n)}{\sigma_b^{\ n+3}}-\frac{n(log\, k-\mu_b)^2}{4\sigma_a^{\ 2}\sigma_b^{\ n+1}} \end{array}\right]$$

5.

$$\frac{\partial\rho}{\partial} L(\mu_a,\mu_b,\sigma_a^{\ 2},\sigma_b^{\ 2},\rho)$$

$$=\left(\frac{1}{2\pi wk\sigma_a\sigma_b}\right)^n \times exp\left\{n\left[-\frac{1}{2(1-\rho^2)}\right]\right\}$$

$$\times\left[\frac{np+n}{\left(\sqrt{1-p^2}\right)^n}-np(1-p)^2\right]$$

$$\times\left[\begin{array}{l} \left(\frac{log\, w-\mu_a}{2\sigma_a}\right)^2-n\rho\left(\frac{log\, w-\mu_a}{\sigma_a}\right)\left(\frac{log\, k-\mu_b}{\sigma_b}\right) \\ +\left(\frac{log\, k-\mu_b}{2\sigma_b}\right)^2 \end{array}\right]$$

Improving the Yield of Glucoamylase and α-amylase in Solid-state Co-culture

Haruka Takefuji[1], Junko Ninomiya[1], Hiroshi Morita[2, a]

[1]*Graduate School of Environmental Engineering ,Kitakyushu University*
Grad. Sch. Univ. kitakyushu
[2]*Faculty of Environment Engineering,Kitakyushu University*
Sch. Univ. kitakyushu

Abstract. Steamed rice inoculated with *Aspergillus oryzae*, called Koji, is an essential ingredient for making *amazake* or brewing *sake*. However, *A. oryzae* usually offers low enzyme yield, especially in case of glucoamylase. Co-culture of *Aspergillus* and *Rhizopus* strains in Koji increased glucoamylase and α-amylase activities. The ratio of initial spore counts of *A. oryzae* and *R. oryzae* influenced the activity of amylolytic enzymes. When this ratio was 1:1, α-amylase showed maximum activity (573 U/g-substrate, 95 h), and when this ratio was 200:1, glucoamylase showed maximum activity (180 U/g-substrate, 95 h).

1 Introduction

Amylases are among the most important enzymes, with great significance in biotechnological applications, constituting around 25% of the enzyme industry [1]. The amylase family has two major classes: glucoamylase and α-amylase.

Glucoamylase (α-1,4-glucan glucohydrolase or amyloglucosidase, EC 3.2.1.3) consecutively hydrolyzes the α-1,4-glycosidic and α-1,6-glycosidic bonds at the non-reducing ends of starch, glycogen, and similar carbohydrates, thereby resulting in glucose production [2].

α-Amylase (α-1,4-glucan 4-glucohydrolase, EC 3.2.1.1) are extracellular endoenzymes that randomly cleave the 1,4-α-d-glucosidic linkages between adjacent glucose units in the linear amylose chain [3].

These amylolytic enzymes are widely used in the Japanese fermentation industry, e.g., for the manufacture of sake, soy sauce, miso, and amazake [4-6].

Amazake, a traditional non-alcoholic beverage in Japan, is produced from a mixture of rice *Koji* and boiled rice that is incubated at 55 °C to 60 °C for 8 h [4]. During this process, the glucoamylase and α-amylase produced by *Aspergillus oryzae*, a fungus found in the rice *Koji*, hydrolyze starch to dextrin and glucose; the chief sweet component in *amazake* is glucose [4, 7, 8].

The surface of rice grains contains many proteins; the proteolytic enzymes from *A. oryzae* hydrolyze these proteins to generate large amounts of amino acids and peptides [7].

Therefore, glucoamylase and α-amylase production by A. oryzae is important for the production of amazake.

However, the production of glucoamylase by the *Aspergillus* species is lower than that of α-amylase.

Many fungal species are capable of producing glucoamylase under different conditions and techniques of fermentation. Among them, *Rhizopus oryzae*, assigned the "generally regarded as safe" (GRAS) status, is commonly used in the industrial production of glucoamylase [9]. Moreover, much work has been undertaken to enhance the activity of the glucoamylase isolated from this fungus, such as optimization of the culture conditions and identification of the genes coding for glucoamylase.

Mertens *et al.* [10] reported that the glucoamylase produced by *Rhizopus* can be divided into two different taxonomic groups, namely, Type-I and Type-II, on the basis of the differences in phenotype (e.g., acid production), rDNA internal transcribed spacer regions, and genomic structure.

Glucoamylase-I can absorb raw starch and has high raw starch-digesting activity, while glucoamylase-II cannot absorb raw starch and has extremely weak raw starch-digesting activity. Glucoamylase-I contains starch-binding sites in the N-terminal region, whereas glucoamylase-II is formed by protease digestion of the N-terminal region of the glucoamylase-I and has reduced ability to digest starch [11, 12].

Solid-state fermentation is mostly undertaken with mixed cultures. Mixed culture fermentations are those in which the inoculum consists of two or more organisms and are widely used in many processes, including the production of antibiotics, enzymes, and several types of fermented food, composting, dairy fermentation, and bioconversion of apple distillery and domestic wastewater sludge [13]. The mixed culture may offer

[a] Corresponding author: morita@kitakyu-u.ac.jp

© The Authors, published by EDP Sciences. This is an open access article distributed under the terms of the Creative Commons Attribution License 4.0 (http://creativecommons.org/licenses/by/4.0/).

internal regulation of growth and product formation, which might be difficult to achieve otherwise. Several reports describe the co-culture of two strains for enhanced enzyme production [14, 15]. For example, the synergistic interaction of *Trichoderma reesei* RUT-C30 and *Aspergillus niger* LMA in a submerged fermentation system allowed more efficient cellulose degradation than in monoculture with either *T. reesei* or *A. niger*, owing to the complementary interactions between the cellulases from *T. reesei* and the glucosidases from *A. niger* that allowed complete cellulose hydrolysis [16]. In addition, Yun-wei and Yan [17] reported that acid protease activity was enhanced by the co-culture of *A. oryzae* AS3042 and *A. niger* SL-09. The inoculation of *A. niger* SL-09 significantly enhanced enzyme production, primarily because of the synergetic interaction between the strains.

Building upon these previous findings, we developed a solid-state co-culture involving two fungi for glucoamylase and α-amylase production. *A. oryzae* and *R. oryzae* were used for α-amylase and glucoamylase production, respectively.

2 Materials and Methods

2.1 Microorganisms

Two fungal strains, namely, *R. oryzae* NBRC 4716 and *A. oryzae* NBRC 5238, were used as producers of glucoamylase and α-amylase, respectively. The strains were maintained on potato dextrose agar (Nissui Pharmaceutical Co., Ltd., Tokyo, Japan) slants at 30 °C for 1 week.

2.2 Rice *Koji* preparation

Material rice (*Yumetukushi*) was obtained from Fukuoka (Japan) and polished to 90%. Steamed rice was inoculated with the spore suspension (final concentration, 2.5×10^4 spores/g-substrate; initial water content, 38%) and cultured for 20, 45, 70, 95, or 120 h. This system was incubated at 31 °C at 90% moisture for the initial 20 h. Then, it was moved to 38 °C at 85% moisture.

2.3 Enzyme extraction from rice *Koji*

The crude enzyme was extracted from 10 g of rice *Koji* by using 50 mL of sterile water for 3 h at room temperature. It was filtered through the filter paper (No. 2; diameter, 125 mm; Tokyo Roshi Kaisya, Ltd., Tokyo, Japan) to remove the *Koji*; the resulting filtrate was used as the crude enzyme solution.

2.4 Co-culture system with pre-culture time

The co-culture system was set up for a controlled culture time. For pre-culture, the spores (final concentration, ~1.25 × 104 spores/g-substrate) of *A. oryzae* or *R. oryzae* were added to steamed rice, and cultured for 20 h (initial water content, 38%). The inoculated rice was incubated at 31 °C at 90% moisture for the initial 20 h. The pre-culture *Koji* was inoculated with the spore suspension (final concentration, 1.25×10^4 spores/g-substrate) of other strains and cultured for 20, 40, 65, 90, 115, and 140 h. This system was incubated at 31 °C at 90% moisture for 20 h. It was subsequently moved to 38 °C at 85% moisture.

2.5 Co-culture system with controlled spore count

The co-culture system was set up with controlled number of spores. *R. oryzae* and *A. oryzae* (from the slants) were separately added to 10 mL of sterile water. Spore suspensions containing different ratios of *R. oryzae* and *A. oryzae* spores (1:200, 1:50, 1:10, 1:1, 10:1, 50:1, or 200:1) were prepared, and the total spore count was adjusted to 2.5×10^4 spores per gram of substrate.

2.6 Determination of glucoamylase, α-amylase, and acidic protease activities

Glucoamylase activity was measured using the method described by Morita et al. [18]. The reaction mixture for glucoamylase activity assay was composed of 1 mL of 2% (w/v) soluble starch (Nacalai Tesque Co., Kyoto, Japan) solution in 0.1M acetate buffer (pH 4.5) and 1 mL of the enzyme solution. The reaction progressed at 40 °C for 20 min and it was stopped by heating the test tubes in a boiling water bath for 10 min. The amount of liberated glucose was determined using the glucose oxidase method [19], by using a commercially available glucose test kit (Wako Pure Chemical Industries, Ltd., Osaka, Japan). One unit of glucoamylase activity was defined as 1 microgram of glucose liberated from soluble starch per minute per milliliter of the enzyme solution.

α-Amylase activity was measured using an α-amylase assay kit (Kikkoman, Ltd., Chiba, Japan), according to the manufacturer's instructions. One unit of α-amylase activity was defined as 1 micromole of 2-chloro-4-nitrophenol liberated from 2-chloro-4-nitrophenyl 65-azide-65-deoxy-β-maltopentaoside per milliliter of the enzyme solution.

A modification of the Anson hemoglobin method [20] was used for determining acidic protease activity. Briefly, 2.5 mL of hemoglobin solution (0.6% in 0.05M acetate buffer, pH 4.5) and 1 mL of the enzyme solution were mixed and incubated at 40 °C for 30 min. The reaction was stopped by adding 5 mL of 5% trichloroacetic acid. After 60-min incubation at 40 °C, the precipitated unhydrolyzed hemoglobin was removed by filtration, and the absorbance of the solution was measured at 280 nm. One unit of the acidic protease was defined as one microgram of tyrosine per gram of solid-substrate per minute.

3 Results and Discussion

3.1. Comparison of enzyme activity in the co-culture and pure culture systems

Figure 1 shows the glucoamylase, α-amylase, and acidic protease activity in *A. oryzae* NBRC 5238 pure culture, *R. oryzae* NBRC 4716 pure culture, and co-culture.

The highest glucoamylase activity was 79 U/g-substrate for 70 h and the highest α-amylase activity was 573 U/g-substrate for 95 h in co-culture. The glucoamylase and α-amylase activities were extremely low in *R. oryzae* pure culture. The acidic protease activity in *A. oryzae* NBRC 5238 pure culture, *R. oryzae* NBRC 4716 pure culture, and co-culture was not significantly different (Fig. 1 C). These results indicated that the highest glucoamylase activity in the co-culture system was lower than that observed in *A. oryzae* NBRC 5238 pure culture (167 U/g-substrate, 95 h). This can be attributed to the fact that it might be difficult to grow *Rhizopus* species on steamed polished rice cultures, because the enzyme produced by *Rhizopus* might not act on the heat-denatured proteins contained in rice. In particular, *Rhizopus* strains possessing low acidic carboxypeptidase activity are incapable of breaking down heat-denatured proteins. Therefore, the growth of *Rhizopus* species is delayed because of the lack of nitrogen sources [21, 22]. Moreover, Harayama *et al.*[23] reported that *R. oryzae* IFO 5418 and *Rhizopus oligosporus* NRRL 2710 showed extremely weak α-amylase and glucoamylase activities in rice and soybean *Koji*. These results were in agreement with those of our study.

A

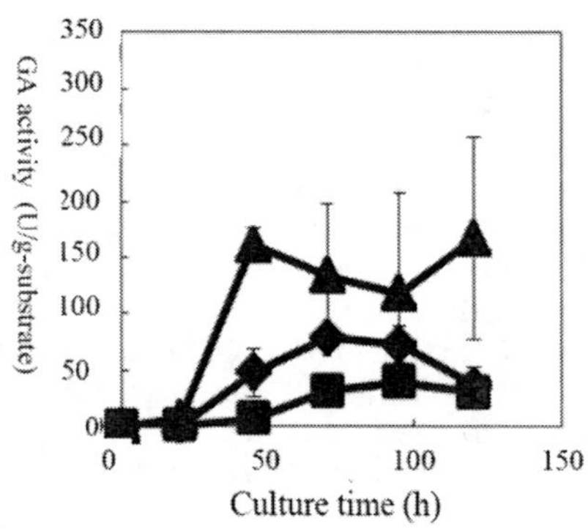

B

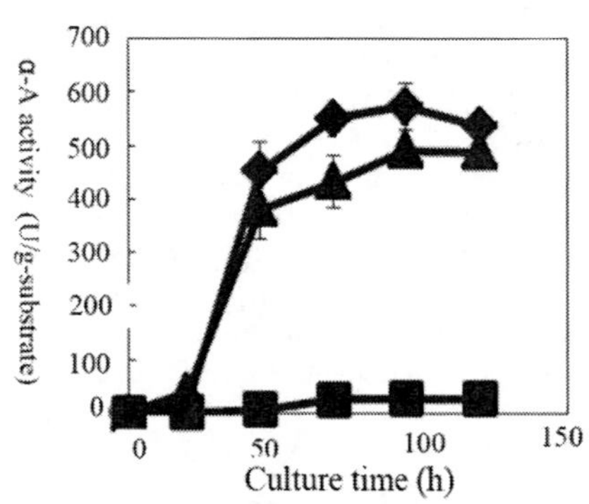

C

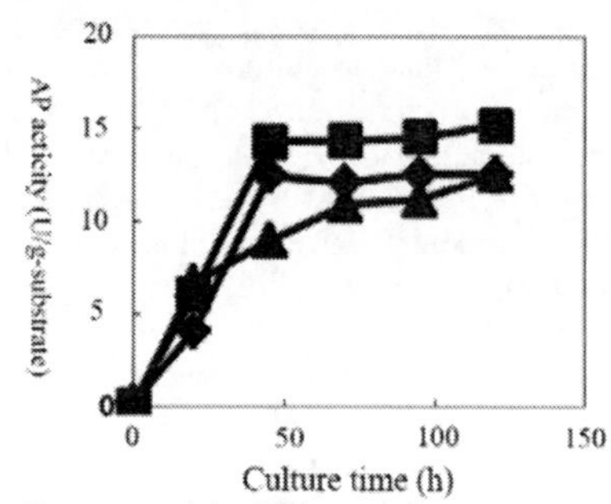

Figure 1. Enzyme activity in the co-culture system using pure cultures of *A. oryzae* NBRC 5238 and *R. oryzae* NBRC 4716
A. Glucoamylase (GA) activity (U/g-substrate),
B. α-Amylase (α-A) activity (U/g-substrate), and
C. Acidic protease (AP) activity (U/g-substrate)
◆: Co-culture involving equal number of *R. oryzae* and *A. oryzae* spores (1:1), ▲: *A. oryzae* pure culture, ■: *R. oryzae* pure culture

3.2 Effects of a co-culture system with pre-culture time on glucoamylase and α-amylase production

Previous studies showed that the glucoamylase activity in the co-culture system was lower than that in the *A. oryzae* NBRC 5238 pure culture system. Therefore, we adopted the co-culture system involving pre-culture time to enhance glucoamylase activity.

A

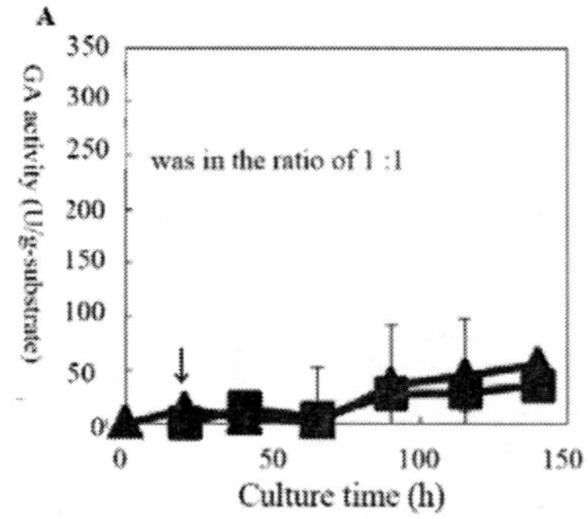

B

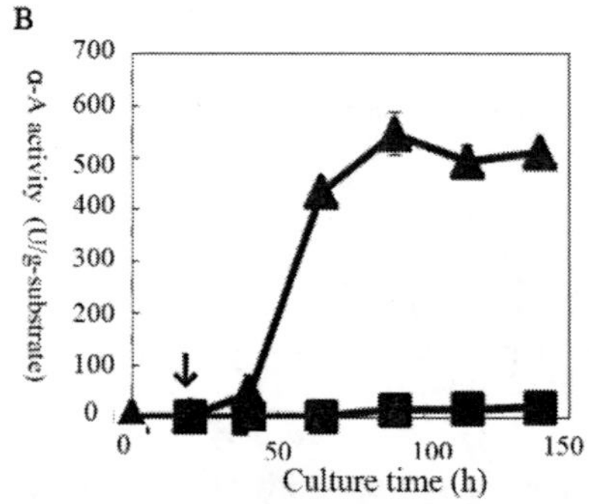

Figure 2 Influence of co-culture system with pre-culture time on glucoamylase and α-amylase production
A. Glucoamylase (GA) activity (U/g-substrate) and
B. α-Amylase (α-A) activity (U/g-substrate)
▲: pre-culture of *A. oryzae*, ■: pre-culture of *R. oryzae*, ↓: co-culture start point

Figure 2 shows the glucoamylase and α-amylase activities of *A. oryzae* or *R. oryzae* during the pre-culture time. High α-amylase production (545 U/g-substrate) was observed at 90 h (Fig. 2B), but glucoamylase production was reduced (Fig. 2A) in the co-culture system with pre-culture time. According to some reports, enzyme production by *A. oryzae* depended on the temperature [24, 25]. Acid protease is well produced at temperatures below 35 °C, but temperatures above 38 °C can decrease the production. The expression of *pepA* (acid protease-encoding gene) is induced at low temperatures [24]. Hisada et al. [25] reported that *glaB* (glucoamylase-encoding gene) promoter is induced by exposure to high temperature (for example, growth at 42 °C).

In this experiment, the temperature was maintained at 31 °C for 40 h during the early stages of cultivation. Thus, we assumed that glucoamylase production by *A. oryzae* was inhibited at low temperature.We found that the co-culture system with pre-culture time was not the optimal culture method.

3.3 Effects of a co-culture system with controlled spore count on glucoamylase and α-amylase productions

The above-described results showed that the glucoamylase activity in the co-culture system with pre-culture time was extremely low. Additionally, α-amylase production by the seed culture of *Rhizopus* in the co-culture system was extremely low. Mikai *et al.*[2] have reported the co-culture of *A. oryzae* and *Rhizopus arrhizus*. Therefore, we tried another co-culture system with controlled spore count.

The graphs in Fig. 3 and 4 show glucoamylase and α-amylase activities in the presence of controlled spore count.

The optimal initial spore count of *A. oryzae* and *R. oryzae* was in the ratio of 200:1 for glucoamylase production (180 U/g-substrate, 95 h, Fig. 3). The highest glucoamylase activity in the co-culture system (spores of *A. oryzae* and *R. oryzae*, 200:1) was higher than that observed in the *A. oryzae* NBRC 5238 pure culture (167 U/g-substrate, 120 h). The optimal initial spore count of *A. oryzae* and *R. oryzae* for α-amylase production (573 U/g-substrate, 95 h) was in the ratio of 1:1 (Fig. 4).

Glucoamylase and α-amylase activities improved on equalizing or increasing the initial spore ratio of *A. oryzae* and *R. oryzae* (Fig. 3A, 4A) and it reduced on increasing the initial spore ratio of *R. oryzae* and *A. oryzae* (Fig. 3B, 4B). Taken together, these data showed that the optimal conditions for the production of both glucoamylase and α-amylase in a co-culture system involve equal spore count of *A. oryzae* and *R. oryzae* (1:1 ratio). Under these conditions, glucoamylase activity (167 U/g-substrate) reached ~47% of that observed in *A. oryzae* pure culture, and α-amylase activity (491 U/g-substrate) reached ~117% of that observed in *A. oryzae* pure culture.

Thus, these results suggest that the production of glucoamylase and α-amylase can be controlled by adjusting the numbers of *A. oryzae* and *R. oryzae* spores. Furthermore, these data imply that increased production of glucoamylase and α-amylase can be obtained by co-culturing *A. oryzae* and *R. oryzae*.

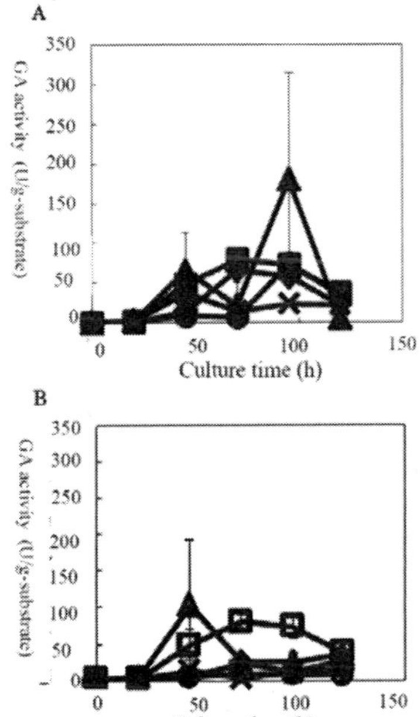

Figure 3 Influence of co-culture system with a controlled spore count on glucoamylase production

A. *A. oryzae*:*R. oryzae*: 1:1, ■: 10:1, ◆: 50:1, ●: 100:1, ×: 200:1,▲

B. *A. oryzae*:*R. oryzae*: 1:1, □: 1:10, ◇: 1:50, ○: 1:100, ×: 1:200,△

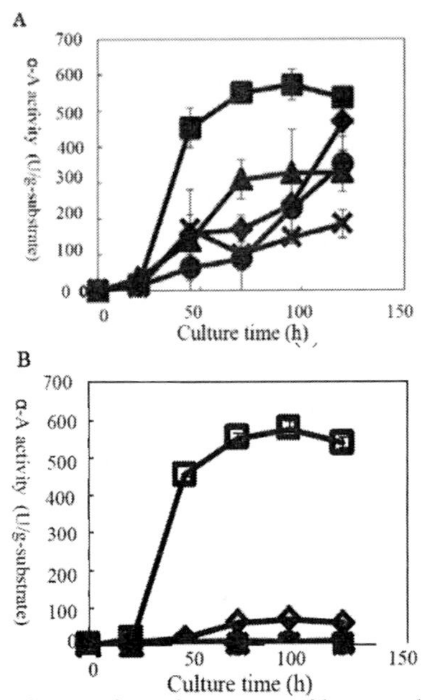

Figure 4 Influence of co-culture system with a controlled spore count on α-amylase production

A. *A. oryzae*:*R. oryzae*: 1:1, ■: 10:1,◆: 50:1, ●: 100:1, ×: 200:1,▲

B. *A. oryzae*:*R. oryzae*: 1:1, □: 1:10, ◇: 1:50, ○: 1:100, ×,1:200 ,△

4 Conclusions

In this study, we investigated the production of glucoamylase and α-amylase by *A. oryzae* and *R. oryzae* in solid-state co-culture.

Glucoamylase activity of co-culture *Koji* was optimized 180 U/g-substrate when the ratio of initial spore counts was 200:1 and α-amylase of co-culture *Koji* was optimized 573 U/g-substrate when the ratio of initial spore counts was 1:1. These results indicated that amylolytic enzymes can be regulated from adjusting the numbers of *A. oryzae* and *R. oryzae* spores.

However, from a perspective in productivity of glucoamylase and α-amylase, our results suggested that the co-culture system with simultaneous inoculation with equal number of spores of *A. oryzae* and *R. oryzae* represents optimal conditions for amylase production.

Further investigations are necessary to improve the production of glucoamylase and α-amylase under optimal conditions (e.g., moisture, temperature) required for the co-culture system.

Acknowledgements

This study was supported in part by the Kitakyushu Foundation for the Advancement of Industry Science and Technology on "Reserch & Development Grant Program for Small and Medium Enterprises based on Collaboration between Industry and Academia." Advice and samples given by Kokurajirushi Co., Ltd has been a great help in this study. Special thanks to Kokurajirushi Co., Ltd.

References

1. M. B. Rao, A. M. Tanksale, M. S. Gathe, V. V. Deshpande; Microbiol. Mol. Biol. Rev., *Molecular and Biotechnological Aspects of Microbial Proteases*. **62**, 597-635 (1998).
2. S. Mikai, N. I. J. Konomi, Y. Sato, M. Era, J. Ninomiya, H. Morita; J. Food Eng., *Simultaneous Increase of Glucoamylase and α-Amylase Production in Submerged Co-culture of Aspergillus and Rhizopus Strains.***16**, 111-121, 123 (2015).
3. D. Raul, T. Biswas, S. Mukhopadhyay, S. K. Das, S. Gupta; Fermentation. Biochem. Res. Int., *Production and Partial Purification of Alpha Amylase from Bacillus subtilis (MTCC 121) Using Solid State Fermentation.***2014**, ID568141,5 (2014)
4. Y. Oda, Y. Ichinose, H. Yamaguchi; Food Sci. Technol. Res., *Utilization of Lactobacillus amylovorus as an Alternative Microorganism for Saccharifying Boiled Rice.* **8** (2), 166-168, (2002)
5. F. C. Pavezzi, E. Gomes, R. da Silva; Brazilian J. Microbiol., *Production and Characterization of Glucoamylase from Fungus Aspergillus awamori Expressed in Yeast Saccharomyces cerevisiae using Different Carbon Sources.* **39**, 108-114 (2008).
6. Y. Hata, H. Ishida, Y. Kojima, E. Ichikawa, A. Kawato, K. Suginami, S. Imayasu; J. Ferment. Bioeng.,*Comparison of Two Glucoamylases Produced by Aspergillus oryzae in Solid-State Culture (Koji) and in Submerged Culture.* **84**, 532-537 (1997).
7. N. Saigusa and R. Ohba; Food Sci. Technol. Res., *Effects of koji Production and Saccharification Time on the Antioxidant Activity of amazake.* **13** (2), 162-165 (2007)
8. S. Yamamoto, Y. Nakashima, J. Yoshikawa, N. Wada, S. Matsugo; Food Sci. Technol. Res., *Radical Scavenging Activity of the Japanese Traditional Food, Amazake.* **17** (3), 209-218 (2011)
9. I. Cantabrana, R. Perise, I. Hernández; Intl. J. Gastronomy Food Sci., *Uses of Rhizopus oryzae in the Kitchen.* **2**, 103-111 (2015)
10. J. A. Mertens and C. D. Skory; Curr. Microbiol., *Isolation and Characterization of Two Genes that Encode Active Glucoamylase without a Starch Binding Domain from Rhizopus oryzae.* **54**, 462-466 (2007)
11. S. Hayasida; Agric. Biol. Chem., *Selective Submerged Production of Three Types of Glucoamylase by a Black-Koji Mold.* **39**, 2093-2099 (1975)
12. S. Hayashida and P. Q. Florr; Agric. Biol. Chem.,*Raw Starch-Digestive Glucoamylase Productivity of Protease-less Mutant from Aspergillus awamori var. kawachi.* **45**, 2675-2681 (1981)
13. M. Z. Alam, F. Razi, S. A. Aziz, A. H. Molla; Water Air Soil Pollut., *Optimization of Compatible Mixed Cultures for Liquid State Bioconversion of Municipal Wastewater Sludge.* **149**, 113-126 (2003)
14. P. Nigam and D. Singh; J. Basic Microbiol.,*Solid-State (Substrate) Fermentation Systems and Their Applications in Biotechnology.* **34**, 405-423 (1994)
15. A. Pandey, P. Selvakumar, C.R. Soccol, P. Nigam; Curr. Sci.,*Solid state Fermentation for the Production of Industrial Enzymes.* **77** (1), 149-162 (1999)
16. A. Ahamed, P. Vermette; *Biochem. Eng. Journal, Enhanced Enzyme Production from Mixed Cultures of Trichoderma reesei RUT-C30 and Aspergillus niger LMA Grown as Fed Batch in a Stirred Tank Bioreactor.* **42**, 41-46 (2008)
17. L. Yun-wei and X. Yan; Afr. J. Biotechnol., *Improvement of Acid Protease Production by a Mixed Culture of Aspergillus niger and Aspergillus oryzae using Solid-State Fermentation Technique.* **10** (35), 6824-6829 (2011)
18. H. Morita, K. Mizuno, M. Matsunaga, Y. Fujio; J. Appl. Glycosci., *Raw Starch-Digesting Glucoamylase Production of Rhizopus sp. MKU 40 using a Metal-Ion Regulated Liquid Medium.* **46**, 15-21 (1999)
19. G. R. Kingseley, G. Getchell; Clin. Chem.,*Direct Ultramicro Glucose Oxidase Method for Determination of Flucose in Biologic Fluids.* **6**, 466-475 (1960)
20. M. L. Anson; J. Gen. Physiol., *The Estimation of Pepsin, Trypsin, Papain, and Cathepsin with Hemoglobin.* **22**, 79-89 (1938)

21. T. Tanaka, N. Okazaki, M. Kitani; J. Brew. Soc. Japan, *Comparison of Growth and Enzyme Production between A. oryzae and Rhizopus sp. Growth of Mold on Uncooked Grains (Part 2)*. **77** (11), 831-835 (1982)
22. H. Murakami; Brew. Soc. Japan, *Koji-gaku. J.* 483 (1986)
23. F. Harayama and H. Yasuhira; J. Brew. Soc. Japan, Comparison between Several Hydrolytic Enzyme Activities of Aspergillus and Those of Rhizopus. **84** (9), 625-629 (1989)
24. H. Kitano, K. Kataoka, K. Furukaawa, S. Hara; J. Biosci. Bioeng.,*Specific Expression and Temperature-Dependent Expression of the Acid Protease-Encoding Gene (pepA) in Aspergillus oryzae in Solid-State Culture (Rice-Koji)*. **93** (6), 563-567 (2007)
25. H. Hisada, M. Sano, H. Ishida3, Y. Hata, M. Machida; Appl. Microbiol. Biotechnol., *Identification of Regulatory Elements in the Glucoamylase-Encoding Gene (gla B) Promoter from Aspergillus oryzae*. **97**, 4951-4956 (2013)

DOI: 10.1051/matecconf/20166203002

Antimicrobial Activity of Fatty Acid Salts Against Microbial in *Koji-Muro*

Aya TANAKA[1], Mariko ERA[1], Takayoshi KAWAHARA[2], Takahide KANYAMA[2], and Hiroshi MORITA[3,a]

[1]*Graduate School of Environmental Engineering, the University of Kitakyushu in Fukuoka, JAPAN*
[2]*Shabondama Soap Co., Ltd. JAPAN*
[3]*Faculty of Environment Engineering, the University of Kitakyushu in Fukuoka, JAPAN*

Abstract. *Aspergillus niger* and *Aspergillus oryzae* are used as *koji* fungi in the spot of the brewing. Since *koji-muro* (room for making *koji*) was a low level of airtightness, microbial contamination has long been a concern to the alcoholic beverage production. Therefore, we focused on the fatty acid salt which is the main component of soap. Fatty acid salts have been reported to show some antibacterial and antifungal activity. This study aimed to find the effectiveness of the fatty acid salt in *koji-muro*. Nine fatty acid salts were tested. The result, C12K was antibacterial effect against *B. subtilis*. C10K and C12K was antifungal effect against *R. oryzae*. These results suggest C12K has potential in the field of *koji-muro*.

1 Introduction

Fungus is ubiquitous in the indoor environment, and breed in the foods such as vegetables and grains, causing corruption and deterioration of these foods in some cases. Bacteria and yeast can be cited as a fellow of other microbial.

Aspergillus species are common contaminants of starchy foods including bread and potatoes. But, *Aspergillus niger* and *Aspergillus oryzae* are used as *koji* fungi in the spot of the brewing. Since *koji-muro* (room for making *koji*) was a low level of airtightness, microbial contamination has long been a concern to the alcoholic beverage production. Typical causes of microbial contamination include *Rhizopus* species in the fungi and *Bacillus* species in the bacteria. *Rhizopus* species are called "*kumonosukabi*" in Japanese, and has been used as *koji* fungi in Asia other than Japan. *Bacillus* species include *Bacillus subtilis* in the typical ones. It is known to shape a durable organ of spores when the growth environment is deteriorated. It has been taken a variety of countermeasure in the current situation to microbial contamination. As an example, the method can be mentioned that using ethanol and invert soap and the fumigation. However, the former is the effect and persistence of low against fungus and spores is a problem. The latter can only be done at the end of the season because it is too powerful.

Therefore, we focused on the fatty acid salt which is the main component of soap. Also, it is an alkali metal salt consisting of alkali and fatty acid (CH_3-(R)-COOH), which is a kind of anionic surfactant. It is used as many industrial material. Fatty acids vary in length and degree of saturation, and naturally occurring fatty acids have a chain length of 4 to 28 carbons, which may be saturated or unsaturated [1]. Moreover, it is known that there is antibacterial activity to bacteria, such as *Staphylococcus aureus* and *Eschrichia coli* [2-4]. So this study examined antimicrobial activities against *Aspergillus oryzae*, *Rhizopus oryzae* and *Bacillus subtilis*.

This study aimed to find the effectiveness of the fatty acid salt in *koji-muro* as antimicrobial agents.

2 Detail Experimental

2.1 Source and preparation of fatty acid salts

Table 1 show the fatty acid salts used in the experiment. Nine fatty acids were tested. Butyric acid (C4:0), caproic acid (C6:0), and linoleic acid (C18:2) were obtained from Wako Pure Chemical Industries, Ltd., Osaka, Japan. Caprylic acid (C8:0), capric acid (C10:0), lauric acid (C12:0), myristic acid (C14:0), oleic acid (C18:1), and linolenic acid (C18:3) were obtained from Tokyo Chemical Industry Co., Ltd.

Hydrated mixtures of fatty acid salts in solution were prepared by gravimetric determination, using fatty acids, KOH: potassium hydroxide pellets (Wako Pure Chemical Industries, Ltd., Osaka, Japan) and water.

Samples were prepared at 350 mM concentrations. The samples were then stirred for 2 h at 75 °C. KOH aq was added to yield theoretical neutralization at pH 10.5 of fatty acid salts. Potassium butyrate (C4K), potassium caproate (C6K), potassium caprylate (C8K), potassium caprate (C10K), potassium laurate (C12K), potassium myristate (C14K), potassium oleate (C18:1K), potassium linoleate (C18:2K), potassium linolenate (C18:3K), and the blank were all adjusted using a KOH pH-adjusted solution (pH 10.5). All fatty acid salts and the KOH pH-

[a] Corresponding author: morita@kitakyu-u.ac.jp

© The Authors, published by EDP Sciences. This is an open access article distributed under the terms of the Creative Commons Attribution License 4.0 (http://creativecommons.org/licenses/by/4.0/).

adjusted solution were filter-sterilized at low temperature (4-6 °C) using a 0.20-µm Millipore filter (Toyo Roshi Kaisya, Ltd., Tokyo, Japan) [5].

Table 1. Samples preparation; FAS

C4K : Potassium btyrate
C6K : Potassium caproate
C8K : Potassium caprylate
C10K : Potassium caprate
C12K : Potassium laurate
C14K : Potassium myristate
C18:1K : Potassium oleate
C18:2K : Potassium linoleate
C18:3K : Potassium linolenate
Control : KOH pH-adjusted solution

2.2 Fungal strains and growth conditions

A. oryzae (*Akita konnno* store, Japan), *R. oryzae* NBRC 4716 and *B. subtilis* NBRC 3335 were obtained from the NBRC (Biological Resource Center, NITE, Tokyo, Japan).

The fungi were initially grown on Potato Dextrose Agar (PDA: *Nissui* Pharmaceutical Co., Ltd., Tokyo, Japan) in the dark and the culture stocks were stored at 4 °C. Plant-pathogenic fungi were routinely sub-cultured on PDA in slants in the dark at 30 °C so that the fresh cultures were available for later use [5].

2.3 Preparation of spore suspensions

Fungi were grown on PDA slants at 30 °C until well sporulated, approximately 10-14 d. Spores were harvested in sterile distilled water using a sterile inoculation loop and gentle agitation; a drop of 0.005 % sulfosuccinic acid bis (2-ethylhexyl) ester sodium salt and 0.9 % NaCl were added to aid wetting of the spores. The spore concentration was determined by counting using a hemacytometer (Thoma, Sunlead Glass Corp., Saitama, Japan). The initial spore concentration was adjusted to 3.0×10^4 spores/mL [5].

2.4 Preparation of cell suspensions

The subcultures were incubated at 30 °C for 1 day. Growth was harvested with autoclaved sterilized water. The bacteria were sedimented by centrifugation at 4200 rpm for 20 min. Pellets were washed three times with 30 mL of autoclaved sterilized water each time. Final pellets were suspended in 20 mL of autoclaved sterilized water. The washed suspensions of *Bacillus* were used on the day of preparation [6].

2.5 Effect of fatty acid salts on fungal spores or bacterial cells

Solutions of 400 µL of fatty acid salts (final concentration of 175 mM in the tubes) and 400 µL of the spore suspension (3.0×10^4 spores/mL or 3.0×10^5 CFU/mL) were prepared in 1.5 mL plastic tubes. Spores mixed with the KOH pH-adjusted solution were used as controls. Final pH of all samples were a range of pH 9.2-10.8. The mixtures were incubated at 25 °C. Samples were counted at 0, 10, 60, and 180 min by plating (100 µL) on PDA or NA. Fungal colonies were counted after incubation for 2 d or 10 d at 30 °C. Viable counts ($\log_{10}$ CFU) of spore was subtracted from the viable count of the control ($\log_{10}$ CFU), and the difference was used as a measure of the antifungal activity. All experiments were performed at least thrice.

2.6 Determination of minimum inhibitory concentrations (MICs)

The MIC is defined as the lowest concentration of drug sufficient for inhibiting visible growth of spores after 10 min of incubation. MICs against fungi were determined using the two-fold dilution method [7, 8].

Table 2 show two-fold dilution method of fatty acid salts. Each fatty acid salt was separately inoculated with 400 µL of *R. oryzae* NBRC 4716 or *B. subtilis* NBRC 3335 at 3.0×10^4 spores/mL or 3.0×10^5 CFU/mL. 1.5 mL plastic tubes containing 400µL of each of fatty acid salts were inoculated separately with 400 µL of the fungi or bacteria. The tubes, each containing a total volume of 800 µL, were incubated at 25°C for each organism for 10 min. After incubation, samples were plating on PDA or NA, incubated at 30°C for 7 d, and then examined for the growth of spores. Following incubation, the end point was visually assessed and expressed in mM. The lowest concentration of the antifungal or antibacterial treatment that inhibited visible growth of the fungi or bacteria after incubation was taken as the MIC of the treatment.

Table 2. Two-fold dilution method of fatty acid salts

Two-fold dilution	concentration [mM]				
	175	87.5	43.8	21.9	10.9
fatty acid salts 175 mM	400 µL	200	200	200	200
KOH pH-adjusted solution	-	200	200	200	200

2.7 Other reagents tested

Linear alkylbenzene sulfonate (LAS, *Shabondama* Soap Co., Ltd., Japan) is an anionic surfactant similar to fatty acid salts, and was used in this experiment for comparison (pH 10.5). LAS was obtained from *Shabondama* Soap Co., Ltd., Fukuoka, Japan.

LAS composed a sodium salt consisting of linear alkyl benzene sulfonic acid, a number of compounds as with soap. 98% of compound elaborated LAS are linear alkyl benzene sulfonic acid sodium of 12 carbon chain, the rest is included other carbon chain length.

3 Results and Discussion

3.1 Antibacterial activity of saturated fatty acid salts against *B. subtilis*

Figure 1 show the antibacterial activity of saturated fatty acid salts against *B. subtilis*. The average initial population of bacteria at 0 min in all samples was approximately 3.0 × 10^5 CFU/mL. Bacteria was incubated for 1 d. Final concentration of saturated fatty acid salts were 175 mM. C12K produced a 5 log-units reduction in the growth of *B. subtilis* NBRC 3335 after incubation for 10 min. Thus, C12K, C18:1K, C18:2K and C18:3K suppressed 99.999% of bacterial growth. Against *B. subtilis* NBRC 3335, C10K produced an antibacterial effect of 5 log-units (suppressing 99.999 % of growth) following incubation for 180 min. C8K and C14K produced an antibacterial effect of 4 log-units (suppressing 99.99 % of growth) following incubation for 180 min. However, C4K, C6K, and pH-adjusted solution (control) were ineffective after 180 min. Saturated fatty acid salts exerted an antibacterial activity, and no effect was produced by the pH-adjusted solution alone.

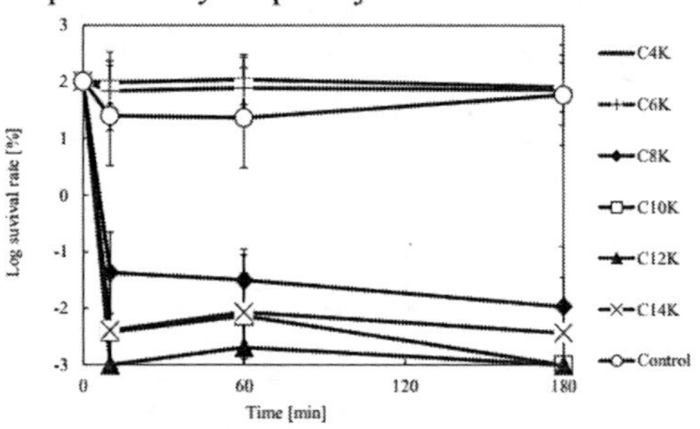

Figure 1. Antibacterial activity of saturated fatty acid salts against *B. subtilis*
Spores were counted at the time of inoculation (0 min) and after 10, 60, 180 min of incubation by means of plating 100 μL portions of the same samples. Log survival rate (expressed log CFU/mL) of the samples stored at 30°C for 1 d were enumerated at the specified time points on NA. Symbols: −, C4K; +, C6K; ◆, C8K; □, C10K; ▲, C12K; ×, C14K; ○, Control (KOH pH-adjusted). Experiments were performed in triplicate, and the error bars represent the standard deviation.

Antibacterial activity of saturated fatty acid salts were found to have defferent intensities of the activity by the defference in the carbon chain. In addition, it was revealed that the antibacterial activity of the relatively long saturated fatty acid salts having a carbon chain is high. Thus, we consider the reason why the difference in antibacterial activity by the difference in the number of carbon chain. Hydrophobic groups of the saturated fatty acid salt is fatty acid. Fatty acid has been considered increasing destabilization of cell membranes by being inserted into the phospholipid bilayer of the cell membrane of the bacteria [9, 10]. Therefore, the saturated fatty acid salts of the short chain saturated fatty acid salts for the hydrophobic group is short, be suspected that did not result in increased destabilization of cell membranes in *B. subtilis*. The saturated fatty acid salt have a medium-chain of hydrophobic group is considered increasing destabilization of cell membranes by a hydrophobic group is inserted into the phospholipid bilayer of the cell membrane of *B. subtilis*, it is believed to have shown an antibacterial effect.

3.2 Antibacterial activity of unsaturated fatty acid salts against *B. subtilis*

Figure 2 show the antibacterial activity of unsaturated fatty acid salts against *B. subtilis*. The average initial population of bacteria at 0 min in all samples was approximately 3.0 × 10^5 CFU/mL. *B. subtilis* was incubated for 1 d. Final concentration of unsaturated fatty acid salts were 175 mM. C18:1K, C18:2K and C18:3K produced a 5 log-units reduction in the growth of *B. subtilis* NBRC 3335 after incubation for 10 min. Thus, C18:1K, C18:2K and C18:3K suppressed 99.999% of bacterial growth. However, pH-adjusted solution (control) were ineffective after 180 min. Unsaturated fatty acid salts exerted an antibacterial activity, and no effect was produced by the pH-adjusted solution alone.

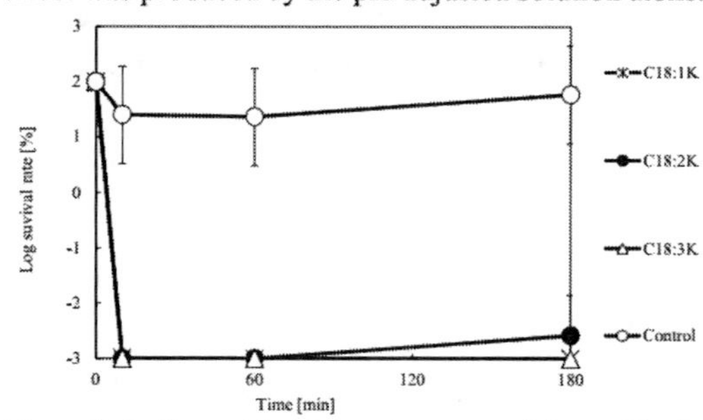

Figure 2. Antibacterial activity of unsaturated fatty acid salts against *B. subtilis*
Spores were counted at the time of inoculation (0 min) and after 10, 60, 180 min of incubation by means of plating 100 μL portions of the same samples. Log survival rate (expressed log CFU/mL) of the samples stored at 30°C for 1 d were enumerated at the specified time points on NA. Symbols: ∗, C18:1K; ●, C18:2K; △, C18:3K; ○, Control (KOH pH-adjusted). Experiments were performed in triplicate, and the error bars represent the standard deviation.

Antibacterial activity of unsaturated fatty acid salts were found to have defferent intensities of the activity by the short saturated fatty acid salts. Thus, we consider the reason why the difference in antibacterial activity by the short saturated fatty acid salts. Hydrophobic groups of the unsaturated fatty acid salt is fatty acid same as saturated fatty acid salts. Fatty acid has been considered increasing destabilization of cell membranes by being inserted into the phospholipid bilayer of the cell membrane of the bacteria [9]. The unsaturated fatty acid salt have long-chain of hydrophobic group. So we considered increasing destabilization of cell membranes by a hydrophobic group is inserted into the phospholipid bilayer of the cell membrane of *B. subtilis*, it is believed to have shown an antibacterial effect [3].

3.3 Antifungal activity of saturated fatty acid salts against *R. oryzae*

DOI: 10.1051/matecconf/20166203002

Figure 3 show the antifungal activity of saturated fatty acid salts against *R. oryzae* NBRC 4716. The average initial population of fungus at 0 min in all samples was approximately 3.0 × 10^4 spores/mL. Fungus was incubated for 1 d. Final concentration of fatty acid salts were 175 mM. C12K produced a 4 log-units reduction in the growth of *R. oryzae* NBRC 4716 after incubation for 10 min. Thus, C12K suppressed 99.99% of fungal growth. Against *R. oryzae* NBRC 4716, C10K produced an antifungal effect of 4 log-units (suppressing 99.99 % of growth) following incubation for 180 min. C14K produced an antifungal effect of 3 log-units (suppressing 99.9 % of growth) following incubation for 180 min. However, C4K, C6K, C8K and pH-adjusted solution (control) were ineffective after 180 min. Saturated fatty acid salts exerted an antifungal effect, and no effect was produced by the pH-adjusted solution alone.

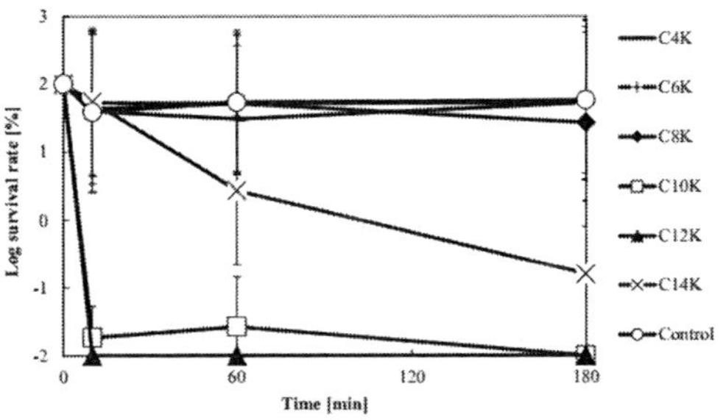

Figure 3. Antifungal activity of saturated fatty acid salts against *R. oryzae*
Spores were counted at the time of inoculation (0 min) and after 10, 60, 180 min of incubation by means of plating 100 μL portions of the same samples. Log survival rate (expressed log CFU/mL) of the samples stored at 30°C for 1 d were enumerated at the specified time points on PDA. Symbols: −, C4K; +, C6K; ◆, C8K; □, C10K; ▲, C12K; ×, C14K; ○, Control (KOH pH-adjusted). Experiments were performed in triplicate, and the error bars represent the standard deviation.

Antifungal activity of saturated fatty acid salts were found to have defferent intensities of the activity by the defference in the carbon chain. The antimicrobial effect of fatty acids, which are the raw material for the production of fatty acid salts, decrease with increasing chain length, and medium-chain fatty acids exhibit stronger activity than longer chain fatty acids [11]. Thus, the reason for the difference in antifungal activity caused by the difference in the number of carbon chain, it is suspected that showed antifungal activity in fungas spores for the same reason as that discussed antibacterial activity of the fatty acid salt against *B. subtilis*. However, some of the saturated fatty acid salt against *R. oryzae* did not show high antifungal activity. *B. subtilis* are classified as Gram-negative bacteria among the bacteria, so this cell walls is thin compared to the fungas [3, 12]. Thus, we considered result did not increase the destabilization of the cell membrane of *R. oryzae* because C8K is not a sufficient length of the hydrophobic group.

3.4 Antifungal activity of unsaturated fatty acid salts against *R. oryzae*

Figure 4 show the antifungal activity of unsaturated fatty acid salts against *R. oryzae* NBRC 4716. The average initial population of fungus at 0 min in all samples was approximately 3.0 × 10^4 spores/mL. Fungus was incubated for 1 d. Final concentration of unsaturated fatty acid salts were 175 mM. C18:1K, C18:2K, C18:3K and pH-adjusted solution (control) were ineffective after 180 min. Unsaturated fatty acid salts exerted an antifungal effect, and no effect was produced by the pH-adjusted solution alone.

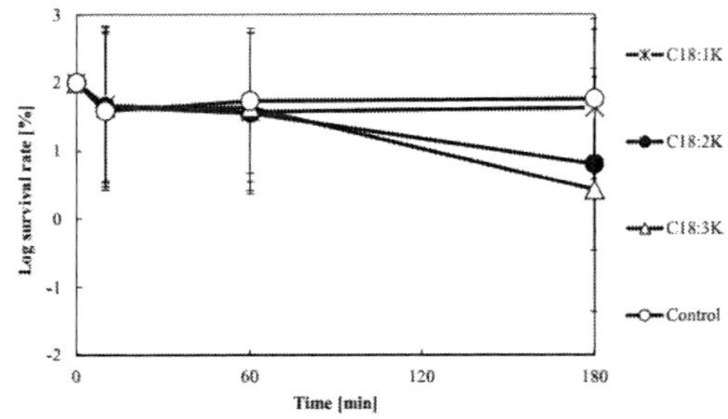

Figure 4. Antifungal activity of saturated fatty acid salts against *R. oryzae*
Spores were counted at the time of inoculation (0 min) and after 10, 60, 180 min of incubation by means of plating 100 μL portions of the same samples. Log survival rate (expressed log CFU/mL) of the samples stored at 30°C for 1 d were enumerated at the specified time points on PDA. Symbols: *, C18:1K; ●, C18:2K; △, C18:3K; ○, Control (KOH pH-adjusted). Experiments were performed in triplicate, and the error bars represent the standard deviation.

Antifungal activity of unsaturated fatty acid salts were found to have different intensities of the activity by the saturated fatty acid salts. However, the reason have not been clarified.

3.5 Antifungal activity of fatty acid salts against *A. oryzae*

Figure 5 show the antifungal activity of fatty acid salts against *A. oryzae*. The average initial population of fungus at 0 min in all samples was approximately 3.0 × 10^4 spores/mL. Fungus was incubated for 2 d. Final concentration of fatty acid salts were 175 mM. No obvious change was observed in tested fatty acid salts against *A. oryzae*. Antifungal effect was not produced by the pH-adjusted solution alone.

So far, fatty acid salts have been observed antimicrobial activity against *R. oryzae* and *B. subtilis* in present study and *Penicillium* [5] and oral bacteria [13] in previous study, but did not observed antifungal activity against *A. oryzae*.

For future work, we would like to examine the reason for fatty acid salt against *A. oryzae* spores did not showed antifungal activity.

DOI: 10.1051/matecconf/20166203002

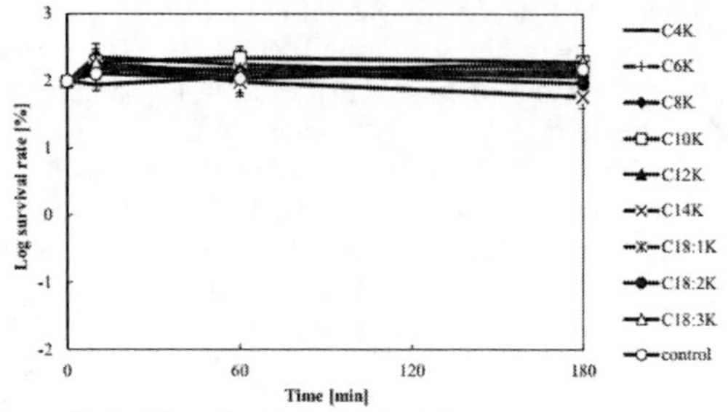

Figure 5. Antifungal activity against *A. oryzae*
Spores were counted at the time of inoculation (0 min) and after 10, 60, 180 min of incubation by means of plating 100 μL portions of the same samples. Log survival rate (expressed log CFU/mL) of the samples stored at 30°C for 2 d were enumerated at the specified time points on PDA. Symbols: −, C4K; +, C6K; ◆, C8K; □, C10K; ▲, C12K; ×, C14K; ∗, C18:1K; ●, C18:2K; △, C18:3K; ○, Control (KOH pH-adjusted). Experiments were performed in triplicate, and the error bars represent the standard deviation.

3.6 Antimicrobial activity of other reagents against *B. subtilis*, *R. oryzae* and *A. oryzae*

Figure 6 show antimicrobial avtivity of LAS against *B. subtilis* NBRC 3335, *R. oryzae* NBRC 4716 and A. oryzae (*Akita konnno* store). The average initial population of bacteria or fungas at 0 min in all samples was approximately 3.0 × 10^5 CFU/mL or 3.0 × 10^4 spore/mL. Bacteria was incubated for 1 d. *R. oryzae* was incubated for 1 d. *A. oryzae* was incubated for 2 d. Final concentration of fatty acid salts and LAS was 175 mM. LAS produced a 5 log-units reduction in the growth of *B. subtilis* NBRC 3335 after incubation for 10 min. Thus, LAS suppressed 99.999 % of bacterial growth. Similar results were obtained when the result of C12K, C18:1K, C18:2K and C18:3K against *B. subtilis* NBRC 3335. We would like to examine the reason for the LAS showed high antibacterial activity. LAS is synthetic detergent so it includes a sulfonate salt having an alkyl group with a carbon chain 10-14. Therefore, it suspected that LAS showed antibacterial activity against *B. subtilis* because it include same carbon chain as fatty acid salts.

LAS produced a 4 log-units reduction in the growth of *R. oryzae* NBRC 4716 after incubation for 10 min. Thus, LAS suppressed 99.99 % of fungal growth. Similar results were obtained when the result of C12K against *R. oryzae* NBRC 4716. We would like to examine the reason for the LAS showed high antifungal activity. The reason for the antifungal activity is suspected same reason as that antibacterial activity of LAS against *B. subtilis*.

No obvious change was observed in tested in LAS against *A. oryzae* (*Akita konnno* store). The reason for the result is suspected same reason as that antifungal activity of fatty acid salts against *A. oryzae* (*Akita konnno* store).

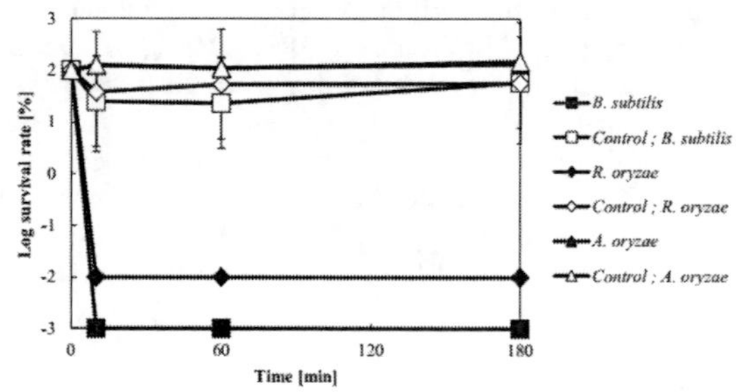

Figure 6. Antimicrovial activity of LAS against *B. subtilis*, *R. oryzae* and *A. oryzae*
Spores were counted at the time of inoculation (0 min) and after 10, 60, 180 min of incubation by means of plating 100 μL portions of the same samples. Log survival rate (expressed log CFU/mL) of the samples stored at 30°C for 1 d or 2 d were enumerated at the specified time points on NA or PDA. Symbols: ■, *B. subtilis*; □, Control of *B. subtilis*; ◆, *R. oryzae*; ◇, Control of R. oryzae; ▲, *A. oryzae*; △, Control of *A. oryzae*. Experiments were performed in triplicate, and the error bars represent the standard deviation.

3.7 MIC of fatty acid salts and LAS

Two-fold dilution samples of the 175 mM solution inoculated with fungus or bacteria were incubated for 10 min, and then applied to the agar medium, and MICs were determined after 7 d of culture. The results are shown in Table 3.

First, we describe the result of MIC of fatty acid salts against *B. subtilis*. All samples of fatty acid salts concentrations > 175 mM were required to inhibit fungal growth. Sample of LAS concentrations > 175 mM were required to inhibit fungal growth.

Next, we describe the result of MIC of fatty acid salts against *R. oryzae*. C10K and C12K at 175 mM inhibited the growth of *R. oryzae* NBRC 4716 for 7 d. However, concentrations > 175 mM of the other samples were required to inhibit fungal growth. LAS at 87.5 mM inhibited the growth of *R. oryzae* NBRC 4716 for 7 d.

Table 3. MIC against *B. subtilis*

FAS	MIC [mM]	
	B. subtilis	*R. oryzae*
C4K : Potassium btyrate	>175	>175
C6K : Potassium caproate	>175	>175
C8K : Potassium caprylate	>175	>175
C10K : Potassium caprate	>175	175
C12K : Potassium laurate	>175	175
C14K : Potassium myristate	>175	>175
C18:1K : Potassium oleate	>175	>175
C18:2K : Potassium linoleate	>175	>175
C18:3K : Potassium linolenate	>175	>175
LAS	>175	87.5

4 Conclusions

In this study, C12K, C18:1K, C18:2K and C18:3K showed the highest antibacterial activity (5 log-units) against *B. subtilis* NBRC 3335, suppressing 99.999% of bacterial growth (Figure 1, 2). A control solution at the same pH as the fatty acid salt solutions did not affect bacterial growth. So we concluded that the antibacterial activity was due to the fatty acid salts themselves, not pH. Also, from the determination of MIC, it was found to be bacteriostatic action. In addition, the antibacterial activity of LAS was same as fatty acid salts.

Next, C12K showed the highest antifungal activity (4 log-units) against *R. oryzae* NBRC 4716, suppressing 99.99 % of fungal growth (Figure 3, 4). A control solution at the same pH as the fatty acid salt solutions did not affect fungal growth. So we concluded that the antifungal activity was due to the fatty acid salts themselves, not pH. Also, the antibacterial activity of LAS was same as fatty acid salts.

No obvious change was observed in tested fatty acid salts against *A. oryzae* (*Akita konnno* store) (Figure 5, 6).

Additional research is required to have high antimicrobial activity with other reagent.

References

1. S. L. Sado-kamdem, L. Vannini, M. E. Guerzoni, Effect of α-linolenic, capric and lauric acid on the fatty acid biosynthesis in *Staphylococcus aureus*, Int. J. Food Microbiol, **129**, 288-294, (2009)
2. S. Khulusi, H. A. Ahmed, P. Patel, M. A. Mendall and T. C. Northfield, The effects of unsaturated fatty acids on *Helicobacter pylori* in vitro, J. Med. Microbiol, **42**, 276-282, (1995)
3. A. P. Desbois; V. J. Smith, Antibacterial free fatty acids : activities, mechanisms of action and biotechnological potential, Applied Microbiology and Biotechnology, **85**, 1629-1642, (2010)
4. S. A. Baskaran, V. Bhattaram, Induupadhyaya, A. Upadhyay, A. Kollanoor-johny, D. Schreiber, JR., and K. Venkitanarayanan, Inactivation of *Escherichia coli* O157:H7 on Cattle Hides by Caprylic Acid and b-Resorcylic Acid, Journal of Food Protection, **76**, 318–322, (2013)
5. M. Era, S. Sakai, A. Tanaka, T. Kawahara, T. Kanyama, H. Morita, Antifungal Activity of Fatty Acid Salts Against *Penicillium pinophilum*, Japan Journal of Food Engineering, **16**, 99-108, (2015)
6. R. Wadowsky, T. M. Wilson, N. J. Kapp, A. J. West, J. M. Kuchta, S. J. States, J. N. Dowling, and R. B. Yee, Multiplication of *Legionella* spp. in tap water containing *Hartmannella vermiformis*, Environmental Microbiology, **57**, 1950-1955, (1991)
7. A. Kollanoor, P. Vasudevan, M. K. M. Nair, T. Hoagland, K. Venkitanarayanan, Inactivation of bacterial fish pathogens by medium chain lipid molecules (caprylic acid, monocaprylin and sodium caprylate), Aquaculture Res., **38**, 1293-1300, (2007)
8. K. Takashima, Mold inspection manual color atlas (Kabi Kensa Manuaru Kara Zufu), Techno System Co., Ltd., 214-215, (2009)
9. T. Kitahara, N. Koyama, J. Matsuda, Y. Aoyama, Y. Hirakata, S. Kamihira, S. Kohno, M. Nakashima, and H. Sasaki, Antimicrobial Activity of Saturated Fatty Acids and Fatty Amines against Methicillin-Resistant *Staphylococcus aureus*, Biol. Pharm. Bull, **27**, 1321—1326, (2004)
10. L. Ababouch, A. Chaibi, F.F. Busta, Inhibition of bacterial spore growth by fatty acids and their sodium salts, J. Food Prot., **55**, 980-984, (1992)
11. L.L. Wang, E. A. Johnson, Inhibition of *Listeria monocytogenes* by fatty acids and monoglycerides, App. Env. Microbiol., **58**, 624-629, (1992)
12. D. Mitoraj, A. Janczyk, M. Strus, H. Kisch, G. Stochel, P. B. Heczko and W. Macyk, Visible light inactivation of bacteria and fungi by modified titanium dioxide, Photochem. Photobiol. Sci., **6**, 642-648, (2007)
13. M. Masuda, M. Era, T. Kawahara, T. Kanyama, H. Morita, Antibacterial Effect of Fatty Acid Salts on Oral Bacteria, Biocontrol Science, **20**, 209-213, (2015)

MATEC Web of Conferences **62**, 03003 (2016) DOI: 10.1051/matecconf/20166203003
ICCFE 2016

Study on the Varying Patterns of Total Phospholipids, Selenium, Phosphorus, Reducing Sugar and Total Sugar, Hydrolyzed Amino Acids in the Velvet Antler of Northeast Sika Deer in Growth Period

Shu-li Wang[1], Yan-mei Wang[1,a]

[1]*Ginseng and Antler Products Testing Center of the Ministry of Agriculture PRC, Jilin Agriculture University, Changchun 130118, China*

Abstract. In this study, the varying patterns of total phospholipids, selenium, phosphorus, reducing sugar and total sugar, hydrolyzed amino acids in the velvet antler of Northeast sika deer in growth period were evaluated. Eighteen Northeast sika deer were allocated into 6 groups according to the antler shedding time. Results indicated that there was significant difference of the selenium content between any two of the six groups ($P<0.05$) except that of Group 1 and Group 2 or Group 5 and Group 6. About the phosphorus there was significant difference between any two of the six groups ($P<0.05$) except that of Group 4 and Group 5 or Group 2 and Group 3 or Group 1 and Group 2. Group 6 had the lowest total Phospholipids content. Both of the reducing sugar and total sugar showed an increasing pattern initially and then decrease gradually.

1 Introduction

The velvet antler is the non-ossification horn of cervidae animals such as Northeast sika deer and red deer. It is a famous traditional Chinese medicine that owns many beneficial effects. Therefore, velvet antler was wildly used in the folk medicine [1]. Velvet antler contains protein, amino acids, peptide, trace elements, phospholipid, fatty acids, and so on [2]. In addition, it also contains some components like chondroitine, acidic polysaccharose, mucopolysaccharides [3, 4]. In this study, the varying patterns of total phospholipids, selenium, phosphorus, reducing sugar and total sugar , hydrolyzed amino acids in the velvet antler of Northeast sika deer in growth period were evaluated. And this may provide some theoretical foundations for the manufacture of velvet antler-health care products. [5, 6]

2. Experimental Section

2.1 Animals and sampling

The velvet antlers were collected from the deer farm of Dongfeng Pharmaceutical factory. Eighteen Northeast sika deer with an average age of four years. Velvet antler with a length of 1 cm was sampled after the animals were anesthetized. The blood in the velvet antler was drained with the vacuum pump [7]. The velvet antler was dried with the vacuum freeze drier and the weight was recorded.

[a] Corresponding author: yanmeimeiw@163.com

2.2 Apparatus and methods

In this study, AFS-933 atomic fluorescence spectrophotometer, DK20 Multi-function digestion device, and Beck-man DU-7500 ultraviolet spectrophotometer were employed. Selenium, phosphorus, total phospholipids, reducing sugar, and total sugar were analyzed according to the followings methods, respectively (Table 1).

Table 1. The content of selenium and phosphorus

Items	Methods
Selenium	GB 5009.93
Phosphorus	GB12393
Total phospholipids	Duan et al., 1988 [8]
Reducing sugar	NY 318 1997-Ginseng products [9]
Total sugar	Lou et al., 2003 [10]
Hydrolyzed amino acids	GB/T5009.124

2.2.1 Selenium

Accurately weigh a certain amount of sample Weigh 0.5g ~ 2g sample, liquid sample lessons 1.00mL~ 10.00mL, placed in the digestion flask, add 10.0mL mixed acid and a few grains of glass beads, covered with a watch glass of cold digestion overnight. The next day on a hot plate heated and timely additional nitric acid. When the solution becomes clear and colorless with white smoke, the heating was continued to a residual volume of about 2mL, must not be evaporated to dryness. Cooling, plus

© The Authors, published by EDP Sciences. This is an open access article distributed under the terms of the Creative Commons Attribution License 4.0 (http://creativecommons.org/licenses/by/4.0/).

5.0mL hydrochloric acid, and heating was continued until the solution becomes clear and colorless with white smoke appears, hexavalent selenium reduced to tetravalent selenium. Cooled and transferred to a 50 mL volumetric flask to volume and mix aside. While doing the blank test.

Instrument Reference conditions: negative high voltage: 340 V; lamp current: 100 m A; atomization temperature: 800℃; furnace height: 8mm; carrier gas flow rate: 500 mL/min; shielding gas flow rate: 1000 mL/min; measurement: standard curve method; reading mode: peak area; delay time: 1s; reading time: 15s; dosing time: 8s; injection volume: 2mL. setting the optimal conditions for the instrument, the furnace temperature was raised gradually to the desired temperature, start measuring stable after 10min~20 min. Continuous series of zero standard tube injection until after the reading has stabilized, the standard series of measurements into the standard curve. Into the sample measurement, the sample and blank samples were measured digestive, before each test different samples should be cleaned injector.

2.2.2 Phosphorus

Accurately weigh a certain amount of sample Weigh 0.1~ 0.5g in the 100m L Kjeldahl flask, 3m L sulfuric acid, 3m L perchloric acid - nitric acid digestion, placed on digestion furnace, beginning to brown-black liquid bottle. When the solution turned to be colorless or slight yellow clear liquid that is completely digested. The solution was allowed to cool, after the transfer plus 20m L, cooled to 100m L flask, washed several times with water Kjeldahl flask washings were combined into the volumetric flask, add water to volume, and mix. This solution is a liquid sample measurement. digested sample taken with the same amount of sulfuric acid, perchloric acid - nitric acid digestion, according to the same method blank solution. phosphorus standard curve Imbibe phosphorus.standard solution 0,0.5,1.0,2.0,3.0,4.0,5.0 mL, were placed 20mL stoppered test tube, followed by adding 2mL molybdic acid solution shake and let stand a few seconds. Join 1m L sodium sulfite solution, 1m L hydroquinone solution shake. Add water to volume, and mix. Standing 0.5h later absorbance was measured at 660nm wavelength spectrophotometer. To measure the absorbance of the standard curve for phosphorus content.

Imbibe the measured liquid 2m blank solution L and the same amount of samples were placed 20m L stoppered test tube, and the remaining steps with the standard curve to measure the absorbance Richard unknown liquid phosphorus content standard curve.

2.2.3 Total phospholipids

Weigh 2g sample in a stoppered Erlenmeyer flask, Folch reagent (chloroform - methanol 2: 1) 20m L, said that given the weight, ultrasound extraction 1h, placed at room temperature, weighed and the weight, plus Folch reagent re-fill, 3000r · min - 1, centrifugal 5min, the supernatant was filtered to 25m L flask, with Folch reagent to the mark, as the test solution.

2.2.4 Reducing sugar

Accurately weighed sample 2g, set in the 100mL beaker, added 85 ring ethanol 50mL, mixing, holding 30 min at 500C in a water bath, filtered, and the residue was then 85% ethanol solution was extracted twice, and the filtrate were combined, the ethanol was evaporated, add a small amount of distilled water and a total volume of 100 mL.

After each tube and mix prepared standard curve method of operating a measured absorbance of each tube, to identify the appropriate reducing sugar content in the standard curve, the samples reducing sugar.

2.2.5 Total sugar

Weigh 1g sample in a test tube, added 10mL, 15mL hydrochloric acid solution and distilled water mixed and heated for 30 min, a solution of iodine with potassium iodide to check the degree of hydrolysis in a boiling water bath, if it has completely hydrolyzed, was not blue color (of starch), cooled I added drops of phenolphthalein indicator, with a 10% sodium hydroxide solution to reddish, filtered and set the volume to 100mL. And then to learn the exact solution 10ml, set people 100mL volumetric flask with distilled water to the mark.

After each tube and mix prepared standard curve method of operating a measured absorbance of each tube, to identify the appropriate reducing sugar content in a standard curve of the samples to the total sugar content.

2.2.6 Hydrolyzed amino acids

Accurately weigh a certain amount of sample to the nearest 0 .0001g, sample weight within 10~20 mg, saying that a good sample put in the hydrolysis tube. Hydrolysis: Add in the hydrolysis tube 6.0 mol/L HCl 15.00mL, was added thioglycolic acid 1.00 ml (for protection), cover stopper with vacuum pump to vacuum, tap into 0 psi, this hydrolysis tube and placed in the oven thermostat seal 110℃ ± 1℃ after 24h of incubation, remove the cooling. Open hydrolysis tube stopper, the hydrolyzate was filtered and washed constant volume to 50.00ml volumetric flask. 1.00 ml draw in the filtrate was 25 ml beaker, evaporated to dryness on a water bath, the residue was dissolved in water 1 ~ 2 ml, then dried up again evaporated to dryness three times, and finally the residue was 0.02mol/L HCl 1.00mL dissolved machine measurement.

3. Results and Analysis

3.1 The contents of selenium and phosphorus in the velvet antler

Results indicated that there was significant difference of the selenium between any two of the six groups ($P<0.05$) except that of Group 1 and Group 2 or Group 5 and

Group 6. About the content of phosphorus there was significant difference between any two of the six groups ($P<0.05$) except that of Group 4 and Group 5 or Group 2 and Group 3 or Group 1 and Group 2 (Table 2).

Table 2. The contents of selenium and phosphorus in the velvet antler

Group	Selenium (ng/g)	Phosphorus (%)
1	211.41a	5.05d
2	221.34a	5.86cd
3	186.24b	6.57c
4	150.44c	7.5b
5	124.01d	7.86b
6	131.31cd	8.81a

Selenium is an important ingredient of glutathione peroxidase and many health care products were wildly consumed because of their high content of this element. As we known, velvet antler was rich in selenium too. Results in this study indicated the highest content of selenium was at the 40 days after the antler shedding. Phosphorus is vital to the life because it is not only the ingredient of nucleic acid and ATP, but also makes up a lot of enzymes with the other elements. Such results Antler Growth Law emergence are inseparable. Deer stag in mid-April after a year in late fall to early May faceplate, antler began to rapidly grow, reached a peak of about 70 d, before the peak of the growth, the growth rate is greater than the speed of ossification, ossification and then faster than the growth rate, and then dried skin off bone horn, bones, horns last fall, the end of an antler growth cycle. Velvet deposited nutrients in addition to the outer part of the minerals by food intake, and thus this body needs to absorb a large number of deer nutrients for the growth of the supply of velvet. Phosphorus involved in energy metabolism, the body if the lack of Phosphorus, the bone marrow, tooth development is not normal. Osteoporosis, softening, easy to fracture or suffer from rickets in children, loss of appetite, muscle weakness.

The content of phosphorus was increased as the shedding time of velvet antler prolonged. Phospholipids can not only facilitate the cell renewal and maintain the metabolism, but also promote the immunity and regenerative capacity of human beings [11, 12].

3.2 The contents of total phospholipids, reducing sugar and total sugar in the velvet antler

Table 3. The contents of total phospholipids, reducing sugar and total sugar in the velvet antler

Groups	Total phospholipids (%)	Reducing sugar (%)	Total sugar (%)
1	2.45ab	0.24d	3.61e
2	2.39ab	0.39c	4.42d
3	2.54a	0.54a	5.04c
4	2.62a	0.50ab	6.15a
5	2.67a	0.47b	5.61b
6	2.18b	0.32c	5.73ab

Generally, Group 5 had the highest total phospholipids content ($P<0.05$) whereas Group 6 had the lowest ($P<0.05$).

The sugar is an important organic compound which is widely distributed in the natural world. It is one of the primary sources for life sustaining. Both of the reducing sugar and total sugar showed an increasing pattern initially and then decrease gradually (Table 3).

The experiment measured plum velvet herbs total phospholipid content, slightly lower than the reported in the literature, may be related to Antler taken herbs processing methods; about Deer Antler total phospholipid content has not been reported. Determination of the total phospholipid content of common mainly colorimetry and HPLC. Phospholipid assay is converted to inorganic phosphorus colorimetrically to calculate the total phosphorus content of the phospholipid content. The HPLC method is generally measured total phospholipid content of each phospholipid component, summing the total phospholipid content of each phospholipid content, this method has the instrument detects a bit fast, but many phospholipid components, each group is not easy to points are measured completely [13], and the price is relatively expensive Ao, measured in terms of total phospholipid higher costs. The colorimetric test solution by a standard phosphorus as a low cost reference and measurement method is simple, the application is also more practical significance to promote [14].

Antler is a valuable traditional Chinese medicine, complex composition, in recent decades, and found a variety of active ingredients, such as amino acids, fatty acids, chondroitin sulfate, hypoxanthine, uracil, and phospholipids polyamine compounds [15]. Wherein the phospholipid is considered to be one of the active ingredients [16], in addition to inhibition of monoamine oxidase activity action outside, stimulate growth and hematopoietic function, enhance the regeneration process and improved neuromuscular function of the system antler etc, velvet phospholipids effect is closely related to. Antler existing literature on total phospholipid segmented detection, content decreasing from top to bottom [17], but did not find them in the diversity of the statistical analysis report.

3.3 Results of hydrolyzed amino acids

Table 4 The content of hydrolyzed amino acids

Group	Amino acid at the middle of velvet antler (%)	Amino acid at the top of velvet antler (%)
1	47.55a	57.15ab
2	45.35ab	59.62a
3	40.25c	50.86c
4	40.13c	50.67c
5	39.05c	51.78bc
6	41.35bc	62.95a

Note: the different superscripts of the data on the same row means significant difference ($P < 0.05$ or $P < 0.01$)

It can be indicated from this Table 4 that among the hydrolyzed amino acids of velvet antler of Northeast Sika Deer in different growth periods, there are significant difference of amino acids content at the middle of velvet antler between group 1 and 3, 4, 5, 6 ($p<0.05$), and between group 2 and 3, 4, 5, 6 ($p<0.05$). There are significant difference of amino acids content at the top of velvet antler between group 1 and 3, 4, 6, between group 2 and 3, 4, 5, and between group 6 and 3, 4, 5 ($p<0.05$).

Amino acids are organic component velvet topped the content of nutrients, the highest glycine; amino acids are the basic components of living organism tissue cells for life events play a pivotal role. If the body lacks any kind of essential amino acids, it can cause physiological dysfunction, affecting the normal antibody metabolism [18], leading to disease. Similarly, if the body lacks certain non-essential amino acids, it will produce antibodies metabolic disorders. Arginine and citrulline formation of urea is very important; cystine intake will cause reduction of insulin, blood sugar. Another example is the post-traumatic cystine and arginine requirements increase [19], such as the lack of adequate energy even if you still can not synthesize proteins. Thus, the presence of amino acids in the human body, not only provides an important raw material for the synthesis of proteins, but also for the promotion of growth and normal metabolism, provides the material basis of life-sustaining

4. Conclusions

Selenium is an important ingredient of glutathione peroxidase and many health care products were wildly consumed because of their high content of this element. As we known, velvet antler was rich in selenium too. Results in this study indicated the highest content of selenium was at the 40 days after the antler shedding. Phosphorus is vital to the life because it is not only the ingredient of nucleic acid and ATP, but also makes up a lot of enzymes with the other elements. The content of phosphorus was increased as the shedding time of velvet antler prolonged. Phospholipids can not only facilitate the cell renewal and maintain the metabolism, but also promote the immunity and regenerative capacity of human beings [20]. Generally, Group 5 had the highest total phospholipids content ($P<0.05$) whereas Group 6 had the lowest ($P<0.05$). The sugar is an important organic compound which is widely distributed in the natural world. It is one of the primary sources for life sustaining [21]. Both of the reducing sugar and total sugar showed an increasing pattern initially and then decrease gradually. These data may provide some theoretical foundations for the manufacture of velvet antler-health care products.

Amino acid is the basic unit composing the protein. Some physiological activities in velvet antler are directly correlated to the amino acid. The amino acid in human body provides the important material for protein synthesis, but also provides the material basis for promoting growth, normal metabolism and maintaining the life [22]. The content of hydrolyzed amino acids increases along with time, while the change is not regular but some groups of hydrolyzed amino acids show significant difference ($p<0.05$), which may be correlated to the accumulation of amino acids at different time. All the results provide the scientific basis for collection and utilizaiton of velvet antler.

5. Summary

These data may provide some theoretical foundations for the manufacture of velvet antler-health care products.

References

1. The pharmacopoeia committee of China. China Pharmacopeia. (People's Medical Publishing House, China 2000).
2. B. Y. Wang and Q L. Zhou: Acta Pharmaceutica Sinica Vol. 26 (1991), p. 714-720.
3. F. Y. Dong, H, Zhang and Y. Chen: Jilin Journal of Traditional Chinese Medicine Vol. 18 (1998), p.62.
4. Wang Quanwei,Dong ZhenWang,Guiwu. Research Progress on Proteins Related to Deer Antler Regeneration, [J] Biotechnology Bulletin, 2015, 1.31.8 (24-29).
5. Zha Enhui, Li Xing xia, Li Dandan, et al.Immunomodulatory ef-fects of a 3.2 k Da polypeptide from velvet antler of Cervus nip-pon Temm inck[J] .International Immunopharm acology , 2013, 16 :210-213.
6. Sui Zhigang, Zhang Lihua, Huo Yushu, et al .Bioactive componentsof velvet antlers and their pharm acological properties[J]. Journalof Pharm aceutical and Biomedical Analysis, 2013(8):1-12.
7. Farberov S, Meidan R. Functions and transcriptional regulation of thrombospondins and their interrelationship with fibroblast growth factor-2 in bovine luteal cells [J]. Biol Reprod, 2014, 91 (3): 58.
8. C. F. Duan, X. H. Gao and J. H. Wang: Special Wild Economic Animal and Plant Research Vol. 4 (1988), p. 41-43.
9. NY318 1997-Ginseng products. (China Standards Press, China 1998).
10. X. H. Lou and Q. F. Wu: China Pharmaceuticals Vol. 12 (2003), p. 57.
11. M. Sadighi, S.R. Haines A. SkottnerA.J. Harris and J.M. Suttie: Journal of Endocrinology Vol. 143 (1994), p. 461-469.
12. R.R. Dalefield and F.W. Oehme: Veterinary and human toxicology Vol. 41 (1999), p. 39-41.
13. John Mikler,Christine Theoret,Jerry C.Effects of Topical Elk Velvet Antler on Cutaneous Wound Healing in Strepto-zotocin-Induced Diabetic Rats High. Alternative and Complementary Medicine. 2004
14. L Eriksson, E Johansson, N Kettaneh-Wold, J Trygg, C Wikstrom, S Wold.Multi- and megavariate data analysis part I: basic principles and applications, second revised and enlarged edition. 2006
15. M. G. Horning, K. L. Knox,C. E. Dalgliesh,E. C. Horning.Gas-liquidchromato-graphic study and estimation of several urinary aromatic acids. Analytical Biochemistry. 1966
16. N. Boussetta, E. Vorobiev, L. H. Le, A. Cordin-Falcimaigne, J.-L. Lanoisellé. Application of electrical treatments in alcoholic solvent for polyphenols extraction from grape seeds [J]. LWT - Food Science and Technology. 2011 (1)

DOI: 10.1051/matecconf/20166203003

17. Zhou Q L, Guo Y J, Wang L J, et al. Velvet antler polypeptides promoted proliferation of chondrocytes and osteoblast precursors and fracture healing. Acta Pharmacological Sinica. 1999
18. R L Garcia, M Sadighi, S M Francis, J M Suttie, J S Fleming.Expression of neurotrophin-3 in the growing velvet antler of the red deer Cervus elaphus. Journal of Molecular Endocrinology . 1997
19. C. Li, J. M. Suttie. Deer antlerogenic periosteum: a piece of postnatally retained embryonic tissue? [J]. Anatomy and Embryology. 2001 (5)
20. J. F. Wang, D. C. Li and Z. Li: Biochemistry. (Chinese Press of Traditional Chinese Medicine, China 2002.
21. Pei Wenxuan, Li, Fei, Dong Ling, Review of Chemical Components and Biological Effect of Pilose Antler, [J] World Science and Technology/Modernization of Traditional Chinese Medicine and Materia Medica, 2012,05:2065-2069.
22. Hong H, Zhou T, Fang S, et al. Pigment epithelium-derived factor（PEDF）inhibits breast cancer metastasis by down-regulating fibronectin [J]. Breast Cancer Res Treat, 2014, 148 (1), 61-72.

Complete mitochondrial DNA and phylogenetic study of qionglai native black chicken

Yaodong Hu, Yun Zhu, Huizhong Pang, Dan Lan

Farm Animal Genetic Resources Exploration and Innovation Key Laboratory of Sichuan Province, Sichuan Agricultural University, Ya'an, Sichuan 625014, China

Abstract. The complete mitochondrial genome sequence of the qionglai black chicken was measured by PCR-based methods, the molecular characterization and phyletic evolution analyzed in detail. Our result showed that the entire mitochondrial genome of the qionglai black chicken is a circular molecule consisting of 16,785 bp (GenBank accession number: KT958484). The contents of A, T, C, and G were 30.25%, 23.74%, 32.54% and 13.48% in the mitochondrial genome, respectively. The complete mitochondrial genome of the qionglai black chicken contains a typical structure, including 13 protein-coding genes, 2 rRNA genes, 22 tRNA genes and 1 control region (D-loop region). The phyletic evolution analysis shown that this chicken was evolution between the red jungle fowl and the special egg chicken white loghorn chicken.This complete mitochondrial genome sequence provides essential information in understanding phylogenetic relationships among Gallus gallus domesticus mitochondrial genomes and the breeding of native chick

1 Introduction

In China, indigenous chickens making up 20% of the poultry market, and the indigenous chicken market is rapidly developing by a rate of 5 to 10% per year[1]. Native black chicken is one of the famous Chinese indigenous chickens in Sichuan province, and it has a large number of breeding in qionglai city which was called as qionglai black chicken. The qionglai black chicken, which has black feathers, black skin and red comb. The qionglai black chicken as the importance indigenous breed and genetic resource was a valuable resource for preserve genetic diversity.

Recently, the chicken market was occupied by specialied commercial populations, which originated in foreign countries. It was the enormous impact for the indigenous chicken market. Qionglai black chicken as the indigenous breeds was a large genetic resousrces as well as having well-adapted and stress toletant as its advantage. While it also have some disadvantage like low production performance. therefore, taking full advantage of this breeds was importance for the indgenous chicken breeding and save the gentic diversity .

Mitochondrial DNA(mtDNA) have the characteristic of quick evolution, matrilineal inherritance and simple molecular structure , which was the most being used as the molecular study marker. Recently, many studies have reported the results of using the fine-gained complete mtDNA to analyses and reconstruct the history of animal domestication, such as horses, pigs and cattle[2-7] .in this study, we research the mtDNA of qionglai black chicken to understanding it characterisric and phylogenetic.

2 Material and Methods

A total of 15 qionglai black chicken were sampled for this study, All birds were reared under the same condition: cage free (density of <35 chicken/100 m 2) and standard conditions of temperature (20° to 22°C), relative humidity (55 to 60%), and ventilation were maintained. The chickens were fed the same professional breeder diet and had free access to feed and water during the entire rearing period. Birds were managed with full consideration of bird welfare. All procedures involving animals were approved by the Institutional Animal Care and Use Committee of the National Institute of Agrobiological Sciences.

2.1 Sample collection

Approximately 2 mL blood was collected from the brachial vein of each individual according to the requirements of animal welfare of Sichuan Agricultural University, and stored at -20°C. The next day, genomic DNA was isolated by phenolic extraction. The quality and integrity of DNA was assessed using the A260/280 ratio and agarose gel electrophoresis.

2.2 Data Analysis

© The Authors, published by EDP Sciences. This is an open access article distributed under the terms of the Creative Commons Attribution License 4.0 (http://creativecommons.org/licenses/by/4.0/).

The complete mitochondrial genome was amplified by 22 pairs of primers designed according to the sequence of the Gallus gallus (GenBank accession number: AP003322). To determine whether the sequences obtained were correct, an NCBI nucleotide BLAST search was conducted (http://blast.ncbi.nlm.nih.gov/). The DNA sequence was analyzed using DNAStar 7.1 software (Madison, MI) and the base composition and distribution of the mitochondrial DNA sequence were analyzed using tRNA Scan-SE1.21 software (http://lowelab.ucsc.edu/tRNAScan-SE/)[8] and DOGMA software (http://dogma.ccbb.utexas.edu/) [9], the phylogenetic were analyzed by mega4.1, respectively.

3 Results and Discussion

3.1 The stucture of the mtDNA

We submission the sequence to the NCBI(http://www.ncbi.nlm.nih.gov/) and the genbank accession number was KT958484.The complete mitochondrial DNA of qionglai native chicken was 16785bp which contain 13 protein-coding genes, 22 transfer RNA genes, 2 ribosomal RNA genes and 1 control region (D-loop region). The gene arrangement and composition were similar to those of other avian species[10, 11]. The overall nucleotide composition of A, T, C and G were found to be 30.28%, 23.74%, 32.50% and 13.48%, respectively, and in the order C>A>T>G. Most of the genes are encoded on the H-strand and they are similar in structure to the typical mitochondrial genome of vertebrates[12].

The molecular characterization of qionglai black chicken mitochondrial genome as followed (Table 1). Firstly, the initiation codon of the protein -coding genes is ATG, except for COX1, which show a GTG initiation codon. Secondly, have nine genes not encoded on the L-stand, including 1 protein-coding gene (ND6) and 8 tRNA genes (tRNA-Gln, tRNA-Ala, tRNA-Asn, tRNA-Cys, tRNA-Tyr, tRNA-Ser, tRNA-Pro and tRNA-Glu). Thirdly, these genes have four different types of termination codons: TAA AGG TAG and T--. Among all gene elements, there were overlaps and spaces. The D-loop contains regulatory elements that control mtDNA replication and transcription, which is a non-coding control region located between the tRNA-Glu and tRNA-Phe genes, and rich in A and T.

3.2 Phylogenetic analysis

We alignment the sequence of qionglai black chicken and other gallus (red jungle fowl, Ceylon jungle fowl, Green jungle fowl, grey jungle fowl, white leghorn), the outgroup were meleagris gallopavo, aix galericulata, alectoris chukar and tadorna ferruginea, the evolution tree shown that the qionglai black chicken phyletic evolution was between the red jungle fowl and white leghorn(Figure 1). As the results shwon, we speculate that the qionglai black chicken was suitable for developent in egg production.

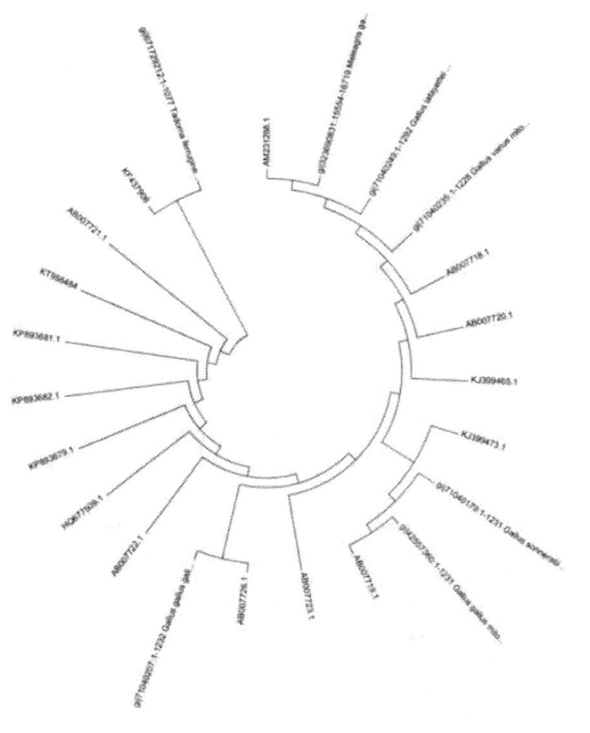

Figure 1. The Unrooted neighbor-joining (NJ) tree of qionglai black chicken and others.

Table 1. The molecular characterization of qionglai black chicken mtDNA

Gene	Position			codon				
	start	end	size	start	stop	anti	stand(H/L)	space(+)/ovrelap(-)
D-loop	1 16716	1162 16785	1232				H	0
tRNA-Phe	1163	1231	69			GAA	H	0
12S rRNA	1232	2207	976				H	0
tRNA-Val	2208	2280	73			GAC	H	0
16S rRNA	2281	3902	1622				H	0
tRNA-Leu	3903	3976	74			GAG	H	0
ND1	3986	4960	975	ATG	TAA		H	9

tRNA-Ile	4961	5032	72			GAT	H	0
tRNA-Gln	5038	5108	71			TTG	L	5
tRNA-Met	5108	5176	69			CAT	H	-1
ND2	5177	6217	1041	ATG	TAG		H	0
tRNA-Trp	6216	6291	76			CCA	H	-2
tRNA-Ala	6298	6366	69			AGC	L	6
tRNA-Asn	6370	6442	73			ATT	L	3
tRNA-Cys	6444	6509	66			GCA	L	1
tRNA-Tyr	6509	6579	71			GTA	L	-1
COI	6581	8131	1551	GTG	AGG		H	1
tRNA-Ser	8123	8197	75			TGA	L	-9
tRNA-Asp	8200	8268	69			GTC	H	2
COII	8270	8953	684	ATG	TAA		H	1
tRNA-Lys	8955	9022	68			CTT	H	1
ATPase 8	9024	9188	165	ATG	TAA		H	1
ATPase 6	9179	9862	684	ATG	TAA		H	-10
COIII	9862	10645	784	ATG	TAA		H	-1
tRNA-Gly	10647	10714	68			ACC	H	1
ND3	10716	11066	351	ATG	TAA		H	1
tRNA-Arg	11068	11135	68			TCT	H	1
ND4L	11136	11432	297	ATG	TAA		H	0
ND4	11426	12803	1378	ATG	TAT		H	-5
tRNA-His	12804	12872	69			ATG	H	0
tRNA-Ser	12874	12938	65			GCT	H	1
tRNA-Leu	12940	13010	71			AAG	H	1
ND5	13011	14828	1818	ATG	TAA		H	0
cytb	14833	15975	1143	ATG	TAA		H	4
tRNA-Thr	15979	16047	69			TGA	H	3
tRNA-Pro	16048	16117	70			TGG	L	0
ND6	16124	16645	522	ATG	TAA	CAT	L	6
tRNA-Glu	16648	16715	68			TTC	L	2

Acknowledgments

This work was supported by the program for the key project from education department of Sichuan Province (14ZA0021), the project of the application basis project from the science and technology department of Sichuan province (2014JY0163), and the project of national natural science foundation of china (31402068).

References

1. Tang, H., et al., *Variation of meat quality traits among five genotypes of chicken.* Poultry science, 2009. **88**(10): p. 2212-2218.
2. Achilli, A., et al., *The multifaceted origin of taurine cattle reflected by the mitochondrial genome.* PloS one, 2009. **4**(6): p. e5753.
3. Achilli, A., et al., *Mitochondrial genomes of extinct aurochs survive in domestic cattle.* Current Biology, 2008. **18**(4): p. R157-R158.
4. Achilli, A., et al., *Mitochondrial genomes from modern horses reveal the major haplogroups that underwent domestication.* Proceedings of the National Academy of Sciences, 2012. **109**(7): p. 2449-2454.
5. Bonfiglio, S., et al., *The enigmatic origin of bovine mtDNA haplogroup R: sporadic interbreeding or an independent event of Bos primigenius domestication in Italy?* PloS one, 2010. **5**(12): p. e15760.
6. Pang, J.-F., et al., *mtDNA data indicate a single origin for dogs south of Yangtze River, less than 16,300 years ago, from numerous wolves.* Molecular biology and evolution, 2009. **26**(12): p. 2849-2864.
7. Wu, G.-S., et al., *Population phylogenomic analysis of mitochondrial DNA in wild boars and domestic pigs revealed multiple domestication events in East Asia.* Genome Biol, 2007. **8**(11): p. R245.
8. Lowe, T.M. and S.R. Eddy, *tRNAscan-SE: a program for improved detection of transfer RNA genes in genomic sequence.* Nucleic acids

DOI: 10.1051/matecconf/20166203004

research, 1997. **25**(5): p. 0955-964.
9. Wyman, S.K., R.K. Jansen, and J.L. Boore, *Automatic annotation of organellar genomes with DOGMA*. Bioinformatics, 2004. **20**(17): p. 3252-3255.
10. Yan, M.-L., et al., *The complete mitochondrial genome sequence of the Daweishan Mini chicken.* Mitochondrial DNA, 2014(0): p. 1-2.
11. Bao, H., et al., *Sequencing and alignment of mitochondrial genomes of Tibetan chicken and two lowland chicken breeds.* Science in China Series C: Life Sciences, 2008. **51**(1): p. 47-51.
12. Boore, J.L., *Animal mitochondrial genomes.* Nucleic acids research, 1999. **27**(8): p. 1767-1780.

Phytochemical Screening, Total Phenolic Content, Antioxidant Activities and Cytotoxicity of *Dendrobium signatum* Leaves

Thitiphan Chimsook[1,a]

[1]Faculty of Science, Maejo University, Chiang Mai, Thailand 50290

Abstract. In the present work, the leaves of *Dendrobium signatum* were extracted with ethanol by maceration; called M; and by sonication-maceration for 30 and 45 mins; called MS30 and MS45. Some biological activities of extracts were investigated. The results showed that the sonication for 30 mins and then maceration 72 hr gave the highest yield at 7.32%. The phytochemical investigation revealed that the bioactive compound from M, MS30 and MS45 were similar which composed of carbohydrates, coumarins, alkaloids, flavonoids, phenolics, sterols and glycosides, respectively. The total phenolic contents were assessed by Folin – Ciocalteau method using gallic acid as the chemical standards. The results showed that MS30 displayed the highest total phenolic contents at the value of 8.83 gGAE/100 g extract. The antioxidant activity was measured using DPPH assay. The results showed that MS30 had the stronger free radical scavenging activity than M and MS45 and had moderate radical scavenging ability compared to ascorbic acid. The cytotoxicity was performed against human cancer cell lines using MTT assay. The results showed that all ethanol extracts displayed weak anticancer activity against MCF-7 and NCI-H187 cell lines, compared to the standard drug doxorubicin. In addition, they showed no activity against KB cell lines.

1 Introduction

The compound which synthesized from the secondary metabolisms was secondary metabolites. Secondary metabolites play a role in reinforcement of many plant tissues. These metabolites are not essential for growth but they produce in order to protect against insects, diseases, and plant regulation (plant hormones). The secondary metabolites compose of a wide range of chemical structures and biological activities. Phytochemicals play an important role for study many bioactive profile of plants and other interesting mircoorganism. Several phytochemical screenings have been carried on for detecting the various groups of naturally occurring phytochemicals. The bioactive compound considers effective in many biological activities A number of phytochemicals including alkaloids, flavonoids, bibenzyl derivatives, phenanthrenes was found in the orchid constituents and suggested various medicinal properties which provided the antimicrobial, antitumor, anti-inflammatory and antiviral activities. Moreover, various other phytochemical constituents have been reported from many orchid species.

In recent decade, the interesting numerous studies have been undertaken to search for the most effective antioxidants because the antioxidants will reduce the oxidative stress in cells and be useful in various treatments of the human diseases [1-3]. Most of plant cells produce the secondary metabolites which have significant potential as antioxidants. Chemical classes of antioxidant secondary metabolites were found such as polyphenol, isoprenoids, tocopherols. Many assay methods were used to estimate the antioxidant content were DPPH method, ABTS method, ORAC assay, PCL assay, β-Carotene linoleic acid bleaching assay, Reducing power assay, Total flavonoid content, Folin-Ciocalteu method [1-3]. Nowaday, the natural antioxidants from plant and other tissues represent the high potentially antioxidant property and less side effect. These can be the alternative for using the natural antioxidants instead of the synthetic antioxidants in the food processing industry and also using in the other pharmacies. Several compound from orchid family were found to have strong antioxidative property. The natural crude obtained from stem, leaves, roots and other parts of orchid which extracted with polar solvents was reported that it presented the significant antioxidant properties. Moreover, several compound were isolated from the Dendrobium orchid showed high and moderate antioxidative properties such as dihydrostilbene isoamoenylin from roots of *Dendrobium amoenum*, Cis- Cismelilotoside, dihydromelilotoside and trans-melilotoside from stems of *Dendrobium aurantiacum*. In addition, The ethanolic extract of *Dendrobium nobile* was found to exhibit antioxidative property equivalent or higher to ascorbic acid. Alkyl ferulates and quercetin were antioxidative compound which were isolated from *Dendrobium monoliforme* and *Dendrobium tosaense* [1-3, 4].

Cancer is the cause of more than 20% of all deaths of one-third of the world's population. Cancer occurs from

[a] Corresponding author: thitiphan.cs@gmail.com

© The Authors, published by EDP Sciences. This is an open access article distributed under the terms of the Creative Commons Attribution License 4.0 (http://creativecommons.org/licenses/by/4.0/).

various evidences such as infection, chemicals, environment factors etc. It is important to investigate new herbal medicine which potential as anticancer or antitumor agents for cancer treatment. Nowadays, the bioactive compounds extracted from the parts of plants and various herb were studied and reported that they had the anticancer activities as called the anticancer plants for example *Panex ginseng*, Paclitaxel (Taxol), Turmeric (Curcumin), Carotenoids, Mushrooms, Green tea, Resveratrol [5]. Moreover, the plants with high anticancer activities are undergoing clinical trial. The species of orchids have been studied and a number of bioactive compound with antitumor activity have been isolated. *Dendrobium nobile*, *Dendrobium chrysanthum*, *Dendrobium fimbriatum* have been a good source of compounds with anti-tumor activity [4-6]. The anticancer or antitumor activity of plant extracts was performed the cytotoxicity using MTT method against various cell lines such as human lung carcinoma, human ovary adenocarcinoma, humanpromyelocytic leukemia cell lines, Hela, MCF-7 cell lines.

Dendrobium, a genus of orchid, has been possessed the useful therapeutic activities including antitumor, anticancer, antimicrobial and antioxidant. In this research, *Dendrobium signatum* was used as the plant study. *D. signatum* was one of the species of genus Dendrobium [4, 6]. The flowers and leaves of *D. signatum* were shown in Figure 1 and Figure 2. The leaves were dark green and the flowers had the white sepals and petals. The lip was white with a yellow center. *D. signatum* bloomed from winter to summer. *D. signatum* can grow in Thailand, Los, Vietnam and India. The information of biological activities of *D. signatum* is not reported. The study of this plant has been focused on the analysis of sequence diversity through internal transcribed spacers to identify various Dendrobium species [7]. Therefore, the objective of this study was to investigate the biological activities of *D. signatum* leaves extracts including phytochemical analysis, antioxidant activities and total phenolic contents and to determine cytotoxic activities of extracts of *D. signatum* leaves.

Figure 1. The flowers of *D. signatum* in Sansai, Chiang Mai, Thailand.

Figure 2. The leaves of *D. signatum* in Sansai, Chiang Mai, Thailand.

2 Materials and Methods

D. signatum was grown and collected from Sansai orchid farm, Chiang Mai Province in April 2015, and authenticated by Faculty of Science, Department of Biotechnology, Maejo University, Chiang Mai, Thailand. The dry leaves in the same size and age were used in these experiments.

2.1 Preparation of crude extraction

The first method of crude extraction was performed under the maceration. The dry leaves of *D. signatum* (100 g) were cut and grind into small pieces and then macerated for 72 hours with 95% ethanol (200 mL) at room temperature under the scheme of extraction shown in Figure 3. The agitation was done every 12 hours for 15 mins. The extraction solution was filtered and concentrated under reduced pressure, lyophilized to yield a residue and stored at −20 °C. All crude ethanol extracts from this method were called "M" and used to determine the phytochemical screening, total phenolic contents, antioxidant potential and *in vitro* cytotoxic activities, respectively.

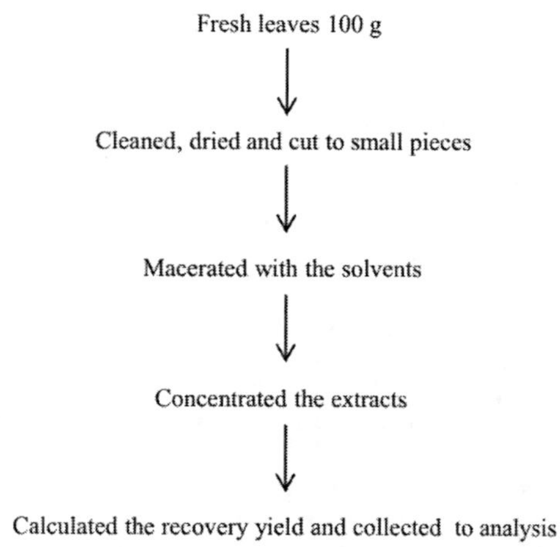

Figure 3. A procedure of the maceration of *D. signatum* leaves extraction.

The second methods of preparation of crude extraction were prepared by using both maceration and sonication. The dry leaves of *D. signatum* (100 g) were cut and grind into small pieces and then sonicated for 30 and 45 mins with 95% ethanol (200 mL) before maceration for 72 hours with the same condition as the first maceration method. The scheme of preparation of the second method was shown in Figure 4. The agitation was done every 12 hours for 15 mins. The extraction solution was filtered and concentrated under reduced pressure, lyophilized and stored at −20 °C. All crude ethanol extracts obtained from the sonication and maceration for 30 or 45 mins were called "MS30 and MS45" and used to determine the phytochemical screening, total phenolic contents, antioxidant potential and *in vitro* cytotoxic activities, respectively.

 DOI: 10.1051/matecconf/20166203005

Fresh leaves 100 g

↓

Cleaned, dried and cut to small pieces

↓

Sonicated with the solvents

↓

Macerated with the solvents

↓

Concentrated the extracts

↓

Calculated the recovery yield and collected to analysis

Figure 4. A procedure of the sonication and maceration of *D. signatum* leaves extraction.

2.2 Phytochemical investigation

Phytochemical investigation was achieved in order to preliminary screen the active compound in the ethanoic of leaves extract of both extraction methods. The dried sample was tested following the standard and applied methods for screening the active compound including the carbohydrates, coumarins, alkaloids, flavonoids, phenolics, sterols, glycosides, tannins, saponins, respectively [8-13].

2.3 Determination of total phenolic contents

The total phenolic contents of ethanol extracts were achieved by Folin–Ciocalteu method. About 0.5 ml of ethanol extracts (10 μg/ml) was added to 2.5 ml of Folin–Ciocalteu solution. After 2 min reaction time, 7% of sodium carbonate solution (2.5 ml) was added to the mixture. Then the mixture was incubated for 30 min at 30 °C to complete the reaction. The absorbance of the resulting mixture was measured at 760 nm against blank. The gallic acid was used as standard. The results were calculated the total contents of phenolic compounds in ethanol extracts in gram gallic acid equivalents (gGAE) to 100 gram of extracts [14, 15].

2.4 Determination of DPPH radical scavenging ability

DPPH is 1, 1-diphenyl-2-picrylhydrazyl. DPPH radical-scavenging ability of ethanol extracts was evaluated. In brief, an aliquot of 1.5 ml of 0.05% DPPH solution in ethanol and 0.1 ml of extract or standard (ascorbic acid) at various concentrations in μg/ml were mixed. The mixture of DPPH and standard or DPPH and extracts was shaken vigorously and incubated for 50 min in the dark at room temperature. Decolourization of DPPH was determined by measuring the absorbance at 517 nm. using a spectrophotometer. Ethanol serves as a blank. The percentage inhibition activity was calculated by using the following formula. [% DPPH radical scavenging activity = [{(Absorbance of control – Absorbance of sample or standard}/ Absorbance of control]*100. The results can express as IC_{50} value in μg/mL [14, 15].

2.5 Cytotoxicity studies

The ethanol extractions were tested for cytotoxic activity against the human oral carcinoma (KB), breast cancer (MCF-7) and human small lung (NCI-H187) lines for the evaluation of growth inhibition. Cell line growth was monitored using the MTT assay as reported [16], and compared with doxorubicin as a positive control. The results were expressed as IC_{50} in μg/mL.

3 Results and Discussion

3.1 Preparation of crude extraction

The ethanol extracts from the leaves of *D. signatum* had the dark green residues. The percentage yield of extracts was presented in Table 1. The results showed that the extraction procedure of sonication and maceration (M30, M45) gave higher yield than that of only maceration (M). These suggested that sonication can help to enhance the extract yields which ethanol as solvent can penetrate and eluted some of the active compound from the small pieces *D. signatum* leaves. It should affect to the qualitative and quantitative of the other biological activities of the extracts.

Table 1. The percentage yield and characteristic of the ethanol extractions (M, MS30, MS45)

Compounds	Percentage yield (%)	Colour
M	5.45±0.79	Dark green
MS30	7.32.±0.43	Dark green
MS45	6.87±0.19	Dark green

The results showed that the extraction procedure of sonication and maceration (M30, M45) gave higher yield than that of only maceration (M). These suggested that sonication can help to enhance the extract yields which ethanol as solvent can penetrate and eluted some of the active compound from the small pieces *D. signatum* leaves. It should affect to the qualitative and quantitative of the other biological activities of the extracts.

3.2 Phytochemical screening

The results of the phytochemical screening of ethanol extracts were shown in Table 2.

Table 2. Phytochemical screening of the ethanol extractions. Meanings are + as "Present" and - as "Absence".

Compounds	Ethanol extracts		
	M	MS30	MS45
Carbohydrates	+	+	+
Coumarins	+	+	+
Alkaloids	+	+	+
Flavonoids	+	+	+
Phenolics	+	+	+
Sterols	+	+	+
Glycosides	+	+	+
Tannins	-	-	-
Saponins	-	-	-

In general, several bioactive compounds including alkaloids, flavonoids, bibenzyl derivatives, phenanthrenes were investigated in the orchid constituent [4, 6]. For Dendrobium species, various organic solvents including petroleum ether, hexane, acetone, ethyl acetate, ethanol and water were used to extract the bioactive substances from the different parts of Dendrobium orchid. In this work, ethanol was used to extract the bioactive compound from *D. signatum* leaves using maceration and sonication – maceration procedures. The M extract was obtained from the maceration. For, MS30 and MS45 were obtained from the sonication – maceration procedures in different sonication time of 30 and 45 mins. Those extracts were tested the phytochemical screening. The results revealed that the bioactive compound from M, MS30 and MS45 were similar which composed of the carbohydrates, coumarins, alkaloids, flavonoids, phenolics, sterols and glycosides, respectively. The tannins and saponins were absent from these ethanol extracts. These phytochemical studies of ethanol extraction from *D. signatum* leaves were the first reported. The phytochemical analyses of Dendrobium have been performed in various parts and organic solvents. Physico-chemical and phytochemical evaluation of *Dendrobium macraei* roots revealed the presence of carbohydrates, coumarins, alkaloids, phytosterols, flavonoids and phenolic compound in the methanol extracts [8]. Qualitative preliminary phytochemical analysis of whole plants of *Dendrobium ovatum* (L.) Kraenzl reported the bioactive compound including flavonoids, alkaloids, triterpenoids, glycosides, steroids and carbohydrates [9]. The phytochemical analysis of *Dendrobium macrostachyum* Lindl revealed the presence of alkaloids, flavonoids, glycosides, sterols, tannins and phenols [10]. Moreover, the qualitative preliminary phytochemical analysis of ethanol extracts of *Dendrobium ochreatum* leaves presented the steroids, alkaloids, flavonoids, saponins, and tannins [11]. In addition, the phytochemical studies of ethanol extract from *D. pandurantum* leaves found that the bioactive compound composed of the tannin, phenols and flavonoids, respectively [12]. From the different types of Dendrobium species which tested the phytochemical screening, it can be concluded that most of all extracts presented flavonoids, alkaloids, sterols, glycosides and tannins, respectively [13].

3.3 Determination of total phenolic contents

The first report of total phenolic contents of three ethanol extracts including M, MS30, MS45 were assessed by Folin – Ciocalteau method using gallic acid as the chemical standard. All of the value showed in Table 3. The results showed that MS30 performed the highest total phenolic contents at 8.83 gGAE/100 g extract. Both MS30 and MS45 had higher total phenolic contents than M. The total phenolic contents of each Dendrobium species were different because of the solvent extraction and also parts of orchids. For example, the previous report of total phenolic contents of the methanol extract of *D. ovatum* (L.) Kraenzl. was 4.51±0.4 g GAE/100 g [9].

Table 3. The total phenolic contents of the ethanol extractions.

Compounds	Total phenolic contents (gGAE/100 g extract)
M	5.52±0.28
MS30	8.83±0.07
MS45	7.43±0.37

This revealed that this species had the high total phenolic contents than all of ethanol extracts (M, MS30 and MS45). Several factors such as the species, growing and the extraction method may be affected to the different of the total phenolic contents in various Dendrobium species. Moreover, a study on the total phenolic content of pollinated and unpollinated flowers of *Dendrobium Sonia* and *Dendrobium Savin* revealed that higher total phenolic contents presented in the unpollinated flowers [17]. The high level of polyphenols and flavonoids was reported in the Chinese medicinal orchid, *D. nobile*, in the methanolic extract of stem and leaves, which was significantly higher than that of chloroform and acetone extracts. Additionally, it was reported that phenolic compound are associated with antioxidant activity and play a significant role in stabilizing lipid peroxidation [18, 19]. The high total phenolic contents and antioxidant activities of different *Dendrobium officinale* extracts were reported in polar solvents [20]. The methanol extracts of stems of *Dendrobium herbaceum* Lindl. were performed the high total phenolic contents using tannic acid, gallic acid as standard and also exhibited the high antioxidant activities [21]. Similar observations were indicated in the methanolic extracts from *Dendrobium denudans* D. Don stem [22]. In this work, the extraction procedures had the influence to the results of total phenolic contents. The sonication of ethanol extract (M30, M45) before the maceration can help to enhance the quantitative of the total phenolic contents which exhibit the value of 8.83 and 7.43 gGAE/100 g extract when sonicating for 30 and 45 mins, respectively. For the extract obtained from the maceration, the lower total phenolic contents presented the value of 5.52 g GAE/100 g extract.

3.4 Determination of DPPH radical scavenging activity

The antioxidant potential of three ethanol leaves extracts including M, MS30 and MS45 were measured using

DPPH assay using ascorbic acid as the chemical standard to identify the DPPH radical scavenging activity. All of the value showed in Table 4.

Table 4. DPPH radical scavenging activity of the ethanol extractions (IC_{50}, μg/mL)

Compounds	IC_{50} (μg/mL)
M	112.35±0.36
MS30	97.21±0.07
MS45	109.25±0.28
Ascorbic acid	21.67±0.11

The highest value of radical scavenging ability was observed from the IC_{50} value of 97.21 μg/mL which was of MS30. For M and M45 showed the lower IC_{50} value of 112.35 μg/mL and 109.25 μg/mL than M30. For, the IC_{50} value of ascorbic acid was determined at 21.67 μg/mL. It can be concluded that all extracts displayed the moderate free radical scavenging activity when compared to standard ascorbic acid. The sonication and sonicated time had effect to the DPPH radical scavenging activity of the extracts. The M30 had stronger free radical scavenging activity than M45 whereas both M30 and M45 exhibited higher antioxidant activity than M. This effect appears the results as same as the total phenolic contents studies. In contrast, the long-time of sonication before maceration may not help to improve the total phenolic contents or bioactive antioxidant compounds from the *D. signatum* leaves extract. In this work, the DPPH radical scavenging activity was studied form the ethanol extract of *D. signatum* leaves. Many Dendrobium species was extracted the bioactive compound from various parts of Dendrobium such as stem, leaves, whole plant and roots and then observed the antioxidant activity using the DPPH assay, ABTS and also ORAC [23-25]. In general, orchids were shown to scavenge DPPH free radicals. The extracts from *D. aqueum* [26], have shown to scavenge DPPH free radicals. The DPPH radical scavenging activities of fractions from *D. aurantiacum* var. *denneanum* showed the moderate antioxidant activity in butanol extract (IC_{50} = 49.50 μg/mL) compared to α-tocopherol (IC_{50} = 25.00 μg/mL) [23]. The phenanthrenes and lignans were isolated from *D. nobile* were determined the scavenging DPPH free radicals. The results revealed that these active substances displayed the strong and moderate antioxidant when compare to ascorbic acid and BHT [25]. In addition, various polysaccarides were extracted and isolated using the hot water or other polar organic solvents and determined the antioxidant activity. For example, it has been reported previously that various polysaccharides isolated from *D. ochreatum* possess the antioxidant properties [27]. The antioxidant activity value was different because of the Dendrobium species, isolated substances or fraction, extraction methods and assay [28-33]. The lower IC_{50} value indicated better scavenging activity when compared to the chemical standard such as α-tocopherol, ascorbic acid, Trolox. In this study, a positive correlation was observed between the total phenolic contents and DPPH radical scavenging activity of ethanol extracts from *D. signatum* leaves. The MS30 possessing high total phenolic contents will exhibit the scavenging of DPPH free radicals. This evidence of such correlations was observed in earlier studies [34]-[36].

3.5 Cytotoxicity studies

The cytotoxicity studies were performed against human cancer (KB, MCF-7 and NCI-H187) cell lines using MTT assay to analyse the cell growth inhibition. The results displayed in Table 5. It revealed that the ethanol extract of M, M30 and M45 displayed weak anticancer activity against MCF-7 cell lines with the IC_{50} value of more than 100, 87.45 and 94.22 μg/mL, compared to the standard drug doxorubicin, which had IC_{50} value of 48.33 μg/mL. These results were similar to the cytotoxicity which tested in NCI-H187 which indicated the IC_{50} value of more than 100, 95.22 and more than 100 μg/mL, compared to the standard drug doxorubicin which had IC_{50} value of 0.10 μg/mL.

Table 5. Cytotoxicity (as the inhibition of cell growth) of the ethanol extractions from *D. signatum* leaves.

Extracts	IC_{50} (μg/mL)		
	KB	MCF-7	NCI-H187
M	-	>100	>100
MS30	-	87.45	95.22
MS45	-	94.22	>100
Doxorubicin	2.55	48.33	0.10

In addition, three extracts showed no activity against KB cell lines in all of ethanol extracts. Nevertheless, it can be indicated that the condition of sonication for 30 mins gave the higher cytotoxicity than the other ones. This can be explained that the optimize condition for bioactive substance extraction from *D. signatum* leaves should combine the two procedures of sonication at the appropriate time before the maceration. However, this work was tested only three human cancer cell lines. However, the other cell lines have to test and evaluate such as the colon carcinoma (SW620), gastric carcinoma (KATO), lung carcinoma (CHAGO), breast carcinoma (BT474) and hepato carcinoma (HEP-G2). The bioactive of antitumor and anticancer of Dendrobium extracts were performed in many preliminary researches such as various substances isolated from *D. nobile* showed cytotoxicity against human lung carcinoma, human ovary adenocarcinoma and leukemia cell lines. Erianin obtained from the stem of *Dendrobium chrysanthum* was found to be a potent inhibitor of proliferation of HL-60 cells. Moreover, the water soluble polysaccharides from *D. denneanum* had strong anti-tumor and immunomodulation abilities. Cytotoxic bibenzyl dimers isolated from the stems of *Dendrobium fimbriatum* Hook exhibited the broad-spectrum and cytotoxic activities with IC_{50} values ranging from 2.2 to 21.2 μM against human cell lines such as HL-60 SMMC-7721, A-549, MCF-7 and SW480, respectively [4, 6, 37-39].

4 Conclusion

In this work, this is the first report of the biological activities of ethanol extracts from *D. signatum* leaves. From the results of the phytochemical analysis and other biological activities such as total phenolic contents, antioxidant activity and cytotoxic activities of ethanolic extracts of *D. signatum* leaves revealed that both antioxidant activity and total phenolic contents of ethanolic extract of *D. signatum* leaves had correlated with the phenolics, flavonoids, coumarins and also phytosterols which found in the phytochemical analysis. Moreover, these bioactive compounds have been reported the high efficiency of antioxidant activity and free radical scavenging. The isolation of these active compounds was the future work in order to characterize the chemical components of ethanolic extract of *D. signatum* leaves and evaluate the other bioactivities such as antibacterial and antifungal activities, respectively.

Acknowledgement

This work was financially supported by Faculty of Science, Maejo University, Chiang Mai, Thailand.

References

1. A. Matkowski, Biotechnol. Adv **26**, 548-560 (2008).
2. D. Krishnaiah, R. Sarbatly, R. Nithyanandam, Food. Bioprod. Process **89**, 217-233 (2011)
3. M. Carocho, I.C.F.R. Ferreira, Food. Chem. Toxicol **51**, 15-25 (2013)
4. S. Singh1, A.K. Singh1, S. Kumar, M. Kumar, P.K. Pandey1 and M.C.K. Singh, Appl. Botany **52**, 11627-11634 (2012)
5. MK. Sarangi, S. Padhi, Int. J. Phytomed **6**, 01-22 (2014)
6. M.M. Hossain, Fitoterapia **82**, 102-140 (2011)
7. Y.T. Liu1, R.K. Chen, S.J. Lin, Y.C. Chen, S.W. Chin, F.C. Chen and C.Y. Lee, Genet. Mol. Res **13**(2), 2709-2717 (2014)
8. N.P. Chetankumar, M.P. Natvarlal, Int. J. Res. Pharm. Bio. Sci **4**(1), 75-80 (2013)
9. S. Ganapaty, M. Ramaiah, K. Yasaswini, V. K. Nuthakki, D. Harikrishnareddy, J. Pharmacogn. Phytochem **2**(3), 113-118 (2013)
10. P.S. Nimisha and Y.R. Hiranmai, Int. J. Pharm. Pharm. Sci **4**(1), 385-386 (2012)
11. J. Banerjee, World. J. Pharm. Sci **4**(8) 1824-1834 (2016)
12. M. Johnson and N. Janakiraman, Indian. J. Nat. Prod. Resour **4**(3), 250-254 (2013)
13. C.B. Singh, M.C. Devi, D.S. Thokchom, M. Sengupta, A.K. Singh, J. Pharmacogn. Phytochem **4**(4), 06-11 (2015)
14. I. Ammara, M. Ennouria, B. Khemakhemc, T. Yanguid, H. Attia, Industrial Crops and Products **37**, 34-40 (2012)
15. A.K. Saha, Md. R. Rahman, M. Shahriar, S.K. Saha, N.A. Azad, S. Das, J. Pharmacogn. Phytochem **2**(2), 181-188 (2013)
16. F.-Q. Xu, F.-C. Xu, B. Hou, W.-W. Fan, C.-T. Zi, Y. Li, F-W. Dong, Y.-Q. Liu, J. Sheng, Z.-L. Zuo, J.-M. Hu, Bioorg. Med. Chem. Lett **24**, 5268-5273 (2014)
17. O.P.S. Rebecca, R. Zuliana, K. Wijenthiran, A.N. Boyce, S. Chandran, Asian. J. Plant. Sci **5**(3), 1-6 (2008)
18. P. Bhattacharyya, S. Kumaria, R. Diengdoh, P. Tandon, Meta Gene **2**, 489-504 (2014)
19. Ö. Beyhan, M. Elmastas, F. Gedikli, J. Med. Plants Res. **4**, 1065-1072 (2010)
20. H. Qin, S. Yangxia, Z. Chengjing, L. Aoxue and F. Yijun, Chin. J. Appl. Environ. Biol **20**(3), 438-442 (2014)
21. L.S. Vattakandya, G. S. Chaudhar, Int. J. Pharm **104**, 302-305 (2013)
22. C.B. Singh, M.C. Devi, D.S. Thokchom, M. Sengupta, A.K. Singh, J. Pharmacogn. Phytochem **4**(4), 06-11 (2015)
23. L. Yang, H. Han, N. Nakamura, M. Hattori, Z. Wang and L. Xu, Phytother. Res **21**, 696-698 (2007)
24. K. Rashmi, S.D. Shweta, C.S. Sudeshna, P.S. Vrushala, T.R. Prashith Kekuda and H.L. Raghavendra, Sci. Technol. Arts Res. J 4(1)160- 164 (2015)
25. X. Zhang, J.K. Xu, N.L. Wang, H. Kurihara and X.S. Sheng Yao, J. Chin. Pharm. Sci **17**, 314-318 (2008)
26. S. Mukherjee, D. Phatak, J. Parikh, S. Jagtap, S. Shaikh, R. Tupe, R. (2012). J. Med. Chem. Drug discov **2**(2), 46-54 (2012)
27. J. Banerjee, B.K. Dey, H. Khanal, B. Dahal, World Journal of Pharmaceutical Research **4**(7), 1824-1834 (2015)
28. S.R. Vandavasi, M. Ramaiah, P.NV. Gopal, J. Pharmacogn. Phytochem **3**(5), 107-111 (2015)
29. L.-Y. Wu1, C.-W. Cheng, Y.C. Tien, C.-L. Kuo, S.-F. Lo, W.-H. Peng, J. Chin. Med **22**(1,2) 47-63, (2011)
30. C.B. Singh, M.C. Devi, D.S. Thokchom, M. Sengupta, A.K. Singh, J. Pharmacogn. Phytochem **4**(4), 06-11 (2015)
31. L.S. Vattakandya, G. S. Chaudhar, Int. J. Pharm **104**, 302-305 (2013)
32. H. Qin, S. Yangxia, Z. Chengjing, L. Aoxue and F. Yijun, Chin. J. Appl. Environ. Biol **20**(3), 438-442 (2014)
33. P. Bhattacharyya, S. Kumaria, R. Diengdoh, P. Tandon, Meta Gene **2**, 489-504 (2014)
34. N. Coruh, A.G.S. Celep, F. Ozgokce, M. Iscan, Food. Chem **100**, 1249-53(2007)
35. J.C. Tilak, S. Adhikari, T.P.A. Devasagayam, Redox. Rep **9**(4), 220-27 (2004)
36. P.T.R. Kekuda, K.S. Vinayaka, D. Swathi, Y. Suchitha, T.M. Venugopal, N. Mallikarjun, E- J. Chem **8**(4), 1886-1894 (2011)
37. Y. Fan and A. Luo, Afr. J. Pharm. Pharmacol **5**(3), 415-420 (2011)
38. F.-Q. Xu, F.-C. Xu, B. Hou, W.-W. Fan, C.-T. Zi, Y. Li, F-W. Dong, Y.-Q. Liu, J. Sheng, Z.-L. Zuo, J.-M. Hu, Bioorg. Med. Chem. Lett **24**, 5268-5273 (2014)
39. T. Phechrmeekha, B. Sritularak, K. Likhitwitayawuid, J. Asian. Nat. Prod. Res **14**(8), 748-754 (2012)

Comprehensive Study on The Solvation of Sr(II) Ion

Ika N. Fitriani[1,a], Wiji Utami[1], Niko Prasetyo[1] and Ria Armunanto[1,2]

[1]*Austria-Indonesia Centre (AIC) for Computational Chemistry, Gadjah Mada University, Indonesia*
[2]*Department of Chemistry, Gadjah Mada University, Indonesia*

Abstract. Solvation of Sr^{2+} ion in liquid ammonia has been studied using the HF, DFT (B3LYP), second-order Møller-Plesset (MP2) and CCSD theory. Single valence basis sets were applied. Total and sequential binding energies are evaluated for all strontium-ammonia clusters containing 1-6 ammonia molecules. Total binding energies and distance calculated using the high level G09 calculations. For each addition of an ammonia molecules, the change of the Sr-N distance in metal-ammonia clusters is the highest at the HF level. HF is the best compromise between computational effort and accuracy.

1 Introduction

Strontium is interest in the field of dentistry. Strontium is also known to inhibit caries formation in teeth and promote apatite formation. In addition, Sr is used as an additive in toothpastes in the form of strontium chloride and strontium acetate to reduce tooth sensitivity. Hence, the incorporation of Sr into bioactive glasses that are currently used as remineralising additives in toothpaste is an attractive proposition [1]. Low doses of strontium are reported to have beneficial effect on bone formation in osteoporotic patients and, in some studies, strontium has been given as a salt [2]. The solvation of strontium is carried over into the solid state with the production of-ammoniates mixed with solid ammonia [3].

When dealing with biological processes, ions constitute an essential foundation of numerous vital processes. As most chemical and virtually all biochemical processes occur in liquid state, solvation of the reaction partners is one of the most prominent topics for the determination of chemical reactivity and reaction mechanisms and for the control of reaction conditions and resulting materials. Besides an exhaustive investigation by various experimental methods [4, 5] theoretical approaches have gained an increasing importance in the treatment of solvation effects [6]. These processes have become more and more the domain of computational simulations and a critical evaluation of the accuracy of simulation methods covering experimentally inaccessible systems is of most importance [7].

Solvation of Sr^{2+} ion has stimulated considerable theoretical interest. Computational chemistry provides information on the arrangements of solvent molecules around metal ions, binding energy, and the vibrational frequencies of the complexes that are often useful for experimentalists. Interplay between theory and experiment is crucial to obtain a molecular level understanding of ion solvation, which will aid to attain correct mechanistic conclusions about biological control of ion movements.

Hybrid *ab initio* Hartree–Fock (HF) simulations proved to achieve the necessary accuracy, although at enormous costs in terms of computational effort. On the other hand, the use of simple density functionals as in the RIDFT method or in Car–Parrinello simulations can lead to first-shell coordination numbers that are lower than almost all experimental and other theoretical data. The more sophisticated B3LYP-DFT formalism leads to a similar degree of accuracy as the *ab initio* Hartree–Fock method, but without saving computer time [8].

Many tests have finally shown that ab initio Hartree–Fock (HF) level is required to produce sufficiently accurate results. Correlated ab initio methods, even Møller-Plesset second order perturbation theory (MP2), would lead to an unaffordable computational effort, but they can be used to estimate the accuracy of HF level by calculations of small solute–solvent clusters at different levels of theory. Unfortunately also DFT level does not appear a suitable solution. QM/MM simulations employing the B3LYP functional, which are almost equally time consuming as ab initio HF calculations, have shown that hydrogen bonds as well as solvation structures are too rigid and thus give a wrong picture of structure and dynamics of solvents and solutions. At this point some misleading nomenclature in publications of simulations has to be clarified. This is indeed not correct as all available density functionals are of a more or less semiempirical nature of the B3LYP functional. Only by ab-initio is also enabled to perform a method inherent control of accuracy and deficiencies by increasing the level of theory from a one determinant alto a multi-determinantal method [9].

[a] Corresponding author: ika.nur.f@mail.ugm.ac.id

© The Authors, published by EDP Sciences. This is an open access article distributed under the terms of the Creative Commons Attribution License 4.0 (http://creativecommons.org/licenses/by/4.0/).

The use of simple density functionals for example, the most commonly used Becke-Lee-Yang-Parr (BLYP) includes further error sources. These functions imply all simplifications of the general gradient approximation formalism and, besides that, the other problems, common to virtually all contemporary density functional theory (DFT) methods, namely, the wrong treatment of kinetic energy, the semiempirical parameterization of some of the terms, and the attempt to compensate errors by empirical formulae, make the interpretation of the results to a certain extent ambiguous [10].

In the present study, ammonia molecules are added to Sr^{2+} ion and the solvated metal ion complexes [$Sr(NH_3)_n$; n = 1-6] is explored using ab initio and density functional theory methods. This study is not only provides insight into the nature of solvation of metal ions but also establishes the method and basis set dependency of this solvation. Because the solvation of metal ions is a topic of great interest, it is necessary to identify theoretical methods that can satisfactorily reproduce experimental results at the lowest computational cost.

2 Method

This method performed cluster calculation for $[Sr\text{-}(NH_3)_{1\text{-}6}]^{2+}$ with different methods including HF, MP2, B3LYP, and CCSD. We have choosen the SV(P) basis set[11] for Sr^{2+} ion and DZP for ammonia molecules[12]. Initially, all strontium-ammonia complexes considered in the present study [Sr(NH3)n; n = 1-6] were explored on their respective potential energy surfaces. Among the various possible conformers, the lowest energy structure were considered for further geometry optimizations at the HF, B3LYP,MP2, and CCSD levels. Besides this, single point calculations were performed at the HF, MP2, B3LYP, and CCSD levels using the geometries optimized at the each levels. Calculations were also performed by the Gaussian-09 (G9) method[13] that employs a sequence of ab initio molecular orbital calculations to estimate the total energy of a given molecular system at a level where it cannot be calculated directly [14].

To calculate the basis set superposition error (BSSE), we have used the counterpoise correction of Boys and Bernardi. BSSE values obtained at the HF, MP2, B3LYP, and CCSD levels are used for BSSE corrections, respectively. The HF, MP2, B3LYP, and CCSD wave function methods with genecp basis set were used to obtain the distance and binding energy. The distance and binding energy were calculated. Total and sequential binding energies were calculated using the following equation:

$$Sr + nNH_3 \rightarrow Sr(NH_3)_n \quad (1)$$

$$Sr(NH_3)_n + NH_3 \rightarrow Sr(NH_3)_{n+1} \quad (2)$$

3 Result and discussion

Tables 1 and 2 list the binding energy as well as the Sr-N distance obtained for geometry optimizations of $[Sr(H2O)n]^{2+}$ (*n*)1-6) clusters at various levels of theory employing the above-mentioned basis sets.

Table 1. Average Binding Energy in kcal/mol for Sr(II)-$(NH_3)_n$ Cluster of Different Size Obtained from HF, MP2, CCSD, and B3LYP calculations

HF	MP2	CCSD	B3LYP
-52.24	-57.82	-54.17	-55.89
-48.81	-54.07	-30.73	-54.91
-48.59	-51.74	-35.57	-52.25
-43.94	-28.54	-36.63	-49.12
-40.75	-45.81	-35.61	-45.27
-38.04	-42.99	-34.29	-77.16

These result also indicate effect of electron correlation to be expected during the simulation. As can be observed from Table 1, the average solvation energy values differ slightly for HF, MP2, and CCSD, but the difference is quite considerable in the case of B3LYP. However, neither MP2 nor CCSD is computationally affordable for simulation at present, so we prefer to use HF as the level of theory for the QM calculation. The relative scale lists the average binding energies per ligand descending from highest to lowest is HF>CCSD>MP2>B3LYP.

Ab initio HF SCF or density functional theory (DFT) are manageable, considering the necessary computational effort. In recent investigations of similar ionic systems, results of HF calculations were in good agreement with experimental data, whereas simple DFT methods always failed, and even the more suitable hybrid B3LYP functional sometimes yielded wrong descriptions or, at best, a proper description of the system without any time saving effect.

Table 2. Average Sr-N Distances in Å for Sr(II)-NH_3 Clusters of Different Size Obtained from HF, MP2, CCSD, and B3LYP Calculations

n	HF	MP2	CCSD	B3LYP
1	2.64	2.44	2.53	2.57
2	2.69	2.44	2.65	2.64
3	2.70	2.44	2.65	2.65
4	2.73	2.44	2.75	2.75
5	2.76	2.44	2.69	2.70
6	2.76	2.44	2.76	2.82

The average bond distances resulting from the different calculations are in good agreement, the B3LYP method showing the largest deviation. Based from the experiment distance from Sr^{2+}-N is 2,62 [15]. HF is in excellent agreement with experimental data. The relative scale of the average Sr^{2+}-N distances from highest to lowest is HF>CCSD>B3LYP>MP2.

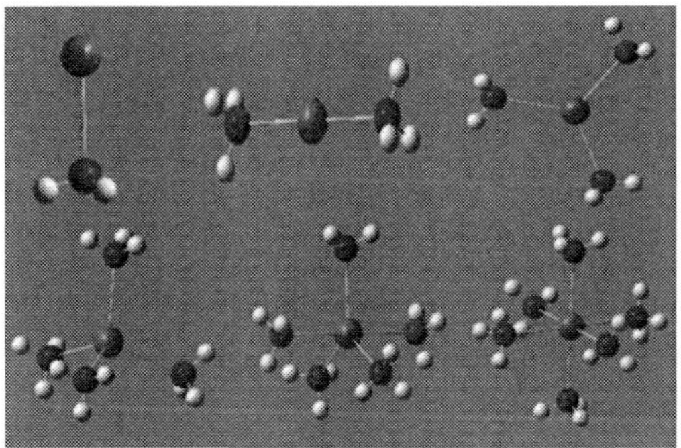

Figure 1. Optimized geometries of $Sr^{2+}(NH_3)_n$ (n=1-6)

Based on the Table 1 and 2, there is slightly difference in energy and distance from several method. The result not depend on correlation interaction because doesn't give significant contribution. MP2, B3LYP, and CCSD have high accuracy for simulation because correlation interaction effect, but HF used for calculation method. HF is the simplest method which spend shorter time than MP2, B3LYP, and CCSD. Optimized geometries can be seen in Figure 1.

Tables 1 and 2 list the binding energy as well as the Sr-N distance obtained for geometry optimization of $[Sr(NH_3)_n]^{2+}$ (n = 1-6) clusters at various levels of theory employing the above mentioned basis sets SV(P). the deviations obtained for the binding energies are in the order of about 1 kcal/mol except for B3LYP, which shows larger values at lower coorfination number. The average bond distances resulting from the different calculation are in good agreement, the B3LYP method showing the largest deviation. The good agreement of the ab initio HF SCF method with correlated methods for the description of solvated ions has been demonstrated recently in similar investigation of the solvated ions Li (I). Based on these investigations and the results of the geometry optimizations the Sr(II) amonia the ab initio HF-SCF method was to be preferred in the construction of the pair and three body interaction potensials as this method is more accurate in tthe case of the mono- and diammine than the DFT method. Due to compatibility, the ab initio HF scheme had to be applied in QM part of the simulation as well. As a general conclusion one can state the influence of electron correlation is minor for such system and that many body effect are less pronounced for monovalent ions than for those with higher charges

HF delivers strontium ion-ammonia distances close to the values derived by CCSD calculations, although the stabilization energy is too small. The difference becomes smaller with increasing number of ligand. Simulation in solution with abundance of ligands is favorable. The MP2 calculations are in good agreement with CCSD ones. The energy of strontium(II) clusters is underestimated to a similar extent as by HF, whereas the binding distances are too small again. The inclusion of dispersion terms (B3LYP) yield better result, but more recent B3LYP approach leads to too rigid hydrogen bonds and make slowing down the dynamics[16].

The estimation of the influence of the basis set superposition error (BSSE) was performed for Sr^{2+}-NH_3 (see Table 3). The BSSE (basis set superposition error) calculations were also performed for the Sr(II)-NH3 different levels of theory and were included in Table 3. The data indicates that BSSE and electron correlation do not have any significant contribution for the system of interest.

With respect to the level of theory for the QM region, either ab initio HF SCF or density functional theory (DFT) are manageable,considering the necessary computational effort. In recent investigations of similar ionic system, result of HF calculation were in good agreement with experimental data, whereas simple DFT method such, B3LYP always failed and even the more suitable hybrid B3LYP functional sometimes yielded wrong descriptions or at best, a proper description of the system without any time saving effect.

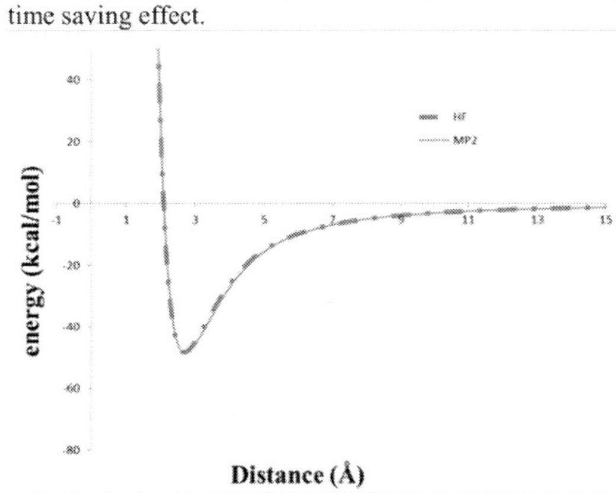

Figure 2. Interaction Energy Curve of Sr^{2+}-NH_3

Correlated method such as full MP2 and CCSF are still computationally too demanding. HF less costly than conventional MP2. MP2 accounts for electron correlation.

An approximate way of assessing BSSe is the counterpoise correction. In this method the BSSE is estimated as the difference between monomer energies as the difference between monomer energies with the regular basis and the energies calculated with the full set of basis functions for the whole complex. Consider two molecules A and B, each having regular nuclear-centred basis sets denoted with subscript a and b, and the complex AB having the combined basis set ab. The geometries of the two isolated molecule and of the complex are first optimized or otherwise assigned. The geometries of the A and B molecules in the complex will usually be slightly different than for the isolated species, and the complex geometry will denoted with a*. The dimer energy minus the monomer is the directly calculated complexation energy.

$$\Delta E_{complexation} = E(AB)^*_{ab} - E(A)_a - E(B)_b \quad (3)$$

To estimate how much of this complexation energy is due to BSSE, four additional energy calculations are needed. Using the a basis set for A and the b basis set for B, the energies of each of the two fragments are calculated with the geometry they have in the complex. Two additional energy calculations of the fragments at the complex geometry are then carried out with the full

[a] Corresponding author: ika.nur.f@mail.ugm.ac.id

ab basis set. This means that the energy of A is calculated in the presence of both the normal a basis functions and with the b basis function of fragment B located at the corresponding nuclear positions, but without the B nuclei present, and vice versa. Such basis functions located at fixed points in space are often referred to as ghost orbital. The fragment energy for A will be lowered due to these ghost fuctions, since the a basis becomes more complete. The counterpoise correction is defined in eq 4.

$$\Delta E_{cp} = E(A)^{*}_{ab} + E(B)^{*}_{ab} - E(A)_{a} - E(B)_{b} \quad (4)$$

The counterpoise-corrected complexation energy is then given in the eq 5.

$$\Delta E_{complexation} - \Delta E_{cp} \quad (5)$$

For regular basis sets, this typically stabilizes at the basis set limiting value much earlier than uncorrected values, but this is not necessarily the case if diffuse function are included in the basis set. Delta ECP is an approximate correction and gives an estimate of the BSSE effect but does not provide either an upper or lower limit [17].

It is usually observed that the counterpoise-correction for methods including electron correlation is larger and more sensitive to the size of the basis set than at the HF or DFT level. This is in line with the fact that the HF wave function converges much faster with respect to the size of the basis set than correlated wave functions.

Table 3. Basis Set Superposition Error According to Boys-Bernardi Procedure for Sr(II)-NH_3 Obtained from HF, MP2, CCSD, and B3LYP calculations

BSSE (in kcal/mol)	
HF	0.53
MP2	3.82
CCSD	3.78
B3LYP	1.03

BSSE HF is lower than MP2, CCSD, and B3LYP. From this BSSE value, HF is reliable because the fault in calculation of interaction energy was small.

The HF and MP2 curve are good in agreement with the Lennard-Jones interaction curve, as can be seen in Fig 2. The value of interaction energy gives positive value in short distance due to repulsive among particles. The interaction energy decreases until optimum distance with negative value. The interaction energy of HF method is good in agreement with MP2 method with slightly different in interaction energy.

4 Conclusion

The paper provides calculation of binding energy, distance, and BSSE at level ab initio used G09 methods for the solvation of Sr^{2+} ion. The present study suggests SV(P) and DZP basis set with HF method is the lowest energy conformer of solvated Sr^{2+} ion in liquid ammonia. HF level of theory promised to yield the best compromise between computational effort and accuracy.

Acknowledgement

Financial support for this work from a M. Sc grant of the Indonesia Endowment Fund for Education Ministry of Finance Republik Indonesia (LPDP Kementrian Keuangan RI) for Ika Nur Fitriani (000363/Te/M/BM-AF-2014) is gratefully acknowledged.

References

1. R. A. Martin, H. L. Twyman, G. J. Rees, E. R. Barney, R. M. Moss, J. M. Smith, et al., J. Mater Chem. **22**, 22212–22223 (2012)
2. R. H. Wasserman, Clin. Chem. **44**, 437-439 (1998)
3. F. A. A. P. W. Atkins, T. L. Overton, J. P. Rourke, M. T. Weller, *Inorganic Chemistry*, (2010)
4. N. Korber, A. Chemie, R. Friedrich-wilhelms-universita, R. V May, Inorg. Chem. **1669**, 4906-4908 (1997)
5. S. Lysgaard, A. L. Ammitzbøll, R. E. Johnsen, P. Norby, U. J. Quaade, T. Vegge, Int. J. Hydrogen Energy. **37**. 18927-18936 (2012)
6. T. S. Hofer, B. R. Randolf, B. M. Rode, J. Phys. Chem. B. **110**, 20409-20417 (2006)
7. D. J. Defrees, A. D. Mclean, J. Chem. Phys. **82**, 333-341 (1985)
8. T. S. Hofer, H. T. Tran, C. F. Schwenk, B. M. Rode, J. Comput. Chem. **25**, 211-217 (2004)
9. B. R. Randolf, Theor. Comput. Inorg. Chem. **62**, 143-175 (2010)
10. B.M. Rode, T.S. Hofer, B.R. Randolf, C.F. Schwenk, D. Xenides, V. Vchirawongkwin, Theor. Chem. Acc. **115**, 77-85 (2005)
11. M. Kaupp, P. R. Schleyer, H. Stoll, H. Preuss, J. Chem. Phys. **94**, 1360–1365 (1991)
12. T. H. Dunning, J. Chem. Phys. **53**, 2823 (1970)
13. M. Caricato, M. J. Frisch, *Gaussian 09 IOps Reference*, (2013)
14. J. S. Rao, T.C. Dinadayalane, J. Leszczynski, G. N. Sastry, J. Phys. Chem. **112**, 12944–12953 (2008)
15. G. Bacg, W. Hall, D. Di Tommaso, N. H. De Leeuw, Cryst. Growth Des. Artic. **10**, 4292–4302 (2010)
16. L. R. Canaval, B. M. Rode, Chem. Phys. Lett. **618**, 78–82 (2015)
17. F. Jensen, *Introduction to Computational Chemistry*, Second Ed, Wiley, (2007)

Fast Biofilm Formation and Its Role on Power Generation in Palm Oil Mill Effluent Fed Microbial Fuel Cell

Maksudur R. Khan[1,a], Baranitharan E[1]., Prasad D. M. R.[2] and Chin K. Cheng[1]

[1]*Chemical and Natural resources engineering Department, University Malaysia Pahang, 26300 Gambang, Pahang, Malaysia*
[2]*Petroleum and Chemical engineering Department, Institut Teknologi Brunei, BE1410 Gadong, Negara Brunei Darussalam, Brunei*

Abstract. In the present study, fast formation and characterization of biofilm and its role on power generation in the microbial fuel cell (MFC) were investigated and the biofilm formation was also correlated with electrochemical behavior of the MFC. MFC was operated with palm oil mill effluent as substrate and carbon cloth as electrode. A biofilm comprising electrochemically active bacteria on the anode surface showed crucial effect to enhance the performance of the MFC. Infrared spectroscopy and thermogravimetric analysis confirmed the presence of biofilm and scanning electron microscopy examined a biofilm and microbial clumps on electrode surface. The current density was directly dependent on the biofilm growth and increased significantly during the initial growth. Electrochemical impedance spectroscopy was done to monitor the progress of the anode colonization by the microorganisms in the MFC. The findings of this study demonstrated that biofilm formation facilitated electron transport as well as decreased the charge transfer resistance of the anode and thus increased the power generation in the cell.

1 Introduction

Microbial fuel cell (MFC) is a bioelectrochemical energy system that opens a new window for eco friendly energy production to meet the growing demand and an economical pathway to a sustainable energy future. MFCs are capable of converting chemical energy available in organic substrates into electrical energy using bacteria as a biocatalyst to oxidize the biodegradable substrates [1]. The factors, such as substrate degradation, electrode material, type of microorganisms, rate of electron transfer from microorganisms to the anode, circuit resistance and proton mass transfer in the liquid mainly affect the MFC performance [2]. Different types of microorganisms have been studied in MFC such as *Clostridium butyricum* [3], *Escherichia coli* [4], *Geobacter sulfurreducens* [5], *Klebsiella pneumoniae* [6], *Pseudomonas aeruginosa* [7], *Rhodoferax ferrireducens* [8], *Shewanella oneidensis* [9], *Shewanella putrefaciens* [10] and so on. Eventhough, some pure cultures (e.g., *Geobacter metallireducens, Shewanella putrefaciens and Pseudomonas aeruginosa*) called electrogens are able to transfer electrons directly and produce current more or less equal to mixed cultures, it cannot utilize mixed or complex substrates. But, most of the industrial and domestic wastewater contains complex substrates and it makes pure culture not suitable for treating wastewater using MFC. Due to this reason, in most of the cases, anaerobic sludge [7] and activated sludge [11] obtained from wastewater treatment plant, have been used as inoclulum for the experiments. Therefore, these microorganisms should have grown and adapted to the condition of the wastewater.

Nevertheless, MFCs show real challenges to generate electric power for practical applications. Kim et al. [12] studied bacterial community structure (BCS), compartmentalization and activity in a MFC for a long period (18 months) and reported that BCS varied significantly in the anode biofilm. The research group as well showed that within 30 days of enrichment of microbial community was generated electricity (0.2 mA), using organic wastewater as the fuel [13]. Recently, Zhang, Zhu, Li, Liao and Ye [14] reported that biofilm is a crucial component in MFC for electrogenesis and it required only 14 days to be completely grown up at the anode compartment. Moreover, there are some barriers which limit the practical applications of MFCs, include ionic transfer restrictions, may increase the internal electrical resistance and oxygen transfer, therefore, decrease the overall cell performance [15]. Various attempts have been taken into account to overcome the electronic barrier. A number of electron-transfer mediators like biofilms have been used to facilitate electron transfer between an electrode and electrochemically active micro-organisms [16]. Anode potential is one of the most important factors at which the MFC is operated due to the control of theoretical energy gain for microorganisms [17]. In addition, the effect of external resistance applied to the electrical circuit also received wild attention and Zhang, Zhu, Li, Liao and Ye [14] reported that MFC performance increases with decreasing the applied external resistance [18] which may

[a] Corresponding author: mrkhan@ump.edu.my

© The Authors, published by EDP Sciences. This is an open access article distributed under the terms of the Creative Commons Attribution License 4.0 (http://creativecommons.org/licenses/by/4.0/).

be due to the effect of the structure of the biofilm at the anode.

Further improvement in power generation from the MFC has been achieved by lowering the polarization resistance of the electrode (anode) which is highly dependent on the nature of the biofilm formed during the operation. Therefore, the study of biofilm formation has been considered as a crucial factor to ameliorate the performance of MFC. The study of rapidly formed biofilm based on palm oil mill anaerobic sludge and electrochemical impedance characteristics towards practical applications still suffers from lackings. In our previous study [19] we have used carbon felt as electrode but due to its large resistance it hindered the power generation. In order to avoid that, carbon cloth was used in this study because it has several advantages such as flexibility, highly porous and provides more surface area for bacterial growth.

In the present study, fast biofilm growth was correlated with electrochemical behavior of the MFC operated with palm oil mill effluent (POME) as substrate and carbon cloth as electrode. Electrochemical impedance spectroscopy (EIS) was employed to investigate the charge transfer resistance in the anode and SEM analysis was used to examine the biofilm.

2 Materials and Methods

2.1 MFC construction

The MFC was made of glass material (borosilicate) and was fabricated in china (Shangai sunny scientific, china). It consisted of two compartments; anode and cathode, with working volume of 450 mL each. The anode chamber of the MFC was filled with 450 mL of POME as substrate and 20 mL of anaerobic sludge (obtained from palm oil anaerobic sludge tank) as inoculum. At the same time, the cathode chamber was filled with 200 mg. L^{-1} of $KMnO_4$ solution. Carbon cloth (CC) was purchased from Shangai Sunny Scientific, China and used as electrode material for all the experiments. Equal size of CC (1.85 cm length and 5.5 cm height) was used in each compartment of dual chamber MFC experiment. The anode and cathode compartments were separated by a Nafion 117 membrane (Dupont Co., USA). Prior to use, CC was washed with water for several times to remove impurities. Nafion membrane was drenched overnight in dilute HCl and subsequently washed with deionized water before use. The anode and cathode electrodes were connected with a resistor by copper wires in order to make a circuit. The fuel cells were operated (batch mode) at ambient temperature of 25-28°C.

2.2 Measurement and analyses

The voltage and current across an external 1 kΩ resistor was measured every 15 min by using digital multimeter with data logger (Fluke 289) which recorded and stored the data [20]. Later, it was retrieved from the multimeter using computer with fluke view forms software [21]. The polarization and power density curves were plotted using the voltage and current data. Power density normalized by volume (P_v) and power density normalized by surface area (P_A) were measured using the following equations:

$$P = VI \tag{1}$$

$$P_{An} = \frac{V^2}{A_{An}R} \tag{2}$$

$$P_v = \frac{V^2}{vR} \tag{3}$$

Where A_{An} = area of anode electrode (m^2), P = power (W), V = potential (V), v = working volume of anode (m^3), R = external resistance (Ω) and I = current flow (A).

2.3 COD removal efficiency

COD was periodically checked by taking small amount of sample from the anode effluent for every 24 hrs. The COD removal efficiency (η) was calculated as described by Baranitharan, Khan, Prasad and Salihon [19].

$$\eta = (COD)_{initial} - COD_{final}/COD_{initial} \text{ X } 100\% \tag{4}$$

Where $COD_{initial}$ = initial COD concentration (mg. L^{-1}) and COD_{final} = final COD concentration (mg. L^{-1}).

2.4 Coulombic efficiency

The CE for complex substrates can be calculated for a fed batch system as described by Baranitharan, Khan, Prasad and Salihon [19].

$$CE = \frac{8\int_0^t Idt}{FV_{An}\Delta COD} \tag{5}$$

Where Δ_{COD} = Change in COD concentration (mg/L), I= current (A), t = change in time (s), V_{An} = volume of liquid in anode compartment (l), F = Faraday's constant (96485 C/mol e^-). In Eq. 5, 8 is a constant, based on M_{O2} = 32 for the molecular weight of oxygen and b = 4 for the number of electrons exchanged per mole of oxygen.

2.5 Sampling

Raw POME and anaerobic sludge were collected from Felda palm oil industries located at Kuantan, Malaysia. The Raw POME was collected from the inlet of the mixing pond and the temperature of raw POME during collection was around 80 to 90°C. Anaerobic sludge was obtained from bottom sampling port of anaerobic treatment plant. The samples were transported to the laboratory in sterile 1000 mL Schott bottles placed in ice and stored at 4°C until use.

2.6 Wastewater and Inoculum characterization

COD of POME was analyzed by standard methods [22]. The predominant microorganisms were characterized using BIOLOG GEN III (Biolog Inc., USA) test. In order to isolate microorganisms from anaerobic sludge, the samples were serially diluted and then pour plate technique was used to get individual colonies from the samples. Serial dilution was done until 10^{-6} dilution then pour plating was performed for the last three dilution of the sample alone (10^{-6}, 10^{-5} and 10^{-4}) to get clear isolated colonies. The colonies were transferred to separate nutrient agar plates. Once the pure cultures from the samples were obtained, BIOLOG was carried out to identify the genus and species of microorganisms by performing carbon consumption tests with the isolated microorganisms.

2.7 Infrared spectroscopy (IR) and Thermogravimetric analysis (TGA)

Samples were prepared for IR test by drying at 120°C overnight to remove the water, and then 1 wt% sample was ground with dry KBr. Samples were pressed into thin discs and analyzed on a Nicolet Avatar 370 DTGS FT-IR spectrophotomer under N_2 atmosphere. TGA samples (5 mg) were analyzed on TA Instruments Q500, thermogravimetric analyzer, at a heating rate of 10°C min^{-1} in N_2 ambient.

2.8 Biofilm examination by SEM

The samples for SEM analysis were prepared by a method described by Chae, Choi, Lee, Kim and Kim [23]. Parts of the CC were cut and removed from the anode chamber then rinsed with a sterile medium, and immediately fixed with an anaerobic solution of 2% glutaraldehyde and 1% formaldehyde. After immersion in 1% osmium tetroxide for 24 h, samples were carefully rinsed three times in a HEPES buffer (pH 6.8) and once in deionized water. Samples were then subjected to a serial dehydration protocol using increased concentrations of ethanol (10, 25, 50, 75, 90, 100, 100, and 100) for 30 min for each stage with gentle periodic agitation and then dried completely at room temperature. Desiccated samples were coated with platinum and observed using a JEOL JSM7800F field emission SEM at 3 kV.

2.9 Electrochemical impedance spectroscopy

Electrochemical impedance spectroscopy (EIS) test was carried out using an electrochemical workstation (AUTOLAB 2273, PAR, USA). The three-electrode mode was employed to analyze the anode electrode. The anode and cathode electrodes were used as working electrode and counter electrode, respectively. The saturated Ag/AgCl electrode was used as the reference electrode and was placed as close as possible to anode electrode [24]. Impedance measurements were conducted at the OCV (Open circuit voltage) of MFCs over a frequency range of 100 kHz to 5 mHz. The Nyquist plots of the impedance spectra were analyzed using NOVA 1.9 (NOVA Software).

3 Results and Discussion

3.1 Performance of MFC with POME

The MFCs were fed with diluted POME with an initial COD concentration of 964 mg. L^{-1} and anaerobic sludge as inoculum for all the experiments as well. The predominant microorganisms present in the anaerobic sludge were identified as *Pseudomonas aeruginosa, Pseudomonas mendocina, Pseudomonas viridiivida, Acetinobacter schindleri, Actinobacillus capsulatus* and *Brevibacterium paucivorans* using the BIOLOG GEN III test. The current generation and COD removals of the palm mill wastewater in the MFC system over time are shown in Figure 1a. After 14 days of operation, the anode effluents showed a maximum COD removal of 63% and coulombic efficiency of about 26%. From Fig. 1a, it was found that the current density tend to increase from day 1 until day 9 and a sharp increase in current generation was evident for 7 days of operation and COD removal efficiency was also increased at the same duration. The current density produced in MFC was directly dependent on the biofilm growth and increased significantly during the initial growth. From day 9 to day 14, the current density was found slightly decreased due to the accumulation of metabolite product such as the successive gathering of H^+ ions depressed the microbial activities [25]. The results showed that current densities and COD removals were proportionally increased with COD removal in MFC up to days 9 but after that period though the COD removal increased, current generation was not increased rather slightly dropped. This COD removal was presumably enriched due to biomass generation, incomplete biodegradation of the substrate, hydrogen production, methanogenesis and aerobic degradation [19]. Figure 1b demonstrates the polarization curves for the MFC operated for 14 days in order to yield anode biofilm with variable physical properties during the period. It is evident from the figure that after 3 days of operation there was little activity on the anode resulting in a low polarization performance. The biofilm development was fast between days 3 and 7 point out a significant enhancement in the cell performance. This improvement may be caused by the effect of microorganisms on the electrode surface amid efficient biofilm formation [20]. The maximum performance was observed on the 7th day of biofilm growth on the anode as shown in Figure 1b. After 7 days, there was a noticeable drop in the polarization performance which might be due to the accumulation of metabolic product such as the accumulation of H^+ ion depressed the microbial activities [26]. The power density profile follows the similar trend as that of the polarization performance with the excellent performance being showed by the MFC on 7th day. A maximum power density of 28 mW. m^{-2} was found on 7th day of operation which is comparatively higher than our previous study using carbon felt (22 mW. m^{-2}) as electrode [19]. Jahim et al. [26] also reported the similar trend studied by using palm oil mill effluent in MFC.

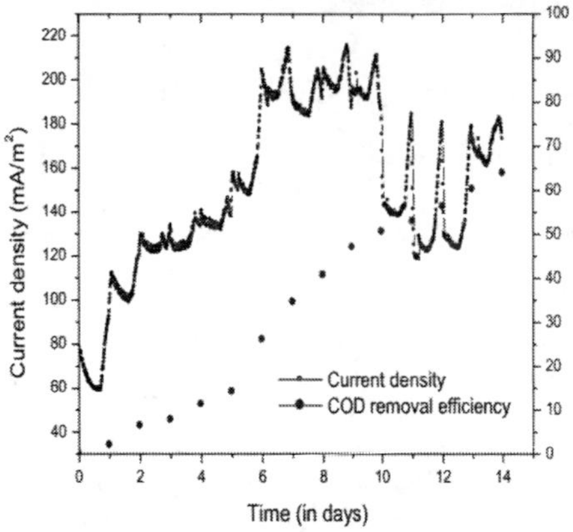

Figure 1a. Profile of current density and COD removal efficiency of carbon cloth MFC with time

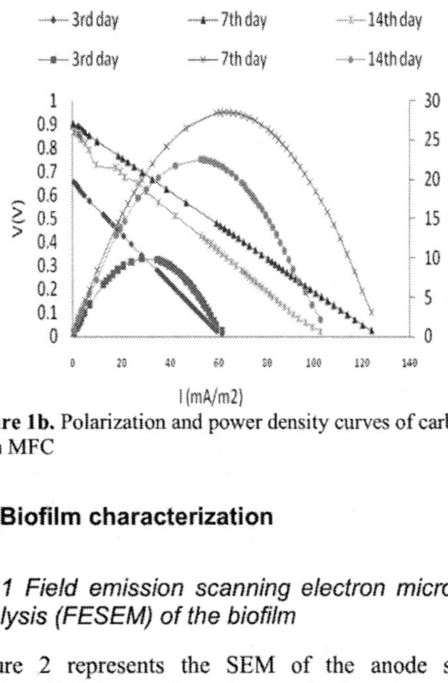

Figure 1b. Polarization and power density curves of carbon cloth MFC

3.2 Biofilm characterization

3.2.1 Field emission scanning electron microscopy analysis (FESEM) of the biofilm

Figure 2 represents the SEM of the anode surface, conducted before (Figure 2a) and after the operation (Figure 2b and 2c) in the MFC in order to confirm the presence of biofilm on the electrode surface. In Figure 2b and 2c, the SEM images showed at different magnification of biofilm growth on CC. Electron micrographs revealed unique biofilm structures and cell shapes. In the case of the higher magnification of SEM images (Figure 2c), the biofilm was formed by an interconnected fibre like structure which covers the electrode surface. Marsili, Rollefson, Baron, Hozalski and Bond [27] reported that the biofilm on an electrode surface contributed the key current through direct electron transfer. Figure 2c shows that biofilms were sparingly distributed on the CC along with highly complicated structures comprised of morphologically different cells. The important component of biofilms known as extracellular polymeric substances (EPS), are high-molecular weight compounds secreted by microorganisms and have a significant influence on the physico-chemical properties including electron transfer. Bacteria of various sizes and shapes were scattered around the electrode, associated with a biofilm formed on its surface. Besides that many rod-shaped cells can also be seen on the biofilm (Figure 2c) including loosely associated microbial clumps present on the electrode surface (CC). Kim et al. [12] have proposed that these microbial clumps consist of bacteria that ferment the complex fuel into simple fermentation products. These products can then be used as substrate for electrochemically active bacteria within the biofilm to generate electricity [13, 28].

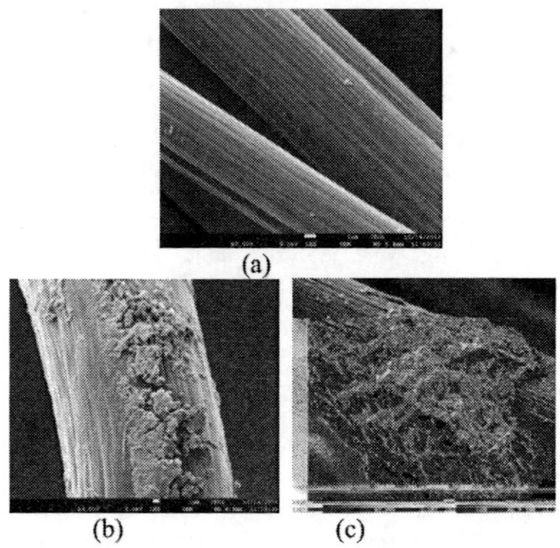

Figure 2. SEM images of (a) new carbon cloth (b and c) bacteria growing on anode carbon cloth in the MFC fed with POME as electron source for 14 days

3.2.2 TGA analysis on biofilm

Anodes were also examined by TGA before and after MFC operation at different intervals (3rd, 7th and 14th day) as shown in Figure 3. Distinct changes in overall weight losses were observed in the conditions the operation carried out. The adhesion of a bacterial biofilm to an MFC anode affects the weight of the anode. Water adhered to the anode surfaces were probably contributed to the weight losses from 25 to 100°C due to evaporation [29]. Therefore, temperatures >100°C up to 900°C was considered remarkable for gravimetric analysis of the anode. A 5% of the total weight loss from 100 to 700°C was observed before anode operation. A slight increase to 9% in anode was observed for the 3rd day while further weight losses were observed to be 11% and 18% corresponding to the 7th and 14th days of the MFC treatment. A 3% change in weight loss was observed between 100°C and 200°C after 14 days of operation but there is almost no loss in anode before operation. This is a temperature range where bacterial organic matter is likely to decompose and burn, and thus this difference is attributed to a bacterial biofilm [29]. It is known that bacterial biofilms commonly form fibrilar and mushroom like structures [30]. The types of structures composing the biofilm and their densities are likely to affect the temperature that weight losses occur [29]. Two distinct weight loss events were observed during anode heating for 3rd, 7th and 14th days of operation. On 3rd day, 2.04% from 100 to 290°C, 3.84% from 300 to 650°C were observed whereas on 7th day, 2.8% from 100 to 290°C and 2.6% from 300 to 650°C were observed and on 14th day, 5% loss from 100 to 290°C and 10% loss from 300 to 550°C were observed. Moreover, at higher temperatures the weight loss showed sharp downward trends may be due to the decomposition of biofilms at the anode surface. The changes in TGA profiles may be attributed to the increased amount of biofilm formation and sustain the effective biofilm as electron carrier on the electrode surface.

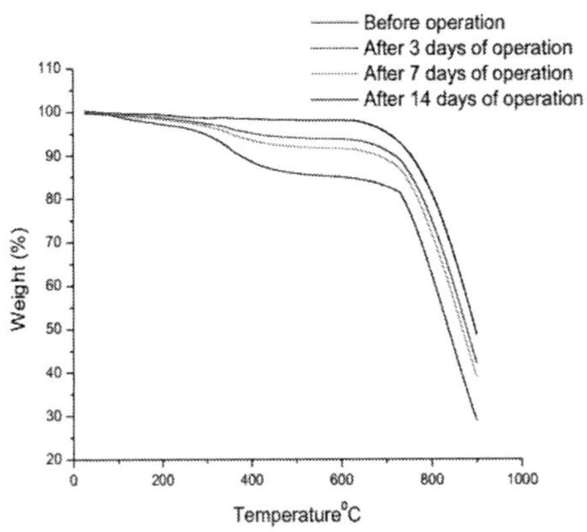

Figure 3. TGA spectrums of CC anodes before and after MFC treatment

3.2.3 FTIR analysis of the biofilm

The anode examined by FT-IR before and after MFC operation at different intervals (3rd, 7th and 14th days) is shown in Figure 4. The broad band observed in the 3500–3300 cm^{-1} range attributed to the free and bound O-H and N-H groups, which could form hydrogen bonding with the carbonyl group of the peptide linkage in the protein. The main bands of peptide linkage were related to C=O stretching at 1643 cm^{-1} (amide I) and N-H banding at 1505 cm^{-1} (amide II) [31]. The bands at 2917 cm^{-1} and 2852 cm^{-1} can be attributed to the functional groups of membrane fatty acids and also by some amino acid side-chain vibrations since here the characteristic C-H stretching vibrations of $-CH_3$ and $=CH_2$ functional groups dominate [32]. The FT-IR spectrum of the anode after MFC operation also revealed that the band observed in the 1200 and 900 cm^{-1} range is due to the symmetric stretching vibration of PO_2^- groups found in nucleic acids and to C-O-C and C-O-P stretching, which exposes the occurrence of carbohydrates and polysaccharides in the cell wall but also the influence of nucleic acids [33]. These results confirm the presence of biofilm on the surface of the electrode.

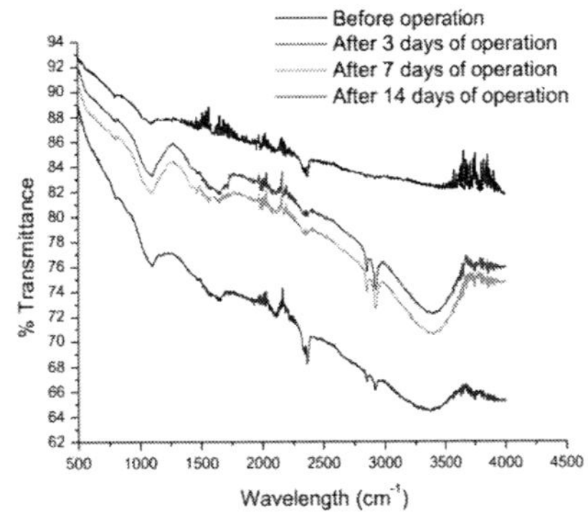

Figure 4. IR spectra of CC anodes before and after MFC treatment

3.3 Electrochemical impedance spectroscopy

Figure 5 shows the Nyquist plots for the anode configuration on 3rd, 7th and 14th days of operation. The real axis (Z) intercept of the Nyquist curves in the high frequency region correspond to the time-independent ohmic resistance in the anode configuration. The magnitude of the Nyquist arc qualitatively yields the electrochemical polarization resistance of the working electrode in anode configuration, while the quantitative values can be obtained by equivalent circuit (EC) fitting of the data [34]. After the reactor startup, the electrochemical impedance spectroscopy of the anode was tested. The impedance spectra of the anode MFC were analyzed by fitting to the equivalent circuit (EC): R(C[RW]) as shown in Figure 5 where R_1, R_2 (R_{ct}), C and W represent the ohmic resistance, charge transfer resistance, capacitance and mass transfer resistance respectively [35]. In Fig. 5, it can be seen that the EIS data were fitted well with the equivalent circuit model and an electric circuit is shown in the upper part of the figure. The R_{ct} values on 7th and 14th days were found to be 22.07 Ω and 20.09 Ω respectively. In Nyquist plot, on 3rd day although the ohmic resistance was low, the charge transfer resistance semicircle was not formed which might be due to the absence of biofilm on electrode surface while on 7th and 14th days semicircle was evident due to the presence of biofilm which causes charge transfer resistance. It is interesting to note that the anode charge transfer resistance (R_{ct}) after 14 days of continuous microbial growth was more or less same as that on day 7. Our previous results [20] showed that stable biofilm can be formed within even 7 days of MFC operation and these results support the same. A biofilm formed on the electrode surface during the MFC operation with anaerobic sludge, was usually enriched with electrochemically active bacteria [13] which might reduce the charge transfer resistance. As a result, the development of the microbial biofilm on the anode was found to be decreased the polarization resistance and facilitates the kinetics of the electrochemical reactions [20]. This investigation confirmed the fact that the anode charge transfer resistance is strongly influenced by the microbes growing on the anode surface.

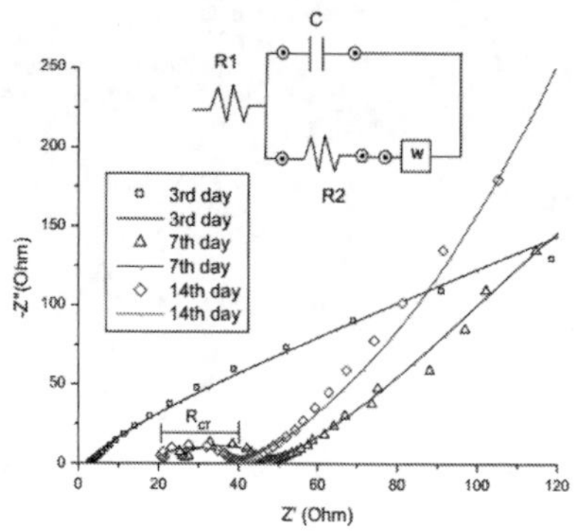

Figure 5. EIS data of MFC with CC after 14 days of operation

4 Conclusion

It can be concluded that the current density produced in MFC was directly dependent on the biofilm growth and increased significantly during the initial growth. FTIR and TGA results confirmed the presence of biofilm formation on anode while FESEM showed the highly complicated structures comprised of morphologically different cells in the biofilm. The maximum power generation and coulombic efficieny achieved in MFC using carbon cloth ($28mW/m^2$, 26%) were comparitvely higher than carbon felt ($22mW/m^2$, 24%). The EIS results confirmed that anodic R_{ct} values from 7th (22.07 Ω) to 14th (20.09 Ω) days were more or less the same, which proves the presence of stable fast biofilm formation on and after 7 days of MFC operation.

References

1. R. Karthikeyan, A. Selvam, K.Y. Cheng, and J.W.-C. Wong, Bioresour. Technol. **200**, 845-852 (2016).
2. S. Velasquez-Orta, I. Head, T. Curtis, and K. Scott, Bioresour. Technol. **102**, 5105-5112 (2011).
3. R.C. Wagner, S. Porter-Gill, and B.E. Logan, AMB express. **2**, 1-6 (2012).
4. T.H. Han, M.H. Cho, and J. Lee, Biotechnol. Bioprocess Eng. **19**, 126-131 (2014).
5. K. Inoue, C. Leang, A.E. Franks, T.L. Woodard, K.P. Nevin, and D.R. Lovley, Environ Microbiol Rep. **3**, 211-217 (2011).
6. L.-Z. Zeng, S.-F. Zhao, and W.-S. Li, Appl Biochem Biotech. **175**, 2637-2646 (2015).
7. M.H.M. Nor, M.F.M. Mubarak, H.S.A. Elmi, N. Ibrahim, M.F.A. Wahab, and Z. Ibrahim, Bioresour. Technol. **190**, 458-465 (2015).
8. S.K. Chaudhuri, and D.R. Lovley, Nature Biotechnol. **21**, 1229-1232 (2003).
9. L.A. Fitzgerald, E.R. Petersen, B.J. Gross, C.M. Soto, B.R. Ringeisen, M.Y. El-Naggar, and J.C. Biffinger, Biosens. Bioelectron. **31**, 492-498 (2012).
10. S. Pandit, S. Khilari, S. Roy, D. Pradhan, and D. Das, Bioresour. Technol. **166**, 451-457 (2014).
11. X. Mei, C. Guo, B. Liu, Y. Tang, and D. Xing, RSC Adv. **5**, 78136-78141 (2015).
12. G. Kim, G. Webster, J. Wimpenny, B. Kim, H. Kim, and A. Weightman, J. Appl Microbiol. **101**, 698-710 (2006).
13. B. Kim, H. Park, H. Kim, G. Kim, I. Chang, J. Lee, and N. Phung, Appl Microbiol Biotechnol. **63**, 672-681 (2004).
14. L. Zhang, X. Zhu, J. Li, Q. Liao, and D. Ye, J. Power Sources. **196**, 6029-6035 (2011).
15. A. Gugliuzza, and A. Basile, *Membranes for clean and renewable power applications*, (Woodhead Publishing, 2014).
16. A.E. Franks, N. Malvankar, and K.P. Nevin, Biofuels. **1**, 589-604 (2010).
17. R.C. Wagner, D.F. Call, and B.E. Logan, Environ. Sci. Technology. **44**, 6036-6041 (2010).
18. A.G. del Campo, P. Cañizares, J. Lobato, M. Rodrigo, and F.F. Morales, (2014).
19. E. Baranitharan, M.R. Khan, D. Prasad, and J.B. Salihon, Water Air Soil Pollut. **224**, 1-11 (2013).
20. E. Baranitharan, M.R. Khan, D. Prasad, W.F.A. Teo, G.Y.A. Tan, and R. Jose, Bioprocess and Biosyst Eng. 1-10 (2014).
21. E. Baranitharan, M.R. Khan, A. Yousuf, W.F.A. Teo, G.Y.A. Tan, and C.K. Cheng, Fuel. **143**, 72-79 (2015).
22. E.W. Rice, L. Bridgewater, and A.P.H. Association, *Standard methods for the examination of water and wastewater*, (American Public Health Association Washington, DC, 2012).
23. K.-J. Chae, M.-J. Choi, J.-W. Lee, K.-Y. Kim, and I.S. Kim, Bioresour. Technol. **100**, 3518-3525 (2009).
24. Z. He, and F. Mansfeld, Energy Environ Sci. **2**, 215-219 (2009).
25. L. SweeSu, J.M. Jahim, W. Wan Ramli, M. Ismail, N. Anuar, S.K. Kamarudin, and S.N. Shari, J Appl Sci. **10**, 3355-3360 (2010).
26. J. Jahim, W. Wan Ramli, M. Ismail, N. Anuar, S.K. Kamarudin, and S.N. Shari, J Appl Sci. **10**, 3355-3360 (2010).
27. E. Marsili, J.B. Rollefson, D.B. Baron, R.M. Hozalski, and D.R. Bond, Appl. Environ. Microbiol. **74**, 7329-7337 (2008).
28. J. Lee, N.T. Phung, I.S. Chang, B.H. Kim, and H.C. Sung, FEMS Microbiol Lett. **223**, 185-191 (2003).
29. J. Kramer, S. Soukiazian, S. Mahoney, and J. Hicks-Garner, J. Power Sources. **210**, 122-128 (2012).
30. B.E. Rittmann, Microbial biofilms. ASM Press, Washington, DC. 359-378 (2004).
31. S. Karnnet, P. Potiyaraj, and V. Pimpan, Polym Degrad Stabil. **90**, 106-110 (2005).
32. P.K. Nanda, K. Krishna Rao, and P.L. Nayak, J. Appl. Polym. Sci. **103**, 3134-3142 (2007a).
33. D. Naumann, Encyclopedia of analytical chemistry. (2000).
34. E. Barsoukov, and J.R. Macdonald, *Impedance spectroscopy: theory, experiment, and applications*, (John Wiley & Sons, New York, 2005).
35. Z. He, N. Wagner, S.D. Minteer, and L.T. Angenent, Environ. Sci. Technol. **40**, 5212-5217 (2006).

MATEC Web of Conferences **62**, 04003 (2016) DOI: 10.1051/matecconf/20166204003
ICCFE 2016

Physical and Electrochemical Characterization of Palm Kernel Shell Biochar (PKSB) as Supercapacitor

Wan Azlina Wan Ab Karim Ghani[1], Praveen Shawn Fernandez[1], Mohamad Qayyum Halele[1], Shafreeza Sobri[1], Jasronita jasni[2]

[1]*Department of Chemical and environmental Engineering , Faculty of Engineering, Universiti Putra Malaysia, Serdang, Selangor, MALAYSIA*
[2]*Department of Electrical and Electronic Engineering, Faculty of Engineering, Universiti Putra Malaysia, Serdang, Selangor, MALAYSIA*

Abstract. A potential low cost and environmentally friendly supercapacitor has been prepared from Palm Kernel Shell Biochar (PKSB). In this study, physical and electrochemical properties of raw, activated and chemical treated (potassium hydroxide (KOH)) as supercapacitors such as high carbon content, high charge storage capacity and stable were evaluated. For physical analyses, the scanning electron microscopy (SEM) was used to study the surface morphology and surface area and porosity were measured using Brunaurer-Emmert-Teller (BET). The chemical treated PKSB shows the highest surface area values of 55.15 m^2/g as compared to raw and activated samples with surface area are 0.17 m^2/g and 19.32 m^2/g, respectively. This is verified by in enhancement of capacitance achieved from 1.76 $x10^{-3}$ Fg^{-1} for the activated biochar and 1.87$x10^{-6}$ Fg^{-1} for untreated PKSB showed by Raman spectroscopy. This enhancement reflected the charge storage capacity is attributed to the creation of broad distribution in pore size and a larger surface area. In addition, this phenomenon also supported by the electrochemical profiles through cyclic voltammogram (CV) measured by Potentiostat-Gavanostat (EIS). CV of the treated PKSB gave better square shape than the activated and raw biochar samples. These characterizations conclude that the raw palm kernel biochar need further treatment to become supercapacitor electrodes to replace activated carbon.

1 Introduction

In response to the changing global landscape, energy has become a primary focus of the major world powers and scientific community. There has been great interest in developing and refining more efficient energy storage devices. One such device, the supercapacitor, has matured significantly over the last decade and emerged with the potential to facilitate major advances in energy storage.

Supercapacitors, also known as ultracapacitors are electrical energy storage devices. Compare to other electrical energy storage devices such as conventional capacitors, batteries, and fuel cells, supercapacitors can store high capacity electrical energy at respectively high power density [1, 2, 3] as illustrated in Figure 1. The charge-storage process for supercapacitor is highly reversible, resulting in longer cycle-life compare to other power storage devices [3]. Advanced carbon materials, such as grapheme, carbon nanotubes and carbon onions from metal carbide have shown high capacitance and high power density in electrochemical supercapacitors and hybrid supercapacitors. However, those advanced carbon materials will be significantly limited by high cost [4]. In addition, most conventional synthesis methods for these innovative carbon materials involve application of toxic strong oxidants, such as chlorine, potassium permanganate and hydrazine [5]. Therefore, hierarchical carbon materials, prepared from sustainable precursors with scalable processes, would be of great benefit to large scale application of supercapacitor in energy storage and conservation.

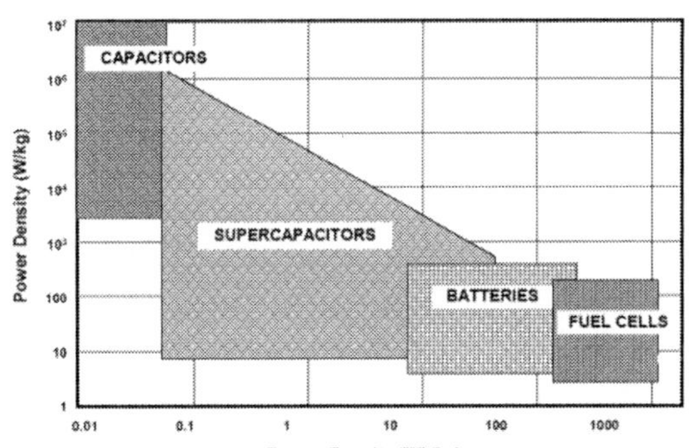

Figure 1. Ragone plot of energy storage devices [6]

The emerging studies shows an interest in agro-based carbon materials to replace the conventional activated carbon or even more, the advanced carbon material like graphene and carbon nanotube. This is because, agro-based materials are much cheaper, sustainable, and readily scalable in the mass market with safer

© The Authors, published by EDP Sciences. This is an open access article distributed under the terms of the Creative Commons Attribution License 4.0 (http://creativecommons.org/licenses/by/4.0/).

 DOI: 10.1051/matecconf/20166204003

synthesizing procedures options [7]. Some agro-based carbon materials (also called Biochar) have been proven to be promising options in producing relatively high specific surface area upon proper activation due to readily available pores in the microstructures. Biochar is a carbon-rich material produced through pyrolysis of biomass, such as wood, manure, and crop residue. Biochar has demonstrated promising potential in the field of supercapacitors. Some of the notable biomass precursors that have been prepared were corn grains [8], cypress [9], distillers dried grains with solubles (DDGS)[10], palm oil empty fruit bunch (EFB) [11], red cedar wood [12], rice husk, silk cocoon [13], and sunflower seed shell [14]. Ultimately, to achieve highly stable reversible electrical energy storage capacity and high power density in supercapacitors, carbon materials to fabricate electrodes, should possess a hierarchical mesoporous structure with high surface, high conductivity, suitable pore size distribution and long-term cyclability.

This paper aimed at comparing physical – electrochemical characterization of palm kernel shell-biochar as supercapacitor electrodes. Malaysia being the second largest oil palm producer in the world with about 5 million hectare of oil palm plantation generates huge amount of oil palm biomass. Oil palm shell shell (PKS) is being generated as by-products from the 426 oil palm mills in Malaysia [15]. Scanning Electron Microscope (SEM),Brunaurer-Emmett-Teller (BET) surface area, Raman spectroscopy, Fourier Transform Infrared (FTIR), and potentiostat-galvanostat (EIS) have been be performed and evaluated to study their performance feasibility as compared to the conventional activated carbon.

2 Materials and Methodologies

For this study, three forms of Biochar samples for each organic materials are prepared, namely; control BC (raw) (not activated and not treated), activated BC (activated but not treated), and treated and activated BC (activated and treated). Biochar was prepared through pyrolysis of a PKS in a Nitrogen gas atmosphere by applying a temperature program to a MTI GSL-1100X tube furnace. The temperature was increased from room temperature to 200∘C and then held at this temperature for1 h to remove moisture. This was followed by increasing temperature to 750∘C at around 3∘C/min and dwelling at this temperature for 1 h to remove volatile organics and to carbonize solid residues. The resulting Biochar chunk was removed from the furnace for characterization and evaluation after the temperature was cooled down to room temperature. The BC chunk was cut into small portions. Those with weights of approximately 1 mg were directly used as supercapacitor electrodes without introducing any organic binder.

2.1 Activation of biochar electrodes

Biochar electrodes were further activated by immersing them into Potassium Hydroxide (KOH) solution at room temperature overnight to increase surface hydrophilicity and the coverage of surface oxy-gen groups. They were rinsed using deionized before use. The activation process was carried out for three hours at 950 °C in a muffle furnace. The rate at which the muffle furnace heating is set is 5°C/min. Once the activation was completed, the activated BC was cooled within the muffle furnace in the same Nitrogen gas environment until the temperature drops to room temperature, which takes approximately nine hours. Then, the activated biochar sample is rinsed with 0.1mol/L hydrochloric acid (HCl) at 100 °C using a spiral condenser. Once that is done, the activated BC is again rinsed this time with deionized water to ph7 (neutral). The reason for using deionized water specifically is to avoid contact of external ions on the surface of the activated BC which might sabotage the experimental results. Lastly, the activated BC sample is dried at 105 °C for the overnight under vacuum.

2.2 Treatment of BC electrodes

The activated BC samples were then further modified with 4M HNO^3. This is to facilitate for the third control sample, which is the activated and treated BC sample. The activated BC samples were left to soak in 20mL of 4 $mol/dm^3 HNO_3$ solution sealed in Polytetrafluoroethylene (PTFE) reactor for 48 hours at 150 °C. Once soaking was done, the treated BC was then cooled and filtered. The BC electrodes were rinsed with deionized water until the filter liquid equaled pH7. Lastly, the filter cake of treated BC was collected and left to dry overnight in a conventional oven at temperature 110 °C.

2.3 Physical characterization

Surface morphology and elemental composition of the electrodes were characterized using a field-emission environmental scanning electron microscope operating at 15 keV equipped with a light-element energy-dispersive spectroscope. BET surface areas of BC samples were measured using a Machine (Micromeritics Gemini VII 2390 V1.03) surface area/pore volume analyzer. Raman measurements were performed using a high-resolution research-grade (Horiba LabRAM HR 3D-capable)Raman spectroscopy imaging system that is optimized for the visible-to-NIR spectral range. A 532 nm wavelength laser was used in this study. DC electrical conductivity of BC was determined from Ohmic resistance, cross-section area, length and porosity of a small Palm Kernel Shell chunk based on Pouillet's law. The Ohmic resistance corresponds to the slope of linear potential–current curve measured by applying potential scan to two ends of the BC electrode chunk

2.4 Electrochemical measurements

Voltammetric and constant current charge–discharge studies of a single BC electrode were carried out in 0.5 mol solution with a coil of 1 mm diameter Platinum wire and a silver/silver chloride (Ag/AgCl) as the counter electrode and the reference electrode, respectively. The

DOI: 10.1051/matecconf/20166204003

electrical contact of a BC electrode was obtained by wrapping fine gold wire around the PKS portion to minimize contact resistance. The geometric surface area is around 0.08 with 2 mm in width and 4 mm in length. The capacitive behavior of a BC supercapacitor constructed by putting two BC electrodes in similar size separated by a sheet of glassy filter-paper in 0.5 mol was evaluated by cyclic voltammetry and chronopotentiometry. Their preliminary durability was evaluated via successive potential cycling.

3 Results and Discussion

3.1 Physical characterization of Palm Kernell Shell biochar.

Figure 2 shows the result SEM of original PKS biochar, activated PKS biochar and treated PKS biochar, respectively. In these images, light region correspond to carbon wall and black region correspond to pores. The PKS biochar developed high porosity, presenting longitudinal pores with sizes ranging from micro to macro pores (10-200μm) [15]. It is clearly shows that the morphology of treated PKS biochar was more porous compared to original PKS biochar. The porous structures of the treated PKS Biochar sample indicate clearly that the potassium hydroxide, KOH activation process did in fact slice through the pores causing a larger pore size structure to be formed, and also the pore distribution is seen to be more frequent in an orderly manner. The positive results obtained from the difference in surface porosity is clear visible here via SEM thus verifying the reasons for samples to undergo Activation process using potassium hydroxide and Treatment using nitric acid, HNO3.

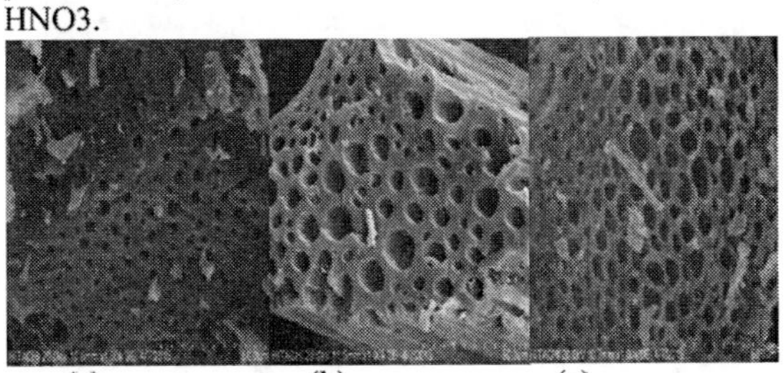

(a) (b) (c)

Figure 2. SEM images of (a) Original PKS Biochar, (b) activated PKS biochar and (c) treated PKS biochar

Table 1 shows the overall Raman spectrum comparison of PKS Biochar samples. The G-band peaks recorded for Original, Activated, and Treated samples are 1583.14 cm-1, 1578.82 cm^{-1}, and 1582.81 cm^{-1} respectively. The peaks that appear for the D-band are 1354.39 cm-1, 1354.06 cm-1 and 1354.72 cm-1 for the same samples of Biochar.

Table 1. Overall Comparison of Biochar Samples

Sample	Original Palm Kernell Shell Biochar	Activated *Palm Kernell Shell* Biochar	Treated *Palm Kernell Shell* Biochar
D-Band (cm-1)	1354.39	1354.06	1354.72
G-Band (cm-1)	1583.14	1578.82	1582.81
(ID/IG) ratio	0.856	0.858	0.857

The ratio between the D-band and G-band intensities (ID/IG) represents the scale of defect concentration. The higher the intensity ratio, the higher the defect concentration. The defect mentioned here corresponds to the proportion of sp^3 carbon formation, which caused disruptions to the original sp^2 bonding, hence attributing to the structural defects being measure. In this case the values of (ID/IG) for Original, Activated and Treated samples are 0.856, 0.858, and 0.857 respectively. The increase in (ID/IG) after the introduction of nitric acid to treat the samples is due to the hydrothermal oxidation caused by nitric acid. This process generates oxygen containing functional groups on the surface of the sample thus increasing the number of sp^3 hybridized carbon, causing increased surface defects.

Figure 3 illustrates the spectrum for raw, treated and activated PKS biochar sample. The peaks recorded for C=O are 1590.71 cm-1, 1591.39 cm-1, and 1591.16 cm-1 respectively which corresponds to the ketones, carboxylic acid, esters and anhydrides, respectively. The stretching that occurred at 1094.38 cm-1, 1018.83 cm-1 and 1243.54 cm-1 can all be assigned to the C-N stretching. The bands around 3700 which can be ascribed to O-H of phenol were not found on the Treated sample but appear on the raw and activated biochar at 3717.22 and 3747.85. This shows that there was a slight shift in Raman bands which was caused by the activation process using potassium hydroxide. Another clear shift in Raman bands occurred in the band range of 1018 – 1243 cm-1. These bands are attributed to the C-N stretching which varied from 1094.38 cm-1 and 1018.83 cm-1 for raw and activated biochar to 1243.54 cm-1 for the treated biochar.

Another obvious change in bands was at band 1362.58 cm-1 which was not present in the Raw biochar, appeared in Activated and Treated biochar although it appears to be shifted slight in the Treated biochar bands at 1243.54 cm -1. This range of bands can be assigned to the C-O stretching and OH bending of alcohol and carboxylic acids.

DOI: 10.1051/matecconf/20166204003

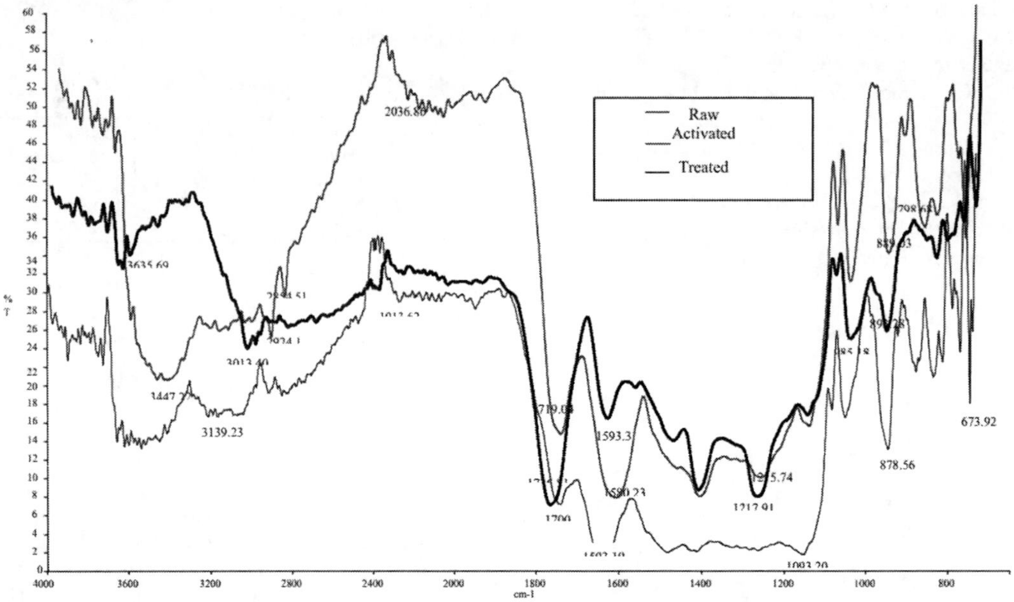

Figure 3. FTIR spectrum raw, activated and treated PKS biochars

In all, similar functional groups were observed on the surface of the activated carbons due to their exposure to similar chemical treatment and narrow variation of temperature of activation and treatment (see Table 2)

Table 2. Peak Assignments for Raw, Activated and Treated Biochar Samples.

	Original	Activated	Treated
Aromatic C-H (cm-1)	3015.23	3012.36	3010.93
Aliphatic C-H (cm-1)	2813.40	2812.74	2815.28
C=O stretching (cm-1)	1704.86	1713.65	1721.86
C=C stretching (cm-1)	1590.71	1591.39	1591.16
O-H stretching (cm-1)	3717.22	3747.85	3648.66
C-N stretching (cm-1)	1094.38	1018.83	1243.54

The values of BET specific surface area are 0.2982 12.9034 and 53.8346 and for original, activated and treated biochar, respectively. The average pore size of the samples also decreased from 338.19797 to 51.78124 nm after potassium hydroxide, KOH activation. The pore size for Treated biochar increased drastically to 7929.64nm. This has been verified by the SEM results obtained that clearly shows the larger pore sizes on the biochar after treatment with nitric acid.

3.2 Cyclic Voltammetry (CV)

The suitability, of the carbon obtained for electrodes in a supercapacitor was evaluated by using an electrochemical technique, cyclic voltammetry (CV). With the prepared Activated biochar as the electrodes, the electrochemical behavior of the biochar was characterized with the CV using 0.5 mol potassium hydroxide as the electrolyte and SCE as reference electrode. The CV response of the electrodes was measured at different scan rates varying from 10 to 200 mV s1 at potentials between 1 and 0 V. Figure 4 shows CV curve of the biochar samples in 0.5 mol potassium hydroxide.

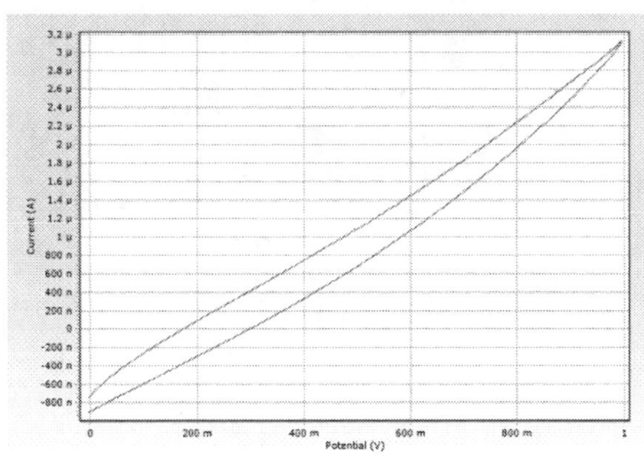

(a)

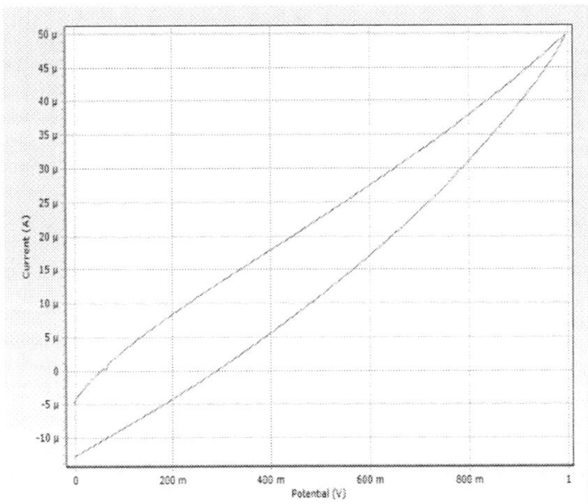

(b)

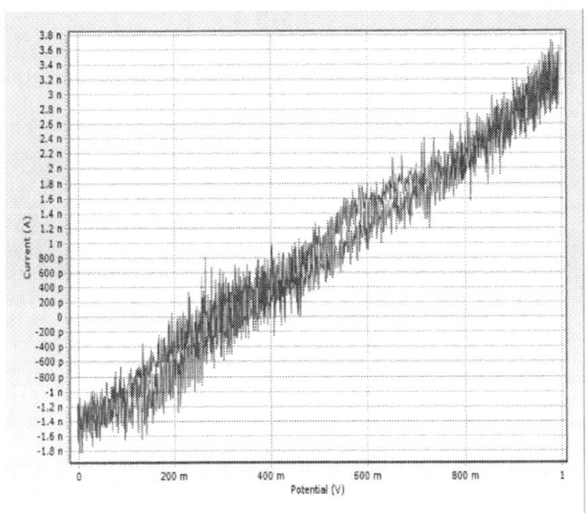

(c)

Figure 4. Cylic Voltammometry (CV) (a) raw, (b) activated and (c) treated PKS biochar electrodes at scanning rate 0.05 mV/s

From these figures, it is observed that the Activated biochar has a better closer resemblance to a square shape compared to the original biochar CV curve. According to this observation, the activated biochar shows better capacitor characteristics as compared to the original biochar. The treated biochar however exhibits a very scattered CV curve due to the high level of noise. This can be attributed to the uneven rough surface of the treated electrode when tested using CV.

4 Conclusion

Sustainable production of activated carbon was produced from Palm Kernel Shell using potassium hydroxide as the activation agent and nitric acid as the treatment medium, was developed and the results reported. A supercapacitor has been developed using 0.5 mol Potassium hydroxide as the electrolyte. Based on the test results albeit not perfect the palm kernel shell based biochar has the potential to be a lower cost substitute for the advanced, more expensive carbon materials currently being commercialized such as metal carbides, graphene or activated graphene. The process of synthesizing the palm kernel shell biochar is without artificial templates hazardous substances.

Acknowledgement

Authors would like to acknowledge the financial support from Ministry of Education (MOE), Malaysia via FRGS grant (Program code:FRGS/1/2014/TK05/UPM/02/8)."

References

1. Xia, K., Gao, Q., Jiang, J. & Hu, J., 2008. Hierarchical porous carbons with controlled micropores and mesopores for supercapacitor electrode materials. Carbon, Volume 46, pp. 1718-1726.A. Mecke, I. Lee, J.R. Baker jr., M.M. Banaszak Holl, B.G. Orr, Eur. Phys. J. E **14**, 7 (2004)
2. Halper, S. M. & Ellenbogen, C. J., 2006. Supercapacitor: A Brief Overview, Virginia: The MITRE Corporation.
3. Pandolfo, A. G. & Hollenkamp, A. F., 2006. Carbon properties and their role in supercapacitors. Journal of Power Sources, 157(1), pp. 11-27.
4. Jin, H., Wang, X., Gu, Z. & Polin, J., 2013. Carbon materials from high ash biochar for supercapacitor and improvement. Journal of Power Sources, Volume 236, pp. 285-292.
5. Junhua Jianga,*, Lei Zhangb, Xinying Wanga, Nancy Holma, Kishore Rajagopalana,Fanglin Chenb, Shuguo Ma, 2013. Highly ordered macroporous woody biochar with ultra-high carbon content as supercapacitor electrodes. Electrochimica Acta, Volume 113, pp. 481-489.
6. Kötz, R. & Carlen, M., 2000. Principles and applications of electrochemical capacitors. Electrochimica Acta, p. 2483–2498.
7. Zhu, Y., 2011. Carbon-Based Supercapacitors Produced by Activation of Graphene. Science, Volume 332, pp. 1537-1539.
8. Balathanigaimani, M.., 2008. Highly porous electrodes from novel corn grains-based activated carbons for electrical double layer capacitors. Electrochemistry Communications, 10(6), pp. 868-871.
9. Ito, E., 2007. Nanoporous carbons from cypress II . Application to electric double layer capacitors. New Carbon Materials, 22(4), pp. 321-326
10. Anjos, D. M., 2013. Pseudocapacitance and performance stability of quinone-coated carbon onions. Nano Energy, 2(5), pp. 702-712
11. Nor, N. S. M., 2015. Influence of gamma irradiation exposure on the performance of supercapacitor electrodes made from oil palm empty fruit bunches. Energy, Volume 79, pp. 183-194.
12. Kuratani, K., 2011. Converting rice husk activated carbon into active material for capacitor using three-dimensional porous current collector. Journal of Power Sources, 196(24), pp. 10788-10790.

DOI: 10.1051/matecconf/20166204003

13. Sahu, V., 2015. Heavily nitrogen doped , graphene supercapacitor from silk cocoon. Electrochimica Acta, Volume 160, pp. 244-253
14. Li, X. et al., 2011. Bioresource Technology Preparation of capacitor's electrode from sunflower seed shell. Bioresource Technology, 102(2), pp. 1118-1123.
15. Wan azlina Wan Ab Karim Ghani, Gabriel da-Silva and azil bahari Alias (2014). Physico-chemical characterizations of sawdust-derived biochar as potential solid fuels. Malaysian journal of analytical science, 18 (3), 724-729.

Semi-mechanistic Model Applied to the Search for Economically Optimal Conditions and Blending of Gasoline Feedstock for Steam-cracking Process

Adam Karaba[1,a], Petr Zámostný[2,a], Tomáš Herink[3] and Vendula Kelbichova[3]

[1]*Unipetrol Centre of Research and Education (UniCRE), Areal Chempark, 436 70 Litvinov, Czech Republic*
[2]*University of Chemistry and Technology Prague, Department of Organic Technology, Technicka 5, 166 28 Prague 6 - Dejvice, Czech Republic*
[3]*Unipetrol RPA, Záluží 1, 436 70 Litvínov, Czech Republic*

Abstract. Steam-cracking is energetically intensive large-scaled process which transforms a wide range of hydrocarbons feedstock to petrochemical products. The dependence of products yields on feedstock composition and reaction conditions has been successfully described by mathematical models which are very useful tools for the optimization of cracker operation. Remaining problem is to formulate objective function for such an optimization. Quantitative criterion based on the process economy is proposed in this paper. Previously developed and verified industrial steam-cracking semi-mechanistic model is utilized as supporting tool for economic evaluation of selected gasoline feedstock. Economic criterion is established as the difference between value of products obtained by cracking of studied feedstock under given conditions and the value of products obtained by cracking of reference feedstock under reference conditions. As an example of method utilization, optimal reaction conditions were searched for each of selected feedstock. Potential benefit of individual cracking and cracking of grouped feedstocks in the contrast to cracking under the middle of optimums is evaluated and also compared to cracking under usual conditions.

1 Scope and Introduction

Steam-cracking is an important industrial process which transforms a wide range of hydrocarbons feedstock to valued petrochemical commodities, olefins and aromatics. Usual types of feedstock include ethane, liquefied petroleum gas and gasoline non-aromatic fractions. Product stream is a wide mixture with major content of light olefins (ethylene, propylene and butadiene), light aromatics (benzene, toluene, xylenes), certain portion of heavy hydrocarbons and many other side-products including hydrogen. The products distribution depends on the feedstock composition and on the cracking conditions as well. Since the process is large-scaled, there is no easy way for direct data transfer from lab-scale to process-scale, therefore mathematical models of cracking are very useful tools for cracking unit operation, control and optimization.

There are numerous papers aimed at modelling of steam-cracking, usually focused to the description of pure substances behaviour [1-3] or the behaviour of simple mixtures. There are also other groups of researches [4-6], mostly focused to experimental investigation of cracking kinetics in lab-scale intended to create mechanistic model. But mechanistic models of industrial steam-cracking of complex mixtures are much less frequent than models developed for lab-scale pyrolysis because the kinetic model is only one part of such a modelling tool. Complicated coil geometry, mixture fluid dynamics including pressure-drop, modelling of the heat transfer in radiant chamber and transfer line exchanger together with mathematical complications with differential-algebraic system and complexity of industrial feedstock makes the creation of end-point-application very extensive and tedious. Therefore, many industrial models are entirely or partially empirical [7-10].

One of the most intended research group, Dente et. al. [11, 12] used mechanistic principles to create a detailed model of steam-cracking using minimum of simplifying assumptions. Their work continued from a simple ethane cracking model and propane-butane mixtures to gasoline fraction cracking. By including more modelling tools (such as coke-formation description, etc.) they created SPYRO® model which is industrially usable.

Another group of researches [13] published their ethane cracking model which is based on detailed and fully mechanistic description of the cracking kinetics using theoretical tools for kinetic parameters estimation.

There are also several studies interested in cracking condition optimization, usually focused to extend the operational period of cracking coil by minimizing of the coke-formation rate [14, 15].

However, there are only several published studies using a mechanistic cracking model for industrial

[a] Corresponding author: adam.karaba@vscht.cz; petr.zamostny@vscht.cz

© The Authors, published by EDP Sciences. This is an open access article distributed under the terms of the Creative Commons Attribution License 4.0 (http://creativecommons.org/licenses/by/4.0/).

application. This study is aimed at the application of previously developed model PYROL to the optimization of cracking unit operation with gasoline feedstock and with respect to market values of cracking products.

2 Model Brief Description

Model PYROL development documentation and technical details including implementation and model parameters tuning has been reported in sufficient details in our previous papers [16-18]. This study is aimed at application only, therefore we offer brief description of the model in this paragraph. The model is based on the combination of semi-mechanistic approach to the description of heavy components decomposition and the semi-empirical model of light components behaviour. Primary reactions are described by radical model considering the steady-state of active radicals and thus pseudo-monomolecular hydrogen abstraction is assumed. Secondary reactions, especially mutual reactions of light species (H_2-C_4), are described by enhanced version of well-established system of formal molecular reactions firstly introduced by Van Damme et. al. [19] as it is also documented in our previous papers.

Reactor is modelled as a plug-flow reactor considering the pressure-drop and it is divided to three parts: cracking coil placed in radiant chamber, adiabatic zone and transfer line exchanger. For the heat transfer modelling, the coil is modelled by a system of surface zones. The radiant chamber is considered to be a system of volumetric zones modelling the flow of flue gas.

Table 1. Model prediction accuracy indicators (average of products composition deviations and average absolute values of ones) for the cracking of gasoline [20].

	Experiment	Balance
N	29	85
COT, °C	812-852	795-845
Component	AVG(Δw$_i$), wt. % AVG(|Δw$_i$|), wt. %	
Hydrogen	-0.31 0.31	-
Methane	0.09 0.45	-
Ethane	0.69 0.69	-
Ethylene	0.53 0.64	-0.04 0.43
Propylene	-0.31 0.52	-0.06 0.60
C4	-0.45 0.81	-
C5	0.07 0.24	-
Benzene	-0.29 0.51	-
C6-C9	-4,57 4,57	-
Oil	1,52 1,52	-

The input of the model is formed by feedstock composition (per component basis) and operational parameters of cracking furnace. Output is the product stream composition, presumed fuel gas consumption and several other parameters.

Model PYROL has been tested in details in our recent paper where it was verified as balance tool [20]. From all reported data, determined deviation in the model prediction is relevant for current work. Table 1 demonstrates determined deviations between model predictions and experimental data (obtained by direct hot gas sampling, for method details see [21]) for the gasoline feedstock and also the comparison to process mass balance results but for all possible types of feedstock together.

3 Economic Criterion: Cracking Value

Cracking products distribution depends on the feedstock composition and reaction conditions. There are many theoretical and experimental studies published that are aimed at searching of ethylene (or other product) maximum yield for some selected feedstock. General rules, about trends in products yields are common knowledge. In this study, we propose relatively simple method for economic evaluation instead technical one.

From a large set of data, provided by our industrial partner, Unipetrol RPA, a very average gasoline, typical cracking feedstock, was chosen as a referential feedstock. A set of conditions typical for gasoline processing on GK6 reactor was chosen as referential. This feedstock provides a typical products distribution under these conditions and the distribution can be priced according to current prices. We propose this cost relativized on a tonne of feedstock as a reference zero point.

Value of any studied feedstock A with composition $\mathbf{w}_A$, cracked under a set of cracking conditions $\mathbf{p}$, can be evaluated according to current prices of products $\mathbf{P}$ using products flow $\mathbf{F} = \mathbf{F}(\mathbf{p},\mathbf{w}_A)$ predicted by simulation. This value in specific form (relativized to the feedstock mass) is referred to the specific value of products obtained by processing of referential feedstock (composition $\mathbf{w}°$) under referential conditions $\mathbf{p}°$.

$$CV_A\,(\mathbf{p},\mathbf{w}_A) = \mathbf{F}(\mathbf{p},\mathbf{w}_A)\cdot\mathbf{P}/\mathrm{FR} - \mathbf{F}(\mathbf{p}°,\mathbf{w}°)\cdot\mathbf{P}/\mathrm{FR}° \quad (1)$$

It is interesting information for industrial producers about the relative profit/loss received from a tonne of studied feedstock cracked under selected conditions $\mathbf{p}$ in comparison to typical feedstock processed under typical conditions $\mathbf{p}°$, but this evaluation is valid only for current prices of products $\mathbf{P}$.

Moreover, the cracking value is significantly depended on the reaction conditions and the construction parameters of reactor coil. Therefore, conditions can be optimized for the maximum of cracking value for given feedstock and given reactor coil but the planner must realize fact, this criterion does not take the heating cost into account. From all operational parameters the coil outlet temperature (COT) has the most significant impact to the products yield. Other parameters, such as feedstock rate (FR), temperature on the outlet of transfer line exchanger (TLXT), are usually determined by facility capacity or had been determined empirically and cannot be easily changed. Dilution of feedstock by steam (SO) is typical for the type of feedstock and the pressure on the outlet of coil (COP) is depended on current state of coil and feedstock rate.

4 Data Needed for Evaluation

In this paragraph, a data needed for evaluation will be shown except product prices because we are not allowed to public wholesale trade prices. Those ones, which were considered for evaluation, were valid for the end of May 2012.

From all historically collected gasoline composition processed or considered to be processed by Unipetrol RPA, referential gasoline feedstock was chosen (see Table 2) and also reaction conditions, typical for the processing of such a feedstock.

Table 2. Reference gasoline feedstock simple characterization and reference conditions on GK6 and SRT III reactor.

Components	Content	Parameter	GK6	SRT III
Paraffins, wt. %	26.9	COT, °C	840.0	833.0
Isoparaffins, wt. %	34.0	COP, kPa	200	198.0
Olefins, wt. %	0.6	FR, kg/h	27.9	26.1
Naphthenes, wt. %	29.2	SO, kg/kg	0.5	0.5
Aromatics, wt. %	9.3	XOT, °C	584.5	600.0
Density, $kg \cdot m^{-3}$	812.0	TLXT, °C	512.7	583.0

Basic characteristics of currently investigated gasoline feedstock are provided in the Table 3. For the mechanistic model, a detailed (per component) composition is required which includes more than 100 individual components. It is not provided for its vastness. For the completeness, a source of each gasoline fraction is provided: A = Light gasoline from the gasoline redestillation, B = Light gasoline from primary distillation, C = Middle fraction from gasoline redestillation, D = Light gasoline separated from hydrocracked vacuum distillates, E = Heavy gasoline from primary distillation, F = Gasoline fraction from hydrocracked products (sharper hydrocracking conditions), G = Heavy gasoline from hydrocracked vacuum distillates, H = Gasoline isolated from the hydotreating of the atmospheric gas oil and kerosene.

Table 3. Feedstocks PIONA (wt. %) and density ($g \cdot l^{-1}$).

Item	P	I	O	N	A	ρ
A	46,1	37,8	0,2	14,5	1,4	666,2
B	43,4	34,9	0,5	18,3	2,9	686,2
C	30,8	35,0	0,6	30,2	3,5	705,1
D	21,2	44,2	0,4	28,2	6,0	692,5
E	27,0	33,6	1,6	27,8	10,0	738,0
F	11,7	37,6	1,7	40,2	8,8	783,2
G	9,3	34,4	2,2	40,5	13,6	768,5
H	18,0	29,7	1,6	30,4	20,3	772,2

The most difficult on this approach is, to determine total prices of individual products, because many cracking products are often transformed to another one in the separation part of cracking unit (selective hydrogenation of acetylene, dealkylation of toluene, etc.).

5 Results and Discussion

In this paragraph, a discussion around the results of CV calculation for selected feedstock and its dependence on reaction conditions will be provided.

5.1 Reaction condition optimization

Current target is to optimize cracking conditions for feedstock which were selected by our industrial partner (see Table 3). Analysis of CV depending on coil outlet temperature with other parameters remaining constant for the reference feedstock is shown on the Figure 1.

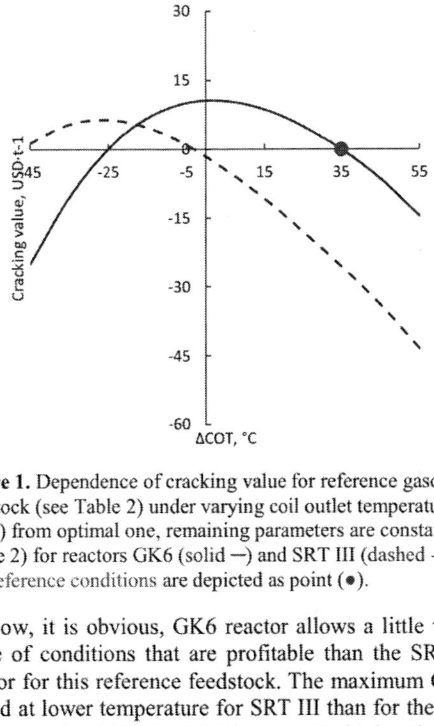

Figure 1. Dependence of cracking value for reference gasoline feedstock (see Table 2) under varying coil outlet temperature (COT) from optimal one, remaining parameters are constant (Table 2) for reactors GK6 (solid —) and SRT III (dashed - - -). The reference conditions are depicted as point (•).

Now, it is obvious, GK6 reactor allows a little wider range of conditions that are profitable than the SRT III reactor for this reference feedstock. The maximum CV is placed at lower temperature for SRT III than for the GK6 but the maximal CV for GK6 is generally higher than the one for SRT III. It is also interesting, reference conditions (and mean of real conditions in the facility as well) are a little sharper than the optimal ones, especially for SRT III reactor, where mean of COT per long period was determined to ΔCOT = +28 °C gives CV = -20.1 $USD \cdot t^{-1}$.

The existence of optimum is logical, because under growing temperature, the most desired product yield (ethylene) grows. But the yields of other important products, such as propylene and C4 hydrocarbons, pass a maximum and fall down. Yield of other desired products also grows (e.g. benzene) but other products (not such desired) grow as well. The yields of important products obtained by cracking of reference gasoline on the GK6 reactor in dependence on coil outlet temperature is shown on the Figure 2.

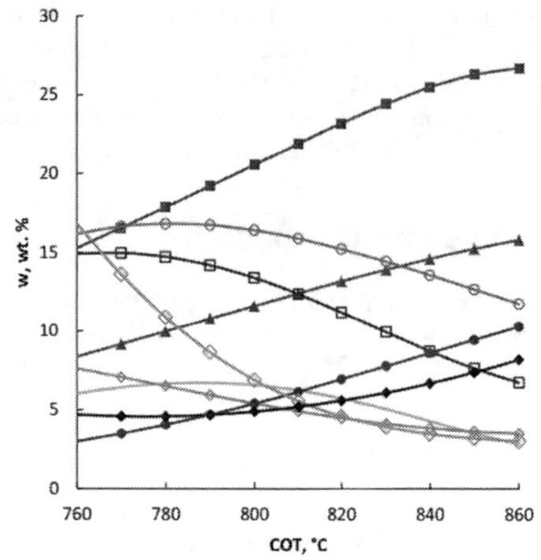

Figure 2. Dependence of products stream composition (simulated) obtained by cracking of reference gasoline feedstock (see Table 2) under varying coil outlet temperature (COT), other parameters remains constant (Table 2), reactor GK6: ethylene (red ■), propylene (green ○), methane (purple ▲), C_4 fraction (navy blue □), sum C_6-C_9 (sky blue ◊), benzene (brown ●), oil (black ♦), ethane (brown line).

The complexity of this dependence is the main reason why cracking value is a very good indicator for evaluation directly in the money per tonne of feedstock.

Similar analysis has been done for selected gasoline feedstock (Table 3) to find the optimal COT for both reactors, GK6 and SRT III. All these data are depicted in the Figure 3.

Figure 4 demonstrates expected difference of cracking value for selected gasoline feedstock separately cracked on individual reactors under the reference (usual) and optimized conditions. These optimized conditions were selected as a mean on previously searched optimal COT for individual feedstock (see Figure 3). As it is obvious, there is an increase of CV between reference (usual) and these optimal conditions for the same set of feedstock.

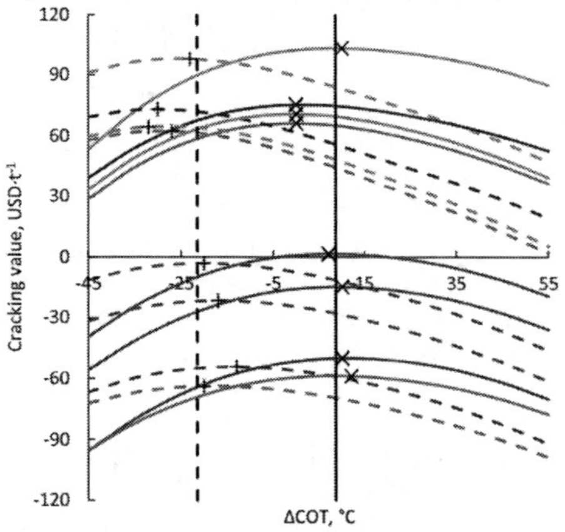

Figure 3. Dependence of cracking value on COT (coil outlet temperature), other parameters remain constant (see Table 2) on both studied reactors (GK6 solid, SRT III dashed) for selected gasoline feedstock: A (orange), B (blue), C (gray), D (pink), E (red), F(purple), G (brown) and H (light green). Optimal temperatures for GK6 (x) and SRT III (+) used for grouping for GK6 (solid black line) and for SRT III (dashed black line).

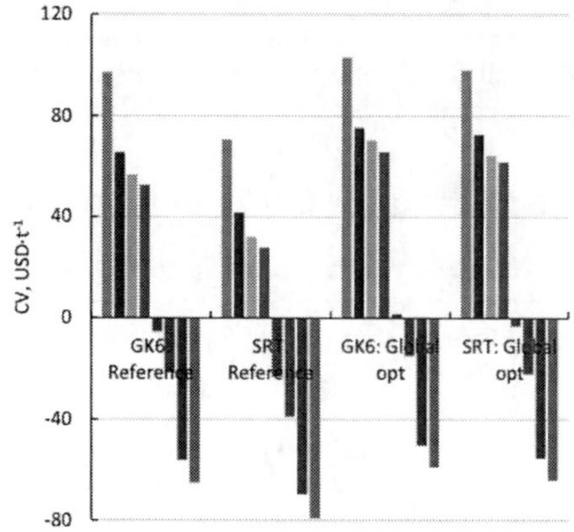

Figure 4. Cracking value of selected gasolines on reactors GK6 and SRT III under reference (usual) and optimal (global optimum for all feedstock) conditions: A (orange), B (blue), C (gray), D (pink), E (red), F(purple), G (brown) and H (light green).

5.2 Optimal feedstock blending

In this paragraph, another type of analysis is provided. If there is a possibility to blend all these gasoline feedstock to two separate tanks in the currently studied facility, it is possible to establish an optimal blending schema to maximize CV. Those two mixtures would be processed separately under conditions specific for resulting mixtures which could provide higher benefit than blending all feedstock together a cracking them at "common optimum" of conditions.

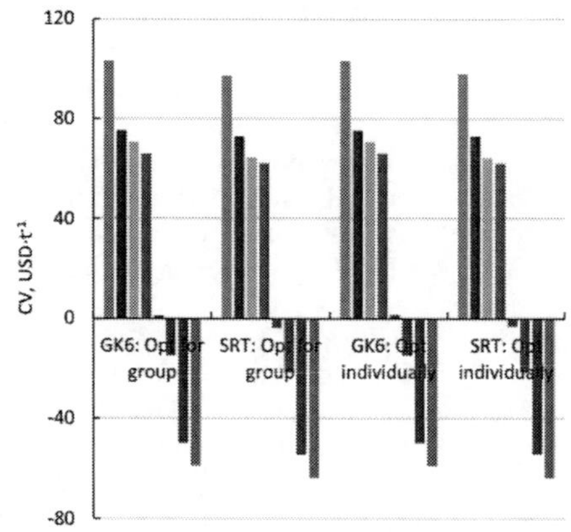

Figure 5. Cracking value of selected gasolines on reactors GK6 and SRT III under conditions optimized for groups and conditions optimized individually: A (orange), B (blue), C (gray), D (pink), E (red), F(purple), G (brown) and H (light green).

As it is shown on the Figure 3, optimal cracking conditions for individual feedstock can be grouped to two groups. These groups are divided by the vertical line (solid for GK6, dashed for SRT III). In this study, grouping of feedstock by the optimal COT on the GK6 reactor is demonstrated. On the Figure 5, cracking value of selected feedstock expected on individual reactors is shown for the case of conditions when these feedstock were grouped by the optimal COT for GK6 rector. COT optimal for group was chosen as arithmetic average of individual optimums. For comparison, cracking values expected under individually optimized conditions for each feedstock on both reactors is shown.

Interesting point is, there is very little potential difference between cracking value obtained by grouped feedstock optimization and individually optimized mode of cracking for each one of feedstock. But it is necessary to remind, there are also another requirement which has to be satisfied in the real facility, such as capacity of cracking coils and non-equal flows of individual feedstocks. Then the optimal blending schema must be evaluated with respect to satisfy optimal capacity of cracking coil and the assignment of individual feedstock to groups can be conditioned right by the satisfying the capacity which has stronger economic impact then optimal products yields (and CV).

As the last comparison, cracking value obtained by processing of all selected feedstock (with assumption of equal flow of these feedstocks) for individual cracking reactors (GK6 and SRT III) is shown on the Figure 6 for individual modes: the individually optimized conditions for each one of feedstocks, those conditions optimized for grouped feedstocks, only one "commonly optimal" conditions as global optimum and reference (usual) conditions.

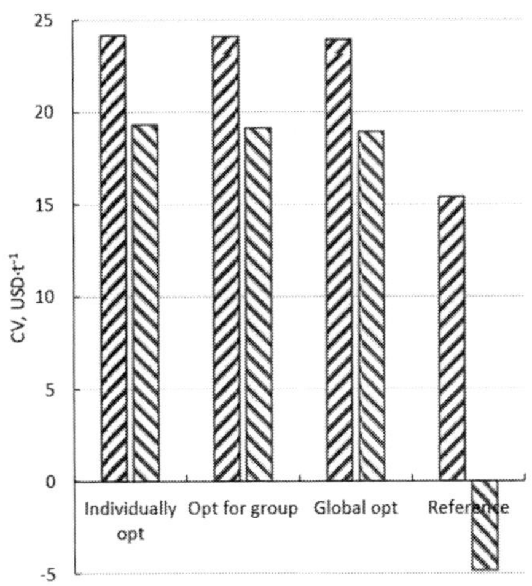

Figure 6. Cracking value expected from the cracking of all feedstocks (with assumption of equal flows) on reactors GK6 (blue, incr. hatched) and SRT III (red, decr. hatched) under individually optimized conditions, conditions optimized for groups, global optimum (for all feedstocks) and reference conditions.

6 Conclusions

An economic criterion CV (cracking value) was proposed in this paper. It is based on the evaluation of expected products yields (determined by previously verified semi-mechanistic model of steam cracking). Total price of products obtained by cracking of given feedstock under given reaction conditions relativized on the feedstock mass is referred (decreased by) the total prise of products obtained by the cracking of reference feedstock under reference conditions.

Using this method, different steam-cracking feedstock can be compared with respect to economic value of obtained products distribution. Optimal cracking conditions can be also determined using CV but these evaluations are valid only for given prices of products and therefore are not constant in time.

Optimal cracking conditions for several selected gasoline feedstocks were searched in this paper. A search for optimal blending scheme was also demonstrated using this criterion. Evaluation of selected 8 available feedstock showed, conditions usual for the processing of gasoline in our industrial partner facility are not the most optimal but a little sharper.

Analysis also reveal, there is a little potential benefit in the possibility to blend selected feedstocks into two separate streams and process them under reaction conditions optimal for resulting mixtures (in comparison with all-over blending). The benefit of individually optimized conditions for each feedstock is negligible.

Acknowledgement

This publication is a result of research project solved in a frame of UniCRE centre whose infrastructure was supported by European Regional Development Fund and the state budget of the Czech Republic. The publication was supported by institutional funds (Ministry of Industry and Trade of the Czech Republic).

References

[1] Van Geem K.M., Reyniers M.-F., Marin G.B., Song J., Green W.H., Matheu D.M.: *AIChE J.*, *52*, 718 (2006).
[2] Xu C., Al S.A.S., Wang C., Carstensen H.-H., Dean A.M.: *J. Phys. Chem. A*, *115*, 10470 (2011).
[3] Herbinet O., Marquaire P.M., Battin-Leclerc F., Fournet R.: *J. Anal. Appl. Pyrol.*, *78*, 419 (2007).
[4] Sabbe M.K., Van G.K.M., Reyniers M.-F., Marin G.B.: *AIChE J.*, *57*, 482 (2011).
[5] Bounaceur R., Burkle-Vitzthum V., Marquaire P.-M., Fusetti L.: *J. Anal. Appl. Pyrolysis*, *103*, 240 (2013).
[6] Raman S., Carstensen H.-H.: *Int. J. Chem. Kinet.*, *44*, 327 (2012).
[7] Villora Cano G., Bodalo Santoyo A., Lopez Cabanes A., Gomez Carrasco J.L.: *An. Quim., Ser. A*, *82*, 137 (1986).
[8] Nabavi R., Niaei A., Salari D., Towfighi J.: *J. Anal. Appl. Pyrol.*, *80*, 175 (2007).
[9] Abghari S.Z., Sadi M.: *J. Taiwan Inst. Chem. Eng.*, *44*, 365 (2013).
[10] Sedighi M., Keyvanloo K., Towfighi J.: *Iran. J. Chem. Chem. Eng. Vol*, *29*, (2010).
[11] Dente M., Ranzi E., Goossens A.G.: *Comput. Chem. Eng.*, *3*, 61 (1979).
[12] van Goethem M.W., Barendregt S., Grievink J., Verheijen P.J., Dente M., Ranzi E.: *Chem. Eng. Res. Des.*, *91*, 1106 (2013).

DOI: 10.1051/matecconf/20166204004

[13] Ranjan P., Kannan P., Al S.A., Srinivasakannan C.: *Chem. Eng. Technol.*, *35*, 1093 (2012).
[14] Edwin E.H., Balchen J.G.: *Chem. Eng. Sci.*, *56*, 989 (2001).
[15] Barazandeh K., Dehghani O., Hamidi M., Aryafard E., Rahimpour M.R.: *Chem. Eng. Res. Des.*, *94*, 307 (2015).
[16] Belohlav Z., Zamostny P., Herink T.: *Chem. Eng. Process.*, *42*, 461 (2003).
[17] Doskocil J., Belohlav Z., Herink T., Zamostny P.: *Chem. Listy*, *97*, 1176 (2003).
[18] Zamostny P., Belohlav Z.: *Pet. Coal*, *45*, 142 (2003).
[19] Van Damme P.S., Narayanan S., Froment G.F.: *AIChE J.*, *21*, 1065 (1975).
[20] Karaba A., Zamostny P., Belohlav Z., Lederer J., Herink T.: *Chem. Eng. Technol.*, *38*, 609 (2015).
[21] Herink T., Fulin P., Lederer J., Belohlav Z.: *Oil Gas J.*, *99*, 50 (2001).

The Research on Analytical Method of Diquat Dibromide in Diquat Technical Concentrates

Bing Gu[1], Hui Ming Xiao[2]

[1]*Pesticides Test Laboratory, ShenYang Research Institute of Chemical Industry, Shenyang, 110000, China*
[2]*Guidaojiaotong Polytechnic Institute, Shenyang, 110000, China*

Abstract. The technical concentrates (TK) is prepared by technical material (TC), the content is lower than TC, it can be used for the preparation of pesticide formulations. Diquat TK is generally brown or dark brown liquid. In recent years, it develops rapidly as herbicides. This paper mainly introduces the research on analyses of the effective components in diquat TK.

1 Introduction

The diquat [1] is light yellow crystal, its ISO name is diquat dibromide, stable in acid and neutral solution, rapidly decomposites in alkaline conditions. Its structural formula is as follows:

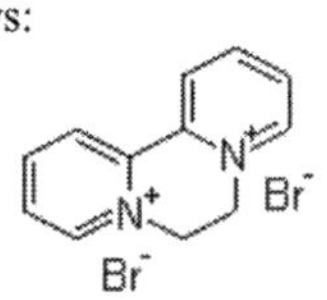

2 Experimental Part

2.1 Qualitative analysis

2.1.1 The qualitative analysis of infrared spectrum for diquat TK

The Qualitative Analysis method [2, 3] of infrared spectrum for Diquat TK is as follows:
-Instrument: Spectrum-100 FT-IR Spectrometers.
-Scan range: 4000 cm^{-1} ~ 400 cm^{-1}.

The diquat dibromide standard samples and diquat TK samples were prepared for the uniform transparent sheet, which will be determined of infrared spectra. The typical infrared spectrum of diquat dibromide is shown in Figure 1.

2.1.2 The qualitative analysis for bromine ion of diquat TK

The reagents and solutions for the qualitative analysis are as follows:
-Water: pure water;
-Potassium bromide standard;

The instruments for the qualitative analysis are as follows:
-Dionex ICS 5000 Ion chromatograph.
-Anion chromatographic column: 250 mm×4.0 mm (i.d.).

The ion chromatography operating conditions are as follows:
-Eluent: KOH solution
-Column temperature: room temperature
-Current velocity (mL/min): 1.2
-Sample volume: 8 L
-Quiescent time: 9.5 min

The determinations of specific steps are as follows:

(1)Preparation of potassium bromide standard solution.

Take an appropriate amount of potassium bromide sample into the volumetric flask, add water to scale, and shake.

(2)Preparation of sample solution

Take an appropriate amount of diquat TK sample into the volumetric flask, add water to scale, and shake.

Under the above mentioned conditions, the retention time of bromine ion in the sample solution of potassium bromide was determined. The typical ion chromatogram for the diquat bromide ion of diquat TK is shown in Figure 2.

© The Authors, published by EDP Sciences. This is an open access article distributed under the terms of the Creative Commons Attribution License 4.0 (http://creativecommons.org/licenses/by/4.0/).

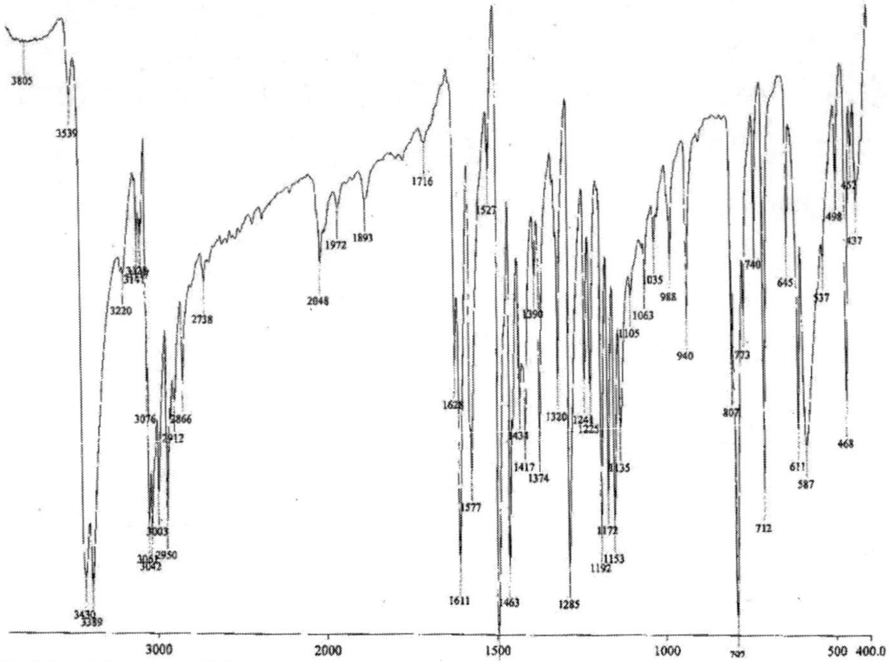

Figure 1. The Infrared Spectrum of Diquat Standard Samples.

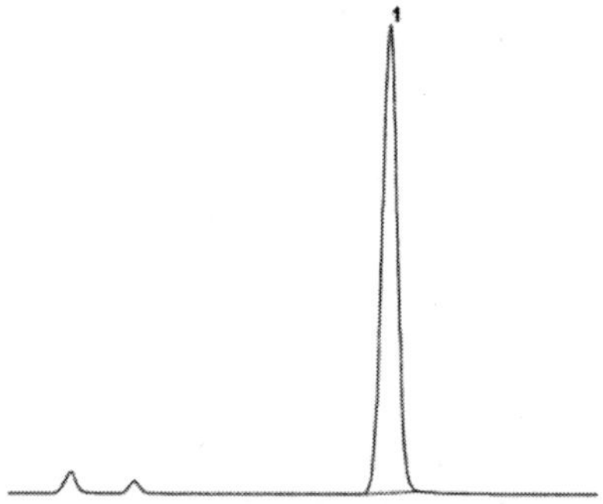

Figure 2. The Ion Chromatogram of Bromine Ion in the Sample.

2.1.3 Liquid chromatography-mass spectrometry

The instrument for Liquid chromatography-mass is as follows [4, 5]:

-Instrument: UPLC Quattro-micro Liquid mass spectrometer.

-Chromatographic column: 150 mm×4.6 mm (i.d.) stainless steel column, filled ZORBAX SB-C18, 5μm filler.

The liquid mass coupling conditions are as follows:

-Mobile phase: ψ (Acetonitrile: 0.1% glacial acetic acid solution) = 5:95.

-Current velocity: 1.0 mL/min.

-Ion source: ESI(+).

-Detection range: m/z 50~1000.

The determinations of specific steps are as follows:

(1) Preparation of standard sample solution

Take an appropriate amount of diquat dibromide samples, which are placed in the volumetric flask, adding acid aqueous solution to the scale, shake.

(2)Preparation of sample solution

Respectively take an appropriate amount of diquat TK samples, which are placed in the volumetric flask, adding acid aqueous solution to the scale, shake.

Mass spectrum determination of diquat dibromide hydrate samples, and the mass spectrum of diquat TK were shown in Figure 3, 4.

From the mass spectrum of diquat TK shown in Figure 3 and Figure 4, we can find that classic date is the same. Some special m/z are 84.7, 157.3, 183.2.

2.1.4 Ultraviolet spectrometry

The instrument for ultraviolet spectrometry [6] is shimadzu UV 2201 ultraviolet spectrophotometer.

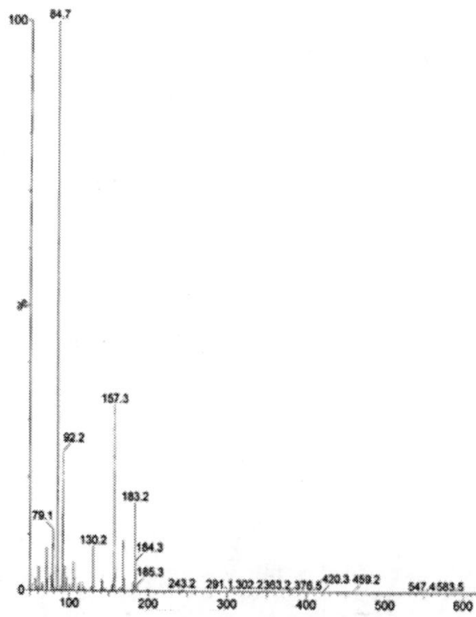

Figure 3. The Mass Spectra of Diquat Samples.

 DOI: 10.1051/matecconf/20166204005

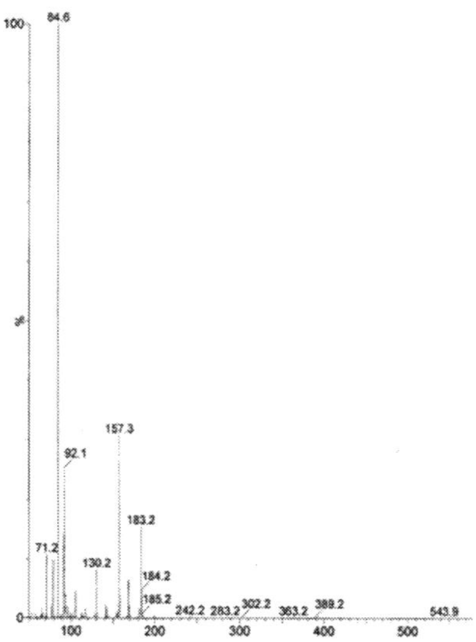

Figure 4. The Mass Spectra of Diquat in Diquat TK.

The ultraviolet spectrum qualitative conditions of diquat standard sample are as follows:
-Light source: deuterium lamp and tungsten lamp.
-Colorimetric pool: 1 cm quartz colorimetric cell.
-Slit: 2.0 nm.
-Scanning range: 400 nm~190 nm.
-Scanning speed: medium.
-Scanning mode: absorbance.
-Solvent: water.
-Reference: water.

Preparation of standard sample solution: take an appropriate amount of diquat standard sample in the sample flask, dissolved with water solution and diluted to the scale, shake, and wait for detection. The typical UV spectra of the diquat standard samples are shown in Figure 5.

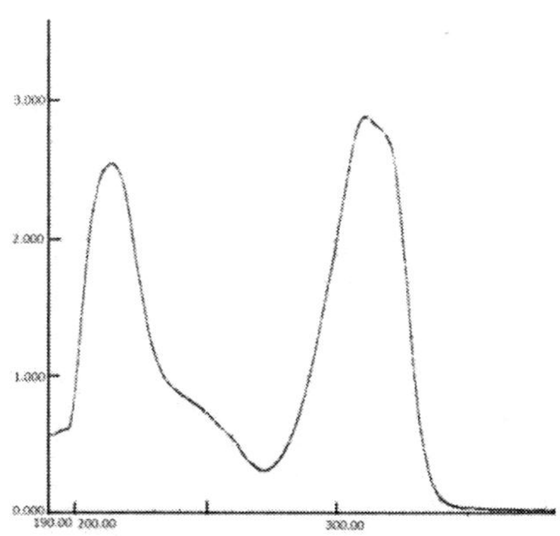

Figure 5. The Ultraviolet Spectra of Diquat Dibromide.

2.2 Quantitative analysis

2.2.1 Reagents and solutions

The specific reagents and solutions [7] are included the following contents:
-Acetonitrile: chromatographic class.
-Phosphoric acid: analytical class.
-Water: ultra pure water.
-Heptanesulfonate: chromatographic pure.
-Triethylamine.
-Phosphoric acid solution: Adjusted with phosphoric acid to pH = 2.1.
-The diquat standard samples.

Heptanesulfonate buffer solution: take an appropriate amount of heptanesulfonate, soluble in 920 mL ultra pure water, adding a certain amount of phosphoric acid, and then mix with triethylamine to a certain pH value, shake.

2.2.2 Method summary

Dissolving the samples in the water, make heptanesulfonate buffer aqueous solution and acetonitrile as mobile phase, stainless steel column and ultraviolet detector with C18 as filler (the maximum absorption wavelength is 310 nm, as shown in Figure 5).

2.2.3 Instrument

Specific instruments include the following contents:
-HPLC instrument with ultraviolet variable wavelength detector.
-Column: 150 mm * 4.6 mm (i.d.) stainless steel column, filled with C18, 5 m filler.
-Quantitative sampling tube: 5 μL.

2.2.4 Operating conditions

The HPLC operating conditions [8] include the following contents:
-Mobile phase: heptanesulfonate aqueous solution acetonitrile = 92:8.
-Flow rate: 1.0 mL/min.
-Column temperature: room temperature.
-Wavelength: 310 nm.
-Injection volume: 5 μL.
-Retention time: the diquat is about 7.8 min.

The high performance liquid chromatography figure of typical diquat TK is shown in Figure 6.

Figure 6. The High Performance Liquid Chromatography Figure of Typical Diquat TK.

2.2.5 Determination

The determinations of specific steps are as follows:

(1) Preparation of standard sample solution

Take an appropriate amount of diquat standard samples into the volumetric flask, add water and shake to be dissolved, dilute with water to the scale, and shake.

(2)Preparation of sample solution

Respectively take an appropriate amount of diquat TK samples into the volumetric flask, add water and shake to be dissolved, dilute with water to the scale, and shake.

2.3 Experimental results and discussion

2.3.1 Linear relationship test

With a certain mass range, the several diquat standard samples were weighed into the different volumetric flasks which have the same volume, dissolved in water, constant volume and shake. According to the operation conditions of the method, the peak area of diquat was determined, and the average results of the two measurements were determined. The standard curve was drew, with the peak area as the horizontal coordinate, the peak area as the vertical coordinateand, the result is shown in Figure 7.

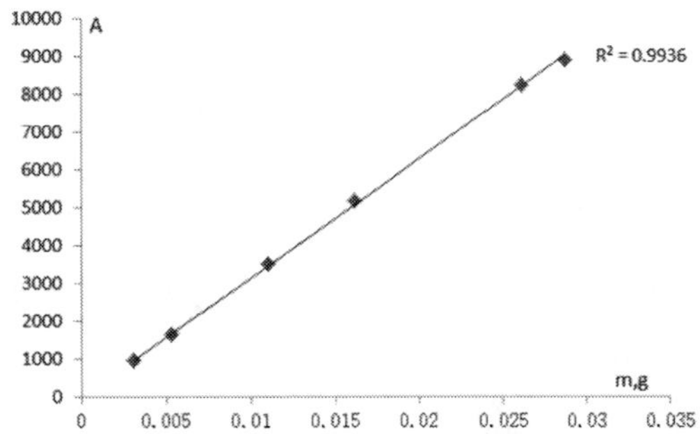

Figure 7. The Relationship Between the Peak Area and Mass Concentration of the Diquat.

From Figure 7, we can see that the diquat dibromide concentration appear a good linear relationship with the corresponding peak area in a certain range, R2 = 0.9936.

2.3.2 Precision test

Select a representative of diquat TK samples (marked 40%), starting from the sample, according to the operation conditions, determination of 6 times, the results are shown in Table 1.

Table 1. The Precision Test results of Diguat TK.

Measure times	**Mass fraction of diguat, %**	**Range,%**	**average value,%**
1	40.46	0.26	40.5
2	40.58		
3	40.71		
4	40.55		
5	40.45		
6	40.51		

From Table 1, it shows the method has good precision, the determination range is 0.26% with the mass fraction of diquat in diquat TK.

2.3.3 Accuracy test

Take a number of diquat TK together with a certain amount of diquat standard sample. According to the method for the determination of the diquat total quality and the recovery rate was calculated, and the results are shown in Table 2.

Table 2. The Diguat TK Recovery Test Results(%).

1#	2#	3#	4#	5#	6#	Average
99.93	100.31	101.02	99.91	99.82	99.73	100.1

3 Conclusion

Firstly, using the infrared chromatography, Liquid Chromatography-Mass Spectrometry, ion chromatography, the determination of bromide ions qualitative methods analysis the diquat TK sample, and then determined by chromatographic methods diquat TK for quantitative analysis.

References

1. Kolberg, D.I.S., MacK, D., Anastassiades, M.; Hetmanski M.T., Fussell R.J., Meijer T., Mol H.G.J., Development and independent laboratory validation of a simple method for the determination of paraquat and diquat in potato [J]. *cereals and pulses Analytical and bioanalytical chemistry*., **404**, 8 (2012).
2. Hao C., Zhao X., Morse D., Yang P., Taguchi V., Morra, F. Optimized liquid chromatography tandem mass spectrometry approach for the determination of diquat and paraquat herbicides[J]. *Journal of chromatography, A: Including electrophoresis and other separation methods*. **1304** (2013).
3. Abu Shawish H.M., Ghalwa N.A., Hamada M.,Basheer A. Modified carbon paste electrode for potentiometric determination of diquat dibromide pesticide in water and urine samples[J]. *Materials science & engineering, C. Biomimetic and supramolecular systems*. **32**, 2 (2012).
4. Xiangbing Mao, Mei Lv, Bing Yu, Jun He, Ping Zheng, Jie Yu, Quyuan Wang, Daiwen Chen. The effect of dietary tryptophan levels on oxidative stress of liver induced by diquat in weaned piglets [J]. *Journal of Animal Science and Biotechnology*. **01** (2015).
5. Norio TESHIMA, Tadao SAKAI. Analysis of Decomposed Species of Diquat using Photocatalytic TiO2 and its Degradation Pathway [C]. *Department of Applied Chemistry Aichi Institute of Technology*, **09**(2006).
6. Peng Xu, Hongmei Li, Shouwei Sun, Xiefei Shi. Separation of Diquat and Paraquat by Chaotropic

DOI: 10.1051/matecconf/20166204005

Effect HPLC [C]. *Department of Applied Chemistry Aichi Institute of Technology*, **09** (2006).

7. Zhang Yan, He Yong,Li Guoao, Chen Xue, Yan Lei. Application of Online Solid-phase Extraction Technology in the Field of Environmental Analysis [J]. *Meteorological and Environmental Research*. **05** (2014).
8. ZHAO Xiu-feng, CHEN Lin, Muhammad I A Rehmani,WANG Qiang-sheng, WANG Shao-hua,HOU Peng-fu, LI Gang-hua, DING Yan-feng. Effect of Nitric Oxide on Alleviating Cadmium Toxicity in Rice [J]. *Journal of Integrative Agriculture*. **09** (2013).

Synthesis of Molecularly Imprinted Polymer for Sterol Separation

Ratanaporn Yuangsawad[1], Nuengruthai Chuayrueng[2], Nattawat Nonthanasin[2], Krittin Binabdullah[2] and Duangkamol Na-Ranong[2,a]

[1]*Division of Chemical Engineering, Faculty of Engineering, Rajamangala University of Technology Krungthep, 2 Nanglinchii, Sathorn, Bangkok 10120, Thailand*
[2]*Department of Chemical Engineering, Faculty of Engineering, King Mongkut's Institute of Technology Ladkrabang, 1 Chalongkrung 1, Ladkrabang, Bangkok 10520, Thailand*

Abstract. Molecular imprinted polymer (MIP) was prepared by bulk polymerization in acetone using acrylamide as a functional monomer, ethylene glycol dimethacrylate as a crosslinker, stigmasterol as a template and benzoyl peroxide as an initiator. The obtained MIPs were characterized using a scanning electron microscope and a fourier transform infrared spectrophotometer. Performance in sterol adsorption of MIPs prepared under various conditions was investigated using a model solution of phytosterols in heptane, comparing with a nonimprinted polymer (NIP). Statistical analysis revealed that the amounts of crosslinker and template strongly affected the performance of MIP while the amount of solvent slightly affected the performance of MIP. MIP synthesized under the optimal condition had adsorption capacity of 1.28 mgsterols/gads which were 1.13 times of NIP.

1 Introduction

Due to high energy consumption and depletion of fossil fuel, demand of biodiesel has been increased for several decades. Palm oil and palm fatty acid distillate (PFAD) are mainly used as raw materials for biodiesel production in many countries, especially in South East Asia [1-4]. However, biodiesel producers are suffering from high cost of production. In order to increase the competency of biodiesel in the market, it is necessary to increase economical value of the overall process including oil refinery and biodiesel production.

PFAD is a by-product from palm oil refinery. It consists of free fatty acids (81.7 %), glycerides (14.4 %) and several high valuable bioactive compounds such as vitamin E, phytosterols and squalene [5]. Phytosterols have several biological activities such as cholesterol lowering effect, anti-inflammatory, anti-atherogenicity, anti-cancer and anti-oxidation. Therefore, applications of phytosterols in pharmaceutical, functional food and cosmetics industries are promising [6].

Recovery of phytosterols from PFAD with reasonable cost should increase economical value of overall process of palm oil refinery and biodiesel production. Several processes including chemical transformation and physical separation steps have been developed mainly to recover phytosterols from soybean oil fatty acid distillate and tall oil [6-9]. These processes usually include the step with high energy consumption such as molecular distillation under high vacuum condition, crystallization at low temperature, recovery of solvent from the solvent extraction and supercritical fluid extraction. Therefore, it is necessary to develop the more environmental friendly and energy saving process for phytosterols recovery.

Adsorption has been widely applied in an industrial process to separate minor contaminants from liquid and gas. It has high potential for application of phytosterols recovery. Since PFAD consists of several types of minor components [5], one of the key factors for the success of the separation of phytosterols by adsorption is selectivity of the adsorbent towards phytosterols.

Molecularly imprinted polymers (MIPs) have been applied in various fields requiring high selectivity to particular molecules such as biosensor development, chromatography and catalysis [10-12]. In this work, MIPs were synthesized from acrylamide (monomer), ethylene glycol dimethacrylate (crosslinker) and stigmasterol (template) by bulk polymerization under various conditions. Amounts of crosslinker, template and solvent were selected as synthesis parameters. Properties of the obtained MIPs were observed using SEM and FT-IR techniques. Batch adsorption was performed to evaluate the performance of the obtained MIPs in sterol adsorption using a model solution of phytosterols in n-heptane. Performance of MIPs was compared with a non-imprinted polymer (NIP).

2 Experimental

2.1 Synthesis of molecularly imprinted polymers

Acrylamide (monomer), ethylene glycol dimethacrylate (EGDMA, crosslinker), stigmasterol (template), benzoyl

[a] Corresponding author: dnaranong@hotmail.com, duangkamol.na@kmitl.ac.th

© The Authors, published by EDP Sciences. This is an open access article distributed under the terms of the Creative Commons Attribution License 4.0 (http://creativecommons.org/licenses/by/4.0/).

peroxide (initiator) and acetone (solvent) were mixed in a 60 ml glass bottle. In order to remove oxygen, nitrogen gas was purged through the solution for 10 min. Then the solution was prepolymerized at 60 °C for 1 h by shaking in an orbital shaker at 220 rpm. After prepolymerization, the temperature in the shaker was raised to 60 °C and the polymerization was allowed to proceed at this condition for 24 h. The obtained solid was washed several times with distilled water to remove all unreacted reagents. To ensure that all the unreacted reagents were completely removed from the obtained solid, water was separated from the solid by centrifugation at 4,000 rpm for 30 min and its absorbance was measured using UV-spectrophotometer. To remove the stigmasterol template from the obtained solid, Soxhlet extraction was performed at 120 °C for 24 h using solution consisting of acetonitrile (85 vol%), water (9.9 vol%), methanol (5 vol%) and acetic acid (0.1 vol%) as extracting solvent. To ensure that the template was completely removed, the obtained MIPs mixed with n-heptane and stirred at 30 °C for 3 h and the solution was taken for analysis of stigmasterol using a gas chromatograph with flame ionized detector (GC-2010plus; Shimadzu).

Different MIPs were prepared by varying ratios of EGDMA, stigmasterol and acetone to acrylamide in the ranges of 2:1 to 4:1 (mol:mol), 0.1:1 to 0.3:1 (mol:mol) and 4:1 to 8:1 (mol:mol), respectively. Amount of benzoyl peroxide was fixed at 0.32 mole per mole of acrylamide. Conditions for the MIPs synthesis were determined using a MINITAB program according to Central Composite Design (CCD) method. The result is summarized in Table 1.

Table 1. Condition for MIP synthesis determined by CCD method. (1 mmol of monomer, initiator 0.32 mmol)

No.	crosslink (mmol)	template (mmol)	solvent (ml)
1	2.5	0.15	5
2	3.5	0.15	5
3	2.5	0.25	5
4	3.5	0.25	5
5	2.5	0.15	7
6	3.5	0.15	7
7	2.5	0.25	7
8	3.5	0.25	7
9	2	0.2	6
10	4	0.2	6
11	3	0.1	6
12	3	0.3	6
13	3	0.2	4
14	3	0.2	8
Std	3	0.2	6

For further discussion, the MIP obtained at the condition of monomer: crosslinker : template : solvent = 1 : 3 : 0.2 : 6 was named as "MIP-STD". The solid sample obtained under this condition before extraction of template was named as "T-MIP-STD". Nonimprinted polymer (NIP) was prepared using this ratio by the same procedure without addition of stigmasterol template.

2.2 Characterization of molecularly imprinted polymers

SEM images were taken to observe morphology of the samples using EVO®MA10 (ZEISS). Extra high tension voltage level was 15.00 kV. Fourier transform infrared spectra were measured using IRPrestige-21 (Shimadzu, Japan) equipped with MIRacle ATR (PIKE Technologies, Inc.) with a resolution of 4 cm^{-1}.

2.3 Adsorption test

Model solution was prepared from sterol mixed, containing campesterol (23.6 wt%), stigmasterol (28.2 wt%) and β-sitosterol (48.2 wt%), and n-heptane. Concentration of total sterols in the model solution was set at 140 $\mu g \cdot g^{-1}$. Batch adsorption was performed as follows. 20 ml of the prepared model solution was mixed with 1 wt% of MIP or NIP and then the mixture was shaken at 220 rpm, 30 °C for 6 h.

Samples (c.a. 0.1 g) before and after adsorption were taken for quantitative analysis of campe-, stigma- and β-sitosterol. Analysis was performed using a gas chromatograph connected with flame ionized detector (GC-2010plus; Shimadzu). Peak separation was achieved using a ZB-5HT capillary column (30 m in length, 0.32 mm in internal diameter, 0.1 μm in film thickness; Phenomenex). Tricaprin was used as an internal standard. Calibration curve of sterols in n-heptane was made in the range of 0 – 120 μg. The obtained calibration curve was expressed as $W_{sterol} = 13.075\ (A_{sterol}/A_{Tricaprin}) + 0.6042$ with $R^2 = 0.9997$.

Performance of the adsorbent was evaluated from percentage of sterol adsorption (%Ads: %, Eq. (1)), adsorption capacity (q_i: $g_{sterols}/g_{ads}$, Eq. (2)) and selectivity based on stigmasterol (S_i: %, Eq. (3)).

$$\%\text{Ads} = \frac{C_0 - C}{C_0} \times 100 \qquad (1)$$

$$q = \frac{(C_0 - C) W_{\text{solution}}}{W_{\text{ads}}} \qquad (2)$$

$$S_i = \left(\frac{W_{i,0} - W_i}{W_{i,0}} \right) \left(\frac{W_{\text{stigma},0}}{W_{\text{stigma},0} - W_{\text{stigma}}} \right) \qquad (3)$$

2.4 Statistical analysis

ANOVA analysis was performed in order to investigate effects of the synthesis parameters on the adsorption performance of the obtained MIP and to predict the optimum condition of MIP synthesis.

3 Results and Discussion

3.1 Polymer characterization

All the synthesized MIPs and NIP were white fine powder and were not dissolved in water and n-heptane. Figures 1(a) and 1(b) show SEM images of MIP-STD with the magnifications of 500 and 20,000, respectively. SEM images reveal that most of the particles had the size smaller than 100 μm and the surface of MIP-STD was not uniformed. Pores with the diameter less than 1 μm and cavities with the size in the order of 2 μm or larger are observed. SEM images of MIP synthesized under other conditions (not shown) had characteristic similar to those of MIP-STP. Figures 1(c) and 1(d) show SEM images of NIP. No large cavity is observed. Therefore, the large cavities in MIP-STP should be considered as a result of existing of stigmasterol during the polymerization and they were formed after removal of the template.

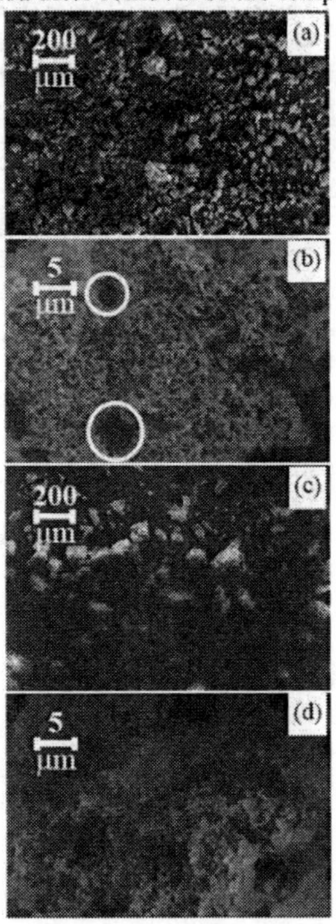

Figure 1. Typical SEM images of MIP-STD (a, b), NIP (c, d).

Figure 2 shows FT-IR spectra of NIP and T-MIP-STD. Identification of absorbed peaks was performed by using database of FT-IR equipment. Characteristics of both spectra were similar. Both spectra had small broad peaks of N-H vibration around 3460 cm^{-1}, small sharp peak of C-H vibration at 2960 cm^{-1}, large sharp peak of C=O vibration at 1730 cm^{-1} and peak of C-N vibration at 1400 cm^{-1}. These peaks were identified as the N-H, C=O, C-N and C-H in the structure of acrylamide-EGDMA polymer. Since no absorption band of C=C (at 1600 cm^{-1}) was observed in these spectrum and both acrylamide monomer and EGDMA have C=C in their structures, this result indicates that unreacted acrylamide and EGDMA were completely removed from the obtained NIP-STD and T-MIP-STD by water washing. Comparison of the spectra of NIP and T-MIP-STD reveals that NIP and T-MIP-STD had almost the same chemical structures. However, the relative intensity of C-H peak to other peaks in the spectrum of NIP was less than the relative intensity observed in the spectrum of T-MIP-STD. This result indicated existing of stigmasterol, containing C-H bond, in T-MIP-STD.

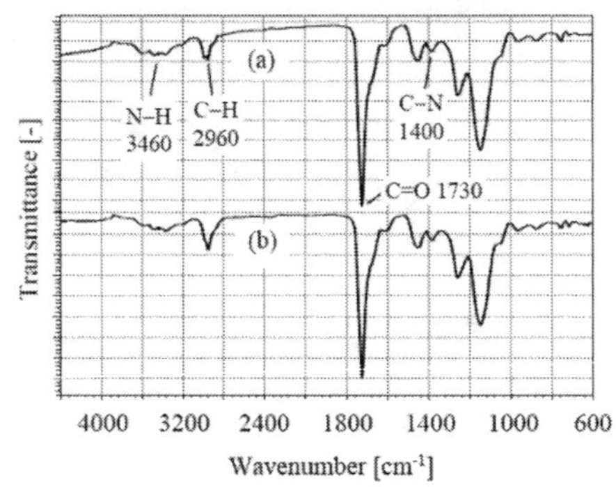

Figure 2. FT-IR spectra of (a) NIP-STD and (b) T-MIP-STD.

3.2 Performance of phytosterols adsorption

Figure 3 shows the performance in sterol adsorption of MIPs comparing with NIP. As shown in the figure, MIPs prepared at the condition number of 2, 4, 14 and STD adsorbed larger amount of sterols comparing with NIP. Moreover, adsorption experiment revealed that all the prepared MIPs adsorbed campe-, stigma- and β-sitosterol with almost the same percentage of the original amounts presenting in the model solution; $S_{campesterol} \approx S_{\beta\text{-sitosterol}} \approx 1$. No specific selectivity towards stigmasterol adsorption should be a result of very similar molecular structures of these sterols. This result should be considered as the possibility of applying MIP in recovery of all types of sterol from fatty acid distillates. Therefore, adsorption capacity was calculated from total amount of campesterol, stigmasterol and β-sitosterol adsorbed on the adsorbents.

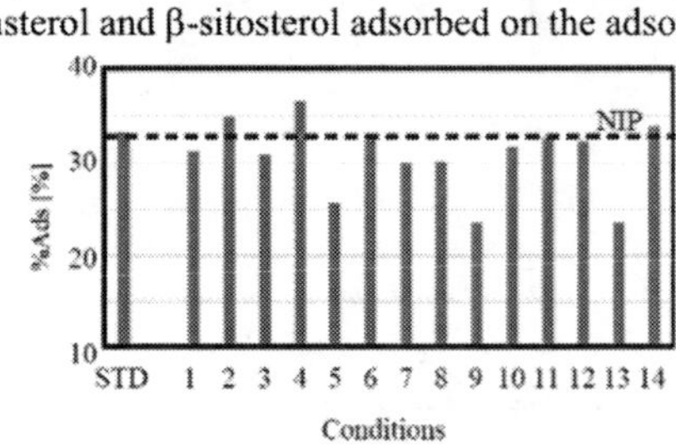

Figure 3. Performance of molecularly imprinted polymers in sterol adsorption at 30 °C, 6 h; (- - -) NIP.

3.3 Effect of synthesis parameters on adsorption performance of MIP

The relation between percentage of sterol adsorption and amounts of crosslinker, template and solvent obtained from statistical analysis are shown in Figure 4(a) – 4(c).

As shown in Figure 4(a), in the investigated range, increasing the amount of crosslink improved the performance in sterol adsorption of the obtained MIPs. The effect was more significant when the amount of template was low. However, increasing the amount of template (0.1 to 0.3 mmol) non-monotonously affected adsorption performance of the obtained MIPs. The lowest efficiency was observed from the MIP synthesized at the template (0.1 to 0.3 mmol) non-monotonously affected adsorption performance of the obtained MIPs. The lowest efficiency was observed from the MIP synthesized at the template (0.1 to 0.3 mmol) non-monotonously affected adsorption performance of the obtained MIPs. The lowest efficiency was observed from the MIP synthesized at the amount of template around 0.15 mmol. Figure 4(b) shows that increase of the amount of crosslink improved the performance in sterol adsorption of the obtained MIPs whereas the amount of solvent insignificantly affected the performance in sterol adsorption of the obtained MIPs. Effect of the amount of template on the performance in sterol adsorption of the obtained MIPs shown in Figure 4(c) agrees well with the same effect shown in Figure 4(a).

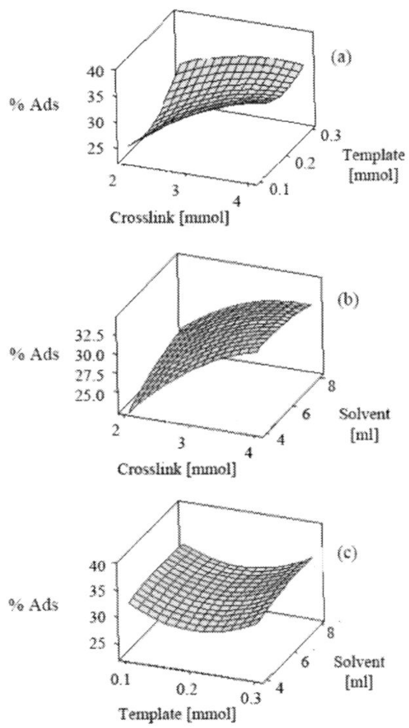

Figure 4. Effects of synthesis parameters on performance of MIPs. Synthesis conditions: (a) volume of solvent = 5 ml, (b) amount of template = 0.20 mmol, amount of crosslink = 3 mmol.

According to response analysis, the optimal amounts of crosslinker, template and solvent were 3.5 mmol, 0.3 mmol and 7 ml, respectively and the MIP synthesized under this optimized condition had adsorption capacity of 1.28 $mg_{sterols}/g_{ads}$ or 1.13 times of NIP.

4 Conclusions

Molecularly imprinted polymer (MIP) using for sterol separation was successfully synthesized by bulk polymerization of acrylamide and EGDMA using stigmasterol as template, acetone as solvent and benzoyl peroxide as initiator. MIP had large cavity formed from stigmasterol in its structure. Performance of MIP in sterol adsorption was significantly affected by the amounts of crosslinker and template but slightly affected by the amount of solvent used in the preparation of MIP. MIP synthesized at the optimal condition had adsorption efficiency of 1.13 times of NIP.

Acknowledgement

The authors would like to acknowledge financial support from National Research Council of Thailand.

Nomenclature

%Ads percentage of sterol adsorption
C total concentration of campesterol, stigmasterol and β-sitosterol in the model solution at 6 h
C_0 total concentration of campesterol, stigmasterol and β-sitosterol in the model solution at 0 h
C_i concentration of species "i" in the model solution
q capacity of sterols adsorption
W_i weight of species "i" in the model solution at 6 h
$W_{i,0}$ weight of species "i" in the model solution at 0 h
W_{sol} weight of model solution

Subscript

i campesterol, stigmasterol or β-sitosterol

References

1. E.N. Ali, C.I. Tay, Procedia Engineering, **53**, 7 (2013)
2. V.V. Kuss, A.V. Kuss, R. G. Rosa, D.A.G. Aranda, Y.R.Cruz, Renew. Sust. Energ. Rev., **50**, 1013 (2015)
3. M.H. Mosarof, M.A.Kalam, H.H. Masjuki, A.M. Ashrafui, M.M. Rashed, H.K. Imdadul, I.M. Monirul, Energ. Convers. Manage., **105**, 617 (2015)
4. S. Mekhilef, S. Siga, R. Saidur, Renew. Sust. Energ. Rev., **15**, 1937 (2011)
5. A.G.M. Top, Lipid Technol., **22**, 11 (2010)
6. P. Fernandes, J.M.S. Cabral, Bioresour. Technol., **98**, 2335 (2007)
7. M.-J. Dumont, S. S. Narine, Food Res. Int., **40**, 957 (2007)
8. H. Yang, F. Yan, D. Wu, M. Huo, J. Li, Y. Cao, Y. Jiang, Bioresour. Technol., **101**, 1471 (2010)
9. C. J. Chang, Y.-F. Chang, H. Lee, J. Lin , P.-W. Yang, Ind. Eng. Chem. Res., **39**, 4521 (2000)
10. W. J. Cheong,, S. H. Yang, F. Ali, J. Sep. Sci., **36**, 609 (2013)
11. N. Verma, A. Bhardwaj, Appl. Biochem. Biotechol., **175**, 3093 (2015)
12. W. Li, S. Li, Adv. Polymer Sci., **206**, 191 (2007)

Solid-liquid separation of ultrafine Calcium carbonate ($CaCO_3$) suspensions by using therm-compression filtration technology

Xiuli Li[a], Li Fan and Mingyuan Zhou

School of Environmental and Materials Engineering, Shanghai Polytechnic University, Shanghai 201209, China

Abstract. The ultrafine $CaCO_3$ was usually produced by wet chemical synthesis technique currently. The solid-liquid separation techniques of mechanical filtration and two-stage thermal drying were needed to separate the ultrafine powder from the $CaCO_3$ slurry. A new technology of thermo-compression filtration was taken to dispose the ultrafine $CaCO_3$ suspensions for purpose of simplifying the traditional producing technique in this paper. The thermo-compression filtration not only combined the process of mechanical filtration and thermal drying on one single filter press, but also saved energy because of its non-phase-change thermo-compression drying characteristic. The external heat energy was used to offer assistant thermo-motive, but not to vaporize all of the moisture in the saturated cake. The experimental results indicated that the thermo-compression filtration was much more effective than mechanical filtration or thermal drying respectively. The moisture content of filter cake after thermo-compression drying was about 18.75%~21.89%, much lower than that after mechanical filtration, about 25.5%~28.5%. Furthermore, to keep a relative higher temperature of drying plates when mechanical filtration process was an available method to obtain the product with lower moisture content, but use a minimum working time.

1 Introduction

Ultrafine calcium carbonate ($CaCO_3$) is an important chemical product which is widely used in the fields of plastics, rubber, paper, paints, adhesive, sealant, oil ink, pharmaceutical, food, and so on[1-4]. Commercial production of ultrafine $CaCO_3$ powder uses a wet chemical precipitation technique. However, the technique causes considerable agglomeration of particles during synthesis, and requires the precise control of a number of operation parameters to obtain a better distribution of nanoparticles[5]. Another $CaCO_3$ nanoparticle synthesis method is through carbonation of lime solution in microemulsion systems. This method again requires the precise control of various operation parameters [6]. Both methods synthesized the $CaCO_3$ as slurry. The Solid-liquid separation process was needed to obtain the $CaCO_3$ powder.

Centrifugal or mechanical filtration dehydration methods were used currently to dehydrate the water of $CaCO_3$ slurry and the moisture content of the product which seen like paste was 55% ~ 65%. Then, two stages of thermal drying operations were taken separately. The first stage of thermal drying operation usually used the drying machineries such as drying chamber, drying conveyor, disc dryer and vacuum rake dryer, et al. The moisture content of the product after this stage was about 20%. The second stage usually used the methods of rotating flash drying or air drying process to drying the product to the moisture content of 0.5%. Those two drying operations would take about 70% of energy consumption of the whole production process, especially the first stage.

A new thermo-compression filtration technology was used in this paper to instead the mechanical filtration and the first stage of drying. The filter cake would send to the second stage of thermal drying directly. This will simplify the manufacturing technique and economize the energy consumption.

2 Thermo-compression filtration technology

The thermo-compression filtration technology which had been used in some areas of coal preparation, ore concentrate, Ti-gypsum and so on[7-9] was also suitable for realizing the solid-liquid separation of ultrafine $CaCO_3$ suspensions. The thermo-compression filtration process

[a] Corresponding author: xlli@sspu.edu.cn

© The Authors, published by EDP Sciences. This is an open access article distributed under the terms of the Creative Commons Attribution License 4.0 (http://creativecommons.org/licenses/by/4.0/).

consists of two stages. The first stage is mechanical filtration producing the saturated cake that is similar to the conventional mechanical pressing. The second stage is thermo-compression drying which starts after the formation of saturated cake in the filter chamber (Fig.1). Therefore, the entire process of mechanical filtration and thermo-compression drying can be achieved continuously on a single filter press [10,11]. In the second stage the heating medium was pumped into the drying plate (as shown in Fig.1). The saturated cake close to the drying plate was heated, and the moisture in those parts of saturated cake was vaporized and expanded. Thus the moisture in another part of saturated cake which far away from the drying plate could be pushed out as liquid phase by the vapor pressure. That meant the heat energy was used to offer assistant thermo-motive but not to vaporize all of the moisture in the saturated cake. Therefore the thermo-compression filtration technology was much more energy conservation than traditional mechanical filtration and thermal drying.

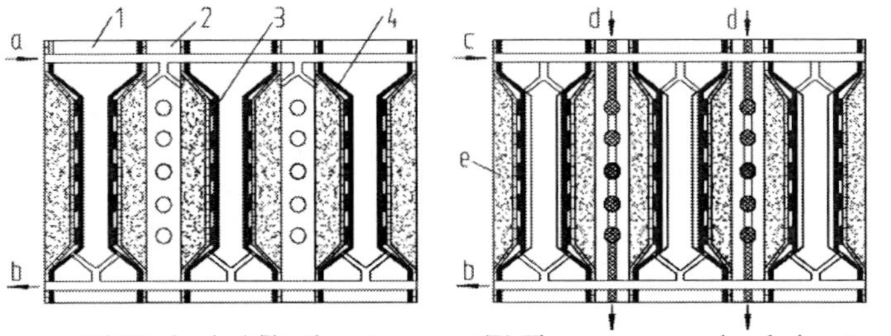

(A) Mechanical filtration stage (B) Thermo-compression drying stage

a – Suspension b – Filtrate c – Compressed air d – Heating medium
e – Filter cake 1 – Recessed filter plate 2 – Drying plate
3 – Squeezing membrane 4 – Filter medium

Figure 1. Schematic Diagram Showing the Operating Principle of the Thermo-compression Filtration System[7]

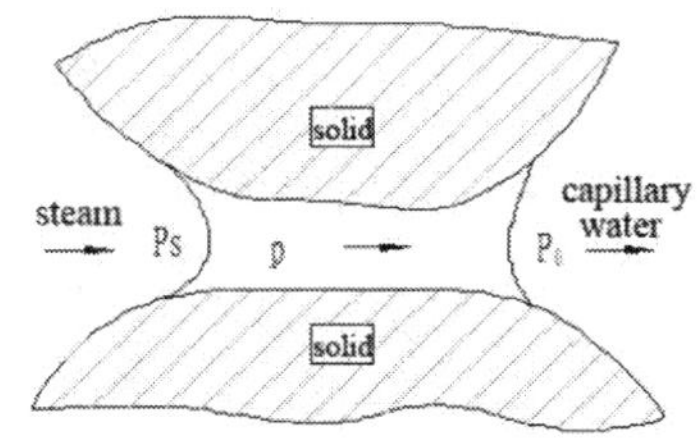

Figure 2. Capillary Water in the Saturated Cakes

Fig. 2 shows the state of the capillary water during the thermo-compression drying process. When the thermo-compression drying process started, the capillary water close to the drying plates was heated and changed from liquid to steam, and then the volume of the steam expanded. The capillary water far away from the drying plates permeated the filter medium and was removed when the saturated vapor pressure (P_S) was higher than the sum of capillary pressure (p) and ambient pressure (P_0) shown by Eq.1 [12]:

$$P_S \geq P_0 + p \geq P_0 + 2\sigma\cos\theta / r \quad (1)$$

where, P_S, P_0, p, and r are the saturated vapor pressure (MPa), the ambient pressure (MPa), the capillary pressure (MPa) and the capillary radius (m) respectively. The capillary pressure (p) and the contact angle (θ) of water depended on the characteristics of the material and the temperature. The filter chamber will be evacuated to reduce P_0 then P_S will decrease as well.

Actually the capillary is not straight but curving and irregularity which will influence its characteristic of heat and mass transfer such as the capillary pressure (p) and the contact angle (θ). Therefore, the tortuosity conception of porous medium was introduced and the thermo-compression filtration kinetics equation was established by combining the Clapeyron-Clausius thermodynamic equation and the material characteristics, as shown by Eq.2.

$$V_m = \frac{d_p^2 \varepsilon^3}{43(1-\varepsilon)^2 \mu L} \cdot \frac{R}{V'' - V'} \ln T + C \quad (2)$$

where, V_m, ε, d_p, L, V'', V', R and T are the average flow velocity in the bent capillary(m/s), porosity of porous medium, granular average effective diameter (m), straight line of both ends of the capillary (thickness of filter cake, m), specific volume of steam (m^3/kg), specific volume of liquid (m^3/kg), latent heat of vaporization (kJ/kg) and the absolute temperature (K). Thus, the thermo-compression solid-liquid separation efficiency has direct relationship with the porosity (ε), granular average effective diameter (d_p), thickness of filter cake (L) and thermo-compression drying temperature (T). When the thickness of filter cake (L) is certainly, the finer the material granularity, the smaller the porosity (ε) is, and the average flow velocity in the bent capillary (V_m) is lower then. The thermo-compression solid-liquid separation efficiency only can be increased by enhancing the thermo-compression drying temperature (T) to enlarge the saturated vapor pressure (Ps).

3 Experimental

3.1 Sample treatment and analysis

The sample ultrafine $CaCO_3$ was produced by Shanghai HuiJing sub-nanometer new material Co., LTD. The ultrafine Calcium carbonate ($CaCO_3$) suspensions with the concentration of 200g/L were prepared. The dispersing agent of sodium dodecyl benzene sulfonate（SDBS）with the content of 0.5~0.8% of $CaCO_3$'s quality was added in the $CaCO_3$ suspensions to increase its stability.

3.2 Experimental method and instrument

Fig. 3 shows the entity laboratory equipment of this system. There were two recessed filter plates with membrane on the surface and one drying plate with snakelike heating pipe that composed to a filter unit. Only one unit was used in this paper to take the experiment. The $CaCO_3$ suspensions were prepared first and then were pumped into the filter chamber by reciprocating diaphragm pump and the filter cake formed gradually. This process was called mechanical filtration. The reciprocating diaphragm pump was not only to transport the material but also proffered the filtration pressure on 0.8MPa. Then the saturated steam of 150℃ which produced by the steam generator was pumped into the drying plate to heating the filter cake. This process was called thermo-compression drying. Those two processes together were called thermo-compression filtration.

There were three experimental methods were taken in this paper as following:

Masked as TC-a: The $CaCO_3$ suspensions were pumped into the filter chamber by reciprocating diaphragm pump. After 80 minutes later the pump stopped, and the filter cake was sampled to test the moisture content, then the thermo-compression dying process started and lasted 30 minutes.

Masked as TC-b: The experimental processes were the same as TC-a, but the mechanical filtration process lasted 25 minutes and the thermo-compression dying process lasted 30 minutes.

Masked as TC-c: The drying plate was kept the temperature of 70~80 ℃ in the mechanical filtration process and the thermo-compression dying process started after 15 minutes when the mechanical filtration process was started. The mechanical filtration process lasted 30 minutes and the thermo-compression dying process lasted 25 minutes. That meant there were 15 minutes working together time of mechanical filtration and thermo-compression and the whole process lasted 40 minutes.

Figure 3. Laboratory equipment of the thermo-compression filtration system

3.3 Analysis method

The moisture content of the filter cake was measured by Rapid Moisture Analyzer (Mettler Toledo HR83). The particle size distribution of the powders of raw material and filter cake were analyzed using particle size distribution analyzer (PSDA, Malvern Zetasizer Nano). 0.5% SDBS was added to improve the dispersion of CaCO3 in distilled water and a total of 90 seconds of ultrasounds were applied to each sample before being analyzed by the PSDA. The scanning electron microscope (SEM) micrographs of raw material and filter cake were obtained using Hitachi S-4800 microscope.

4 Results and discussion

The experimental results of those three different experimental methods were shown in Table 1. It was found that the moisture content of filter cake after thermo-compression drying, about 18.75%~21.89%, was much lower than that after mechanical filtration, about 25.54%~28.49%. TC-a took more long time than TC-b, but the solid content of filter cakes that they obtained were basically the same, and the moisture content of filter cake (TC-a) was just a little lower than TC-b. So the TC-a experimental method could receive a relatively better thermo-compression drying effectiveness, but the economical efficiency was lower than TC-b according to contrast the working time. The dehydrating amounts of those two thermo-compression drying processes were 0.154kg and 0.156kg, separately. That water was removed as liquid phase but not gaseous phase. That meant a large number of energy which was used to evaporate the water was saved and just a fraction of the energy was worked to heat the water close to the drying plates. If we take no account of that fraction energy, the thermo-compression drying processes had the level of energy-efficient as 252.8kJ/kg(dry basis) of TC-a and 266.1kJ/kg(dry basis) of TC-b . The level of energy-efficient was calculated on the basis of the heat of evaporation of 100°C water as 2258.77kJ/kg.

The TC-c method was much similar to the industrial processes as a semi-batch processing. The moisture content of filter cake (TC-c) was lower than TC-b, and the solid content of filter cake that TC-c obtained was much more than TC-a and TC-b, although TC-c's working time was the least. That meant to keep a relatively higher temperature of drying plates when mechanical filtration process was beneficial to feed because high feeding temperature could reduce the viscosity of suspensions. Thus the TC-c method was a most efficient method. The moisture content of the filter cake was lower than 20%, and the product could send to the second stage of thermal drying directly to remove the remainder moisture.

Fig. 4 shows the SEM images of raw material and filter cake. Fig.5 shows the particles size distribution of raw material and filter cake. It was found that the $CaCO_3$ powder of raw material and filter cake had the similar morphology by SEM images, but their particle size had some difference. The particle size distribution of raw material was wide than filter cake, and there was a little more fine particles of raw material than filter cake. That meant the process of thermo-compression filtration would

 DOI: 10.1051/matecconf/20166204007

make the $CaCO_3$ powder had a bit of aggregation but it will not make great effect.

Table 1. experimental results of different experimental method

Experimental method	TC-a	TC-b	TC-c
Mechanical filtration time(min)	80	25	25
Thermal-compression drying time(min)	30	30	25
Total time (min)	110	55	40
Moisture content of filter cake (A)(%)	25.54	28.49	—
Moisture content of filtercake (B)(%)	18.75	21.89	19.87
Quality of filter cake(kg)	1.694	1.695	1.970
Solid content of filter cake(kg)	1.376	1.324	1.579
Dehydrating amount of thermo-compression drying（kg）	0.154	0.156	—
Level of energy-efficient (kJ/kg(dry basis))	252.8	266.1	—

Notice: Moisture content of filter cake (A) indicated the filter cake after mechanical filtration, and (B) indicated the filter cake after thermo-compression drying.

Raw material

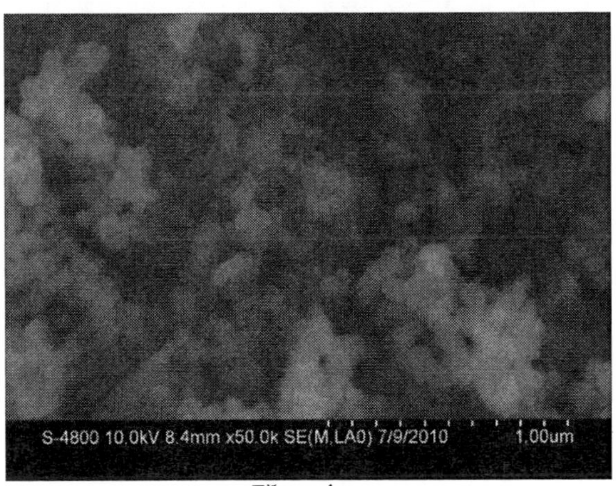

Filter cake

Figure 4. SEM images of raw material and filter cake

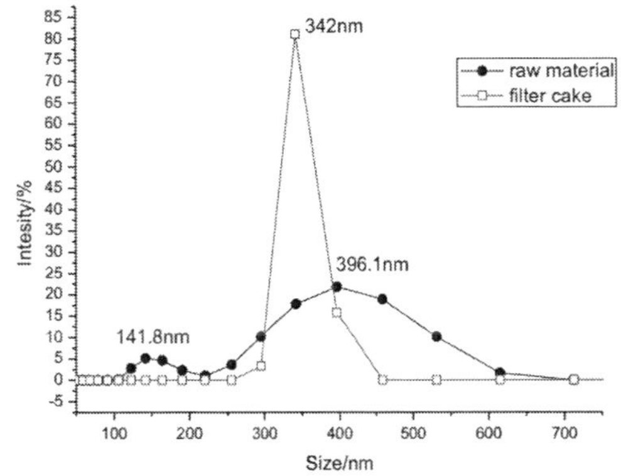

Figure 5. Particle size distribution of raw material and filter cake

5 Conclusion

The thermo-compression filtration technology was proved to be highly energy efficient and high in performance because of its non-phase- change thermo-compression drying process. Under the effect of external heat energy, the remaining moisture of the saturated cakes after mechanical filtration was removed by means of the thrust force produced by the vaporization of capillary water inside the cakes. The ultrafine $CaCO_3$ suspensions could be separated into clear water and dry $CaCO_3$ powder with moisture content lower than 20% using this technology. The products then would send to the second stage of thermal drying directlyn which will simplify the manufacturing technique and economize the energy consumption. The technology was also sutiable to other fine powder slurry such as graphene primary product which was under experiment at present.

Acknowledgment

This work was supported by Leading Academic Discipline Project of Shanghai Municipal Education Commission (No.J51803) and Cultivation subject construction of SSPU(environmental engineering) (XXKPY1303) .

References

1. S.C. Tjong: Materials Science and Engineering Vol.**53** (2006) **,p.**73
2. Y.S. Thio, A.S. Argon, R.E. Cohen and M.Weinberg: Polymer Vol.**43** (2002) , p.3661-3674
3. Qing-Xin Zhang, Zhong-Zhen Yu and Xiao-Lin Xie, et al: Polymer Vol.**45** (2004) , p.5985-5994
4. Chi-Ming Chan, Jing-shenWu and Jian-Xiong Li, et al: Polymer Vol.**43** (2002) , p.2981-2992
5. T. Tsuzuki, K. Pethick and P.G. McCormick: Journal of Nanoparticle Research Vol.**2** (2000) , p.375-380
6. A. Sugih, D. Shukla, H. Heeres and A. Mehra: Nanotechnology Vol.**18** (2007) , p.35

7. Mingyuan Zhou, Li Fan, Xiuli Li and Changwen Ma: Advanced Materials Research Vol.**361** (2012) , p.337-342
8. Mingyuan Zhou, Weigen Zhao, Jie Guan and Jianfeng Bai, in: *The XXIV International Mineral Processing Congress*, Beijing China (2008)
9. Hao Yuan, Mingyuan Zhou, Jie Guan and Li Fan: Applied Mechanics and Materials Vol.**448** (2014) , p.746-751
10. Mingyuan Zhou: Journal of China Coal Society Vol.**24** (1999) , p.108-112
11. Peuker U A, Stahl W: Drying Technology Vol.**19** (2001) , p.807-848
12. Mingyuan Zhou, Jie Guan: Journal of China Coal Society Vol. 35 (2010) , p.472-475

DOI: 10.1051/matecconf/20166205001

Design of an Air-Sparged Tubular Photocatalytic Reactor for the Degradation of Methylene Blue: Mass-Transfer Limitation Studies

Patrick Ramoso[1,a] and Maria Lourdes Dalida [1]

[1]*Department of Chemical Engineering, University of the Philippines Diliman, Quezon City, Philippines*

Abstract. An alternative process for the removal of organic pollutants in aqueous systems is photocatalysis. The challenges hindering its industrial use are electron-hole recombination and mass-transfer limitations. In order to address these problems, the objective of this study is to introduce air by sparging, and design an air-sparged photocatalytic reactor using titanium dioxide immobilized on borosilicate glass. The performance of the reactor on the removal of the model pollutant, methylene blue (MB), was evaluated and compared against the reactor operated without sparging. The effect of mass-transfer limitations on reactor performance was also investigated by regression using a Langmuir-type model equation. The sparged photocatalytic reactor was able to degrade 57% MB in 2 hours, an improvement of 40% compared to no sparging, and is comparable to similar reactors in literature, but with the advantage of using less expensive materials of construction and simpler immobilization technique. Mass-transfer limitation studies showed a good fitting of the initial reaction rate r, with $r = 0.1399Q / (0.6120 + Q)$ for the sparged operation, and Q is the volumetric flowrate of water (L/min). The model also shows that the reactor operates near the reaction-limited regime, and that the extent of mass-transfer limitation effects was reduced by the present reactor.

1 Introduction

The development of advanced wastewater treatment processes has been well studied over the years in light of the detrimental effects of water pollution, particularly organic compounds coming from industrial effluents. These treatment processes are classified under physical, chemical or biological [1, 2]. These treatment processes, however, have their own limitations. For example, adsorption processes have a problem with the disposal of spent adsorbent; chemical treatment processes may cause the formation of harmful byproducts; and biological processes are feasible only for contaminants that are not toxic to the microbes present in the system.

An alternative to these processes is photocatalysis, which uses ultraviolet or visible light to initiate the degradation of organic pollutants using a photocatalyst, resulting to complete mineralization [2-5]. Photocatalysis offers an alternative because the degradation of pollutants is complete – there are no unwanted byproducts; there is no transfer of pollutants from one phase to another; and its application is a wide range of pollutants.

Photocatalytic reactors may be classified either as suspended or immobilized systems. Suspended systems, although they have higher activity, have a problem of requiring post-process separation since the photocatalyst is finely dispersed within the water being treated. This lowers the economic viability of these systems [3, 5]. Post-process separation is not required in immobilized systems, as the photocatalyst is anchored on a support. Immobilized systems, however, also have problems of their own, such as mass-transfer limitations and electron-hole recombination, with the latter covering both immobilized and suspended systems. High mass transfer resistances result to a reactant concentration gradient between the bulk fluid and photocatalyst surface, resulting to lower rates of reaction [6]. Electron-hole recombination occurs as the excited electron in the conduction band returns and fills up the positive hole in the valence band. The series of reactions ceases, and photocatalytic activity decreases [3].

Mass-transfer limitations are manifested by the dependence of the reaction rate r, or the apparent rate constant k_{app}, on the liquid velocity. Various studies have identified the existence of mass-transfer limitations in many immobilized systems, such as tubular reactors [7-8], annular reactors [9-11], rotating disc reactors [9, 12], and packed-bed reactors [7, 13], among others [14]. A commonly used model equation derived from first principles is a Langmuir-type equation based on volumetric flowrate Q [15-16]:

$$r \text{ (or } k_{app}) = abQ/(1+aQ) \quad (1)$$

These two challenges are to be addressed by introducing air into the system by sparging. Adding air bubbles lessens the effect of mass-transfer limitations by promoting mixing of the liquid phase. Introducing air into the system adds an electron-scavenger into the system, with oxygen accepting excited electrons to form the superoxide ion. This leaves the holes available for

[a] Corresponding author: pdramoso@up.edu.ph

© The Authors, published by EDP Sciences. This is an open access article distributed under the terms of the Creative Commons Attribution License 4.0 (http://creativecommons.org/licenses/by/4.0/).

 DOI: 10.1051/matecconf/20166205001

mineralization to take place. The objective of this study is to design an air-sparged, tubular photocatalytic reactor using TiO_2 and study the effect of mass-transfer limitations on its performance.

2 Materials and Methods

2.1. Reactor design

The reactor consists of five tubes radially arranged with a total reactor volume of 600 mL, and a 10W UVA lamp placed at the center of the reactor, as shown in Figures 1-2. The tubes are made of borosilicate glass, and are fabricated with internal protrusions to promote turbulence and provide additional surface area for the catalyst [17]. The catalyst, Aeroxide® P25 TiO_2, was immobilized onto the tubes by spray coating with ethanol as carrier, and the tubes were dried and calcined in a muffle furnace. Air is sparged through nozzles directed at each tube such that the bubbles rise along the tubes as the water descends.

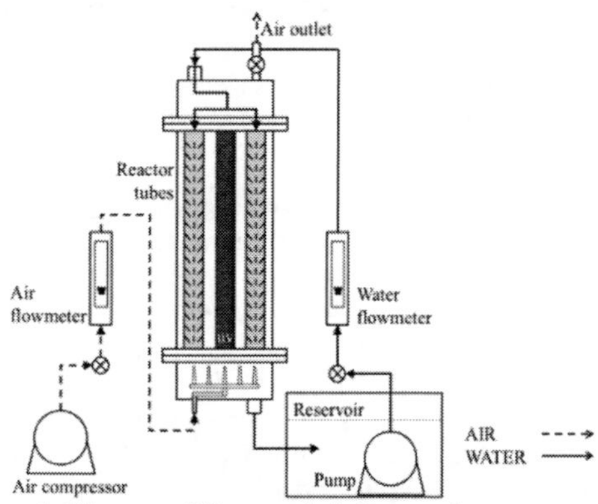

Figure 1. Diagram of Photocatalytic Reactor System

Figure 2. Actual Photocatalytic Reactor Set-up

2.2 Catalyst characterization

Samples of the immobilized photocatalyst were obtained by the destruction of one reactor tube after its calcination in order to characterize its surface. This was done by scanning electron microscopy (SEM) using a Hitachi S3400-N Scanning Electron Microscope. Images of surface were obtained at 300x and 1000x magnification [18].

2.3 Reactor performance evaluation

The performance of the reactor, measured as percent removal of methylene blue (MB), under sparged and non-sparged operation was compared at the following process parameters: 2.69 ppm initial MB concentration, pH 7, 2.8 liters per minute (L/min) water flowrate [17], and 2.0 L/min air sparging rate, at 30, 60, 90 and 120 minutes. Adsorption (dark) and blank (no photocatalyst) runs were done as control [18]. Methylene blue concentration was analyzed using a PerkinElmer Lambda 850 UV/VIS Spectrometer at a peak wavelength of 665 nm.

2.4 Mass-transfer limitation studies

The effect of mass-transfer limitations on the reaction rate was modeled for the sparged and non-sparged operation of the photocatalytic reactor using 2.69 ppm initial MB concentration and pH 7, with a sparging rate of 2.0 LPM. To determine the effect of mass-transfer limitations, the water flowrate was varied to Q = 1.2, 1.6, 2.0, 2.4 and 2.8 LPM. The initial rate of reaction, r, was determined based on the difference between initial and final MB concentrations for a period of 20 minutes [18]. A Langmuir-type model equation was used, particularly was used to relate the initial reaction rate to the volumetric flowrate. The model constants a and b were determined by regression analysis.

$$r = aQ / (b + Q) \quad (2)$$

3 Results and Discussion

3.1 Catalyst characterization

SEM images of the reactor tube surface are presented in Figures 3-4. It could be seen that the catalyst was immobilized throughout the surface of the glass; however, the coating is not uniform. The presence of wavelike patterns indicates a high surface area provided by the roughness of the immobilized surface; this morphology has been reported to increase photocatalytic activity [19]. In addition, the high magnification image in Figure 4 illustrates the roughness of the surface, which may have resulted from the agglomeration of P25 TiO_2 particles upon spray coating.

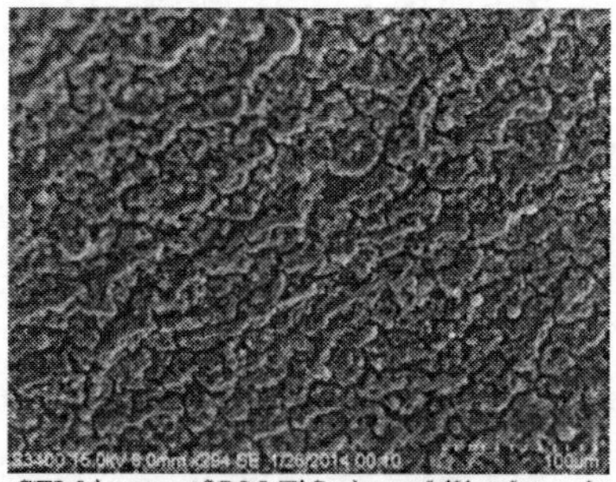

Figure 3. SEM image of P25 TiO_2 immobilized on glass at 300x magnification

DOI: 10.1051/matecconf/20166205001

Figure 4. SEM image of P25 TiO_2 immobilized on glass at 1000x magnification

It is also noted that catalyst entrainment cannot be assessed by surface characterization, as this would require the destruction of another reactor tube. Visual inspection of effluent samples, as well as reactor wash water, did not indicate any presence of entrained TiO_2 particles. The small, if not negligible, extent of catalyst entrainment may be explained by the strong adhesion of the P25 TiO_2 with the borosilicate glass, brought about by the use of ethanol as a carrier of the TiO_2 suspension [18]. Due to the hydrophilic property of P25 TiO_2 [20], the photocatalyst strongly adhered to the surface of the borosilicate glass, aided by the ethanol carrier.

3.2 Reactor performance evaluation

The performance of the reactor in the removal of methylene blue is shown in Figure 5. It confirms that the dominant mechanism for the removal of methylene blue is photocatalysis, with 57% MB removal after 120 minutes. The contribution of adsorption is less than 20%. The negligible extent of MB adsorption on TiO_2 is consistent with other studies [9, 21].

The control runs, which include adsorption on the reactor walls, tubings, reservoir walls, and pump internals, account for an average of 12% MB removal. It is also noted that the control runs are already at steady state as early as 30 minutes, implying that the adsorption of methylene blue on internal surfaces is fast, but contributes to a small extent only.

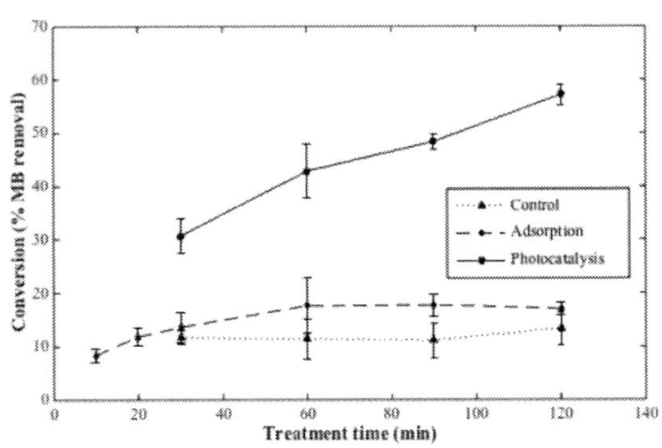

Figure 5. Photocatalytic reactor performance

The effect of sparging on reactor performance is illustrated in Figure 6. The highest conversion of the air-sparged reactor is 57% at 120 minutes, higher compared to the 40% of the non-sparged reactor, or greater by a factor of 1.4 (40%). This validates the use of air sparging to improve reactor performance. The sparging of air into the system resulted to higher dissolved oxygen concentrations in the system. Oxygen serves as an electron-scavenger, taking in the excited electrons, thereby lessening electron-hole recombination. In addition, the rising air bubbles also facilitate better mixing inside the reactor tubes, further increasing the removal of methylene blue.

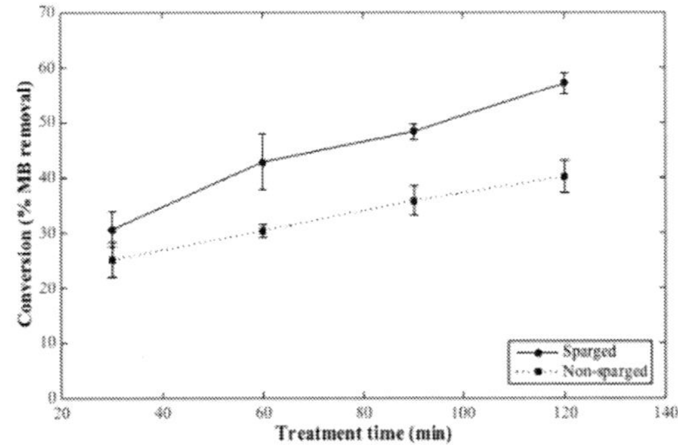

Figure 6. Effect of sparging on reactor performance

The reactor performs better than other reactors in related studies [21-23] and may be considered at par with some reactors [9], but with the advantage of using a simpler immobilization method (spray coating) and the use of less expensive materials of construction (borosilicate glass) [18].

3.3 Mass-transfer limitation studies

The equations for the initial reaction rate, r, in μmol/L-min, in terms of water flowrate, Q, for sparged and non-sparged operations are shown in equations (3) and (4), respectively. The rate is also plotted in Figure 7 below.

$$r = 0.1399Q / (0.6120 + Q) \quad (3)$$

$$r = 0.1024Q / (0.4822 + Q) \quad (4)$$

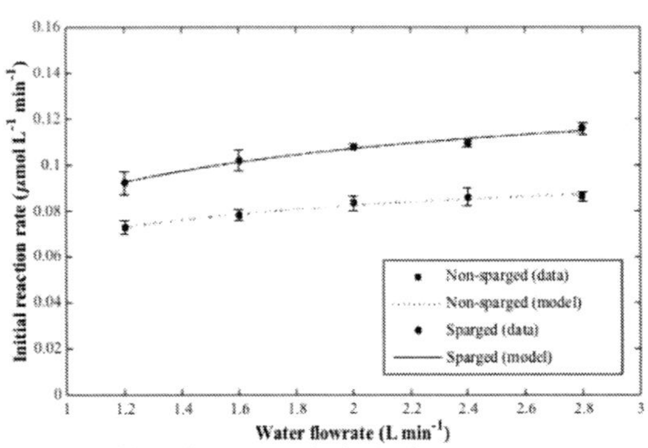

Figure 7. Effect of mass-transfer limitations on reactor performance

The experimental results gave rise to a good fit for both non-sparged and sparged reactor operation models, having R^2 values of 0.9825 and 0.9793, respectively. This confirms a good fit to the Langmuir-type equation relating initial reaction rate with water flowrate. The small slopes seen in both models suggest that the studied range of flowrates is near the reaction-limited regime, and far from the diffusion-limited regime, which is characterized by relatively steep slopes. This observation becomes more obvious when the models are extrapolated to higher and lower flowrates, as seen in Figure 8. With this, it could be concluded that at the range of flowrates studied, particularly at the higher end of the range, the problem of mass-transfer limitations has been addressed by the present reactor [18].

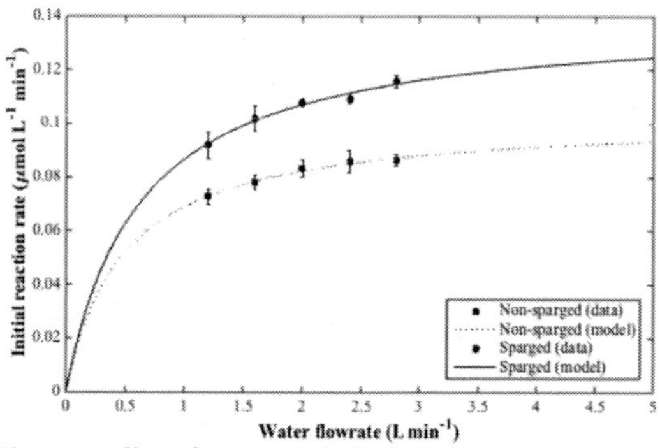

Figure 8. Effect of mass-transfer limitations on reactor performance *(model extrapolated)*

It is also noted from Figures 7-8 that the air-sparged reactor operates at higher rates of reaction than the non-sparged reactor, with maximum initial rates of 0.1157 and 0.0864 μmol L^{-1} min^{-1}, respectively. This further supports the result that air sparging improved the performance of the reactor, as explained in the previous subsection.

4 Conclusions

An air-sparged tubular photocatalytic reactor with immobilized P25 TiO_2 photocatalyst was designed to lessen electron-hole recombination and promote mixing. Surface characterization showed a good immobilization of TiO_2 by spray coating and calcination; no observable catalyst entrainment was found by visual inspection of reactor effluents.

The dominant mechanism for methylene blue removal is photocatalysis, with negligible contribution of adsorption on TiO_2 and other internal surfaces of the reactor. Air sparging improved the performance of the tubular photocatalytic reactor by a factor of 40%. The reactor is comparable to the performance of other similar reactors in literature. However, the reactor of the present study offers an advantage of using a simpler immobilization technique and a less expensive material of construction.

Mass-transfer limitation studies indicate a good fit to a Langmuir-type equation in relating the initial reaction rate with the liquid volumetric flowrate. It also confirms that the present reactor operates closer to the reaction-limited regime, and that the extent of mass-transfer limitation effects has been lessened in this reactor.

Acknowledgments

The authors would like to acknowledge the Department of Science and Technology (Philippines) Engineering Research and Development for Technology (ERDT) Program for the financial assistance, and the University of the Philippines Department of Chemical Engineering for the use of laboratory and analytical equipment.

References

1. M.R. Hoffmann, S.T. Martin, W. Choi, D.W. Bahnemann, Chem. Rev., **95,** 69–96 (1995)
2. R. Vinu, G. Madras, *"Photocatalytic degradation of water pollutants using Nano-TiO₂,"* in *Energy Efficiency and Renewable Energy through Nanotechnology*, L. Zang, Ed. London: Springer London, 625–677 (2011)
3. H. de Lasa, B. Serrano, M. Salaices, *Photocatalytic Reaction Engineering.* New York: Springer Science+Business Media, LLC (2005)
4. A. Mills, S. Le Hunte, J. Photochem. Photobiol. A Chem., **108,** 1–35 (1997)
5. D. F. Ollis, E. Pelizzetti, N. Serpone, Environ. Sci. Technol., **25**, *9*, 1522–1529 (1991)
6. D. Chen, F. Li, A. K. Ray, Catal. Today, **66**, *2–4*, 475–485 (2001)
7. M. F. J. Dijkstra, E. C. B. Koerts, A. A. C. M. Beenackers, J. A. Wesselingh, AIChE J., **49**, *3*, 734–744 (2003)
8. M. F. J. Dijkstra, H. Buwalda, A. W. F. De Jong, A. Michorius, J. G. M. Winkelman, A. A. C. M. Beenackers, Chem. Eng. Sci., **56**, 547–555 (2001)
9. I. Boiarkina, S. Norris, D. Alec, Chem. Eng. J., **225**, 752–765, (2013)
10. D. Chen, F. Li, A. K. Ray, AIChE J., **46**, *5*, 1034–1045, (2000)
11. M. F. J. Dijkstra, H. J. Panneman, J. G. M. Winkelman, J. J. Kelly, A. A. C. M. Beenackers, Chem. Eng. Sci., **57**, *22–23*, 4895–4907 (2002)
12. D. Dionysiou, M. T. Suidan, I. Baudin, J.-M. Laine, Appl. Catal. B Environ., **38**, 1–16 (2002)
13. Y. Hama, M. Itamochi, T. Horikawa, M. Katoh, T. Tomida, J. Chem. Eng. Japan, **42**, *7*, 502–507 (2009)
14. B. Serrano, H. de Lasa, Ind. Eng. Chem. Res., 36, 4705–4711 (1997)
15. D. Chen, A. K. Ray, Appl. Catal. B Environ., **23**, 143–157 (1999)
16. R. W. Matthews, J. Phys. Chem., **91**, *25*, 3328–3333 (1987)
17. A. H. Lim, K. B. Bautista, R. I. Sato, M. L. P. Dalida, *Design and Performance of an Immobilized Photocatalytic Reactor for Water Treatment*, B.S. thesis. Dept. Chem. Eng., Univ. of the Philippines Diliman, Quezon City, Philippines (2013)
18. P. D. Ramoso, M. L. P. Dalida, *Design and Optimization of an Air-Sparged, TiO₂-Immobilized Tubular Photocatalytic Reactor for the Degradation of Methylene Blue*, M.S. thesis. Dept. Chem. Eng., Univ. of the Philippines Diliman, Quezon City, Philippines (2014)

19. M. Uzunova-Bujnova, R. Todorovska, M. Milanova, R. Kralchevska, D. Todorovsky, Appl. Surf. Sci., **256**, *3*, 830–837 (2009)
20. V. Bolis, C. Busco, M. Ciarletta, C. Distasi, J. Erriquez, I. Fenoglio, S. Livraghi, S. Morel, J. Colloid Interface Sci., **369**, *1*, 28–39 (2012)
21. K. Natarajan, T. S. Natarajan, H. C. Bajaj, R. J. Tayade, Chem. Eng. J., **178**, 40–49 (2011)
22. S. Mozia, M. Toyoda, T. Tsumura, M. Inagaki, A. W. Morawski, Desalination, **212**, *1–3*, 141–151 (2007)
23. B. I. Stefanov, N. V. Kaneva, G. L. Puma, C. D. Dushkin, Colloids Surf. A Physicochem. Eng. Asp., **382**, *1–3*, 219–225 (2011)

MATEC Web of Conferences **62**, 05002 (2016) DOI: 10.1051/matecconf/20166205002
ICCFE 2016

Effect of the Bromine-Based Flame Retardant Plastic Pyrolysis of Hydrotalcite

N. Morita[1], Y. Kawabata[1], T. Wajima[2], A. T. Saito[2], and H. Nakagome[2]

[1]*Department of Urban Environment System at Chiba University, Japan.*
[2]*Chiba University, Japan.*

Abstract. In this study, a method is presented to decrease halogen compounds in the product oil from thermolysis of polystyrene and polypropylene mixed plastic spiked with tetrabromobisphenol A. A mixture of hydrotalcite and plastic was pyrolyzed in a glass reactor at 400 °C under a nitrogen atmosphere. Bromine compounds in the residual substances were measured. The yield of product oil increased using hydrotalcite as an additive. The bromine compounds that were the major ingredients in the oil after thermolysis at 400 °C from the mixed plastic, which also included toluene, ethyl benzene, styrene, and 1-methylethyl benzene, were 2-bromohexane, 3-bromo-1-propenyl benzene, 4,5-dibromodecane, 1-bromomethylbenzene, 3-bromophenol, and 4-bromo-2,6-dimethylbenzaniline. However, bromine compounds were not detected in the product oil, residue, or gas when hydrotalcite was added. After the thermolysis of the plastic, bromine compounds in the product oil may decrease because bromine was captured by the added hydrotalcite.

1 Introduction

Plastics originate from fossil fuels and are becoming a problem because of the large quantity of waste. Methods for recycling plastic include thermal, chemical, and material recycling. Chemical recycling by pyrolysis is capable of returning plastics to their raw materials, and has attracted attention for making plastic derived fuel. There is an ever increasing volume of electronic products and associated waste electrical and electronic equipment (WEEE). Moreover, new models of electronic devices are replacing the old at a higher rate than ever before. WEEE includes computers, cellphones, hair dryers, and refrigerators, as well as the cathode-ray tube television [1]. However, the flame retardant materials added in these plastics form a product oil that includes halogen compounds. There are many recyclable materials, e.g., glass and metal; however, the energy consumption of the recycling process is problematic. Protecting the environment is desirable, but there are now more than 15 types of plastic commonly used in electronics [2]. The quantity of WEEE per person in a member nation of the 27 European Union (EU27) countries was estimated as 17 kg per person in 2005 [3]. The EU has introduced laws for the reuse and recycling of plastics to reduce landfill, and control WEEE destruction by thermal decomposition [4]. Plastics can be converted into gas for use as chemical raw materials, and fuel. Oil and residual carbon produced by pyrolysis can also be useful [5]. There has been considerable research on pyrolysis using catalysts and the chemical recycling of plastic [6-17]. For example, the yield of the product oil can be increased using ZSM-5 zeolite red mud mesoporous molecular sieve in the pyrolysis of plastics [18, 19]. However, the toxic brominated flame retardant is a problem for plastic recycling of household electrical appliances [20-24]. The generation of harmful byproducts has been studied to prevent the release of halogen gas, and additives have been used to promote recycling [22]. For example, polybrominated compounds mixed with polypropylene (PP), polyethylene, polystyrene (PS), and polyvinyl chloride have been decomposed safely with a carbon complex of calcium carbonate [25]. Additionally, the quality of the pyrolysis oil has been improved using the commercial hydrogenation catalyst DHC-8 [26]. The yield of product oil was improved to 96% from 93% when the zeolite catalysts, H-USY or H-Beta, were used in a steam contact process for pyrolysis [27]. The content of the bromine compounds in the product oil was decreased by adding hydrotalcite (HT) in the pyrolysis of brominated plastic [28-30]. HT is composed of layers of hydroxides that can form a metal complex hydroxide, containing anionic hosts and Mg^{2+} and Al^{3+} interlamellar guests [31, 32]. HT is an ion exchanger [31-34] and a catalyst [35] for gas removal [36, 37], and these applications have been receiving increasing attention in recent years [38].

The high anion exchange ability of HT, which is similar to organic ion exchange resins, has attracted attention as an adsorbent for the removal of various pollutants from water. The anionic adsorption properties include substitution reactions with chemicals such as

[a] Corresponding author: Naoyuki_Morita@chiba-u.jp

© The Authors, published by EDP Sciences. This is an open access article distributed under the terms of the Creative Commons Attribution License 4.0 (http://creativecommons.org/licenses/by/4.0/).

borate, nitrate, fluoride, phosphate, sulfate, chromate, arsenic acid, and selenic acid. The anionic adsorption mechanism of HT has been proposed as an interlamellar complex formed of anionic exchange reactions [39]. In addition, HT has attracted attention as a catalyst, and reports have shown a yield increase to 77% from 62% in biodiesel fuel generation [40, 41].

In this study, pyrolysis at 400°C of PS and PP plastic mixtures spiked with tetrabromobisphenol A (TBBA) is examined using HT. The effects of additives on the product oil and the behavior of bromine compounds is evaluated.

2 Materials and Methods

2.1 Materials

Plastic samples containing bromine were prepared by mixing 10 g of PSJ PS) (Asahi Kasei Chemicals Corporation Co., Ltd.) and 10 g of PP (Teijin Co., Ltd.) with 2 g of TBBA (Tokyo Chemical Industry Co., Ltd., Tokyo, Japan). To reduce bromine compounds in the product oil, synthetic HT, KW-1000, (Kyowa Chemical Industry Co., Ltd.), sea sand (Wako Pure Chemical Industries, Ltd.), molecular sieve 4A (Wako Pure Chemical Industries, Ltd.), and molecular sieve 13X (Wako Pure Chemical Industries, Ltd.) were used as additives. The experimental conditions are listed in Table 1.

Table 1: Experimental conditions used in this study

Condition	Contents of the sample
I	Polystyrene (PS) +Polypropylene(PP)+ Tetrabromobisphenol A (TBBA)
II	PS + PP + TBBA + Sea sand
III	PS + PP + TBBA + Molecular sieves 4A
IV	PS + PP + TBBA + Molecular sieves 13X
V	PS + PP + TBBA + Hydrotalcite

2.2 Experimental apparatus and procedures

The experimental apparatus used in this study is shown in Figure 1. A 22-g quantity of PS+PP+TBBA, or mixtures with an additive (20 g) was added to a glass reactor. The reactor was purged with nitrogen gas at a flow rate of 50 mL/min for 60 min to remove oxygen. After oxygen substitution, the flow of nitrogen gas was stopped and the pyrolysis experiment was performed. The temperature of the plastic was measured as the initial temperature, and the reactor temperature was increased to the decomposition temperature (400 °C) at a heating rate of 5 °C/min. The gases produced by pyrolysis were condensed and recovered. Non-condensable gases, which pass through the condenser, were collected in a gas pack via bubbling in an alkaline aqueous solution. Residue remained in the reactor after pyrolysis. The mass balance of the pyrolysis products (oil, gas, and residue) was calculated after the experiment. The pyrolysis products obtained after experiments under each condition were analyzed.

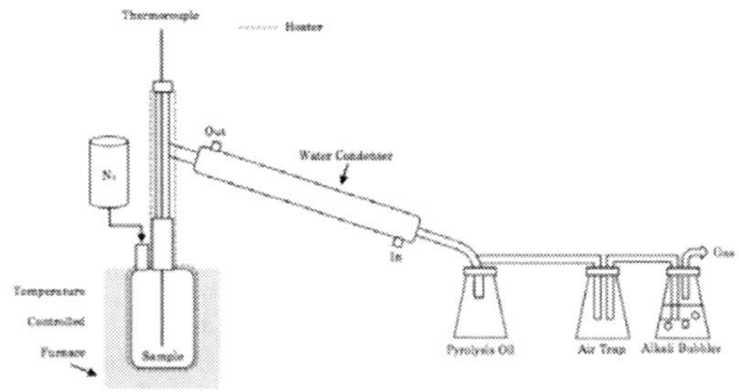

Figure 1. Experimental apparatus

2.3 Analysis

The oils were analyzed by gas chromatography/mass spectrometry (Shimadzu GC-MS-QP2010 Ultra) using a RTX-624 column (thickness: 1.8 μm, inner diameter: 60 mm, length: 0.32 mm) with pure helium as the carrier gas. The main compounds in the oil were identified using a gas chromatography/mass spectrometry spectral library. The analyses were performed at a temperature of 40 °C for 30 min, and then the temperature was increased to 230 °C at a rate of 5 °C/min. The structural change in the sample before and after the experiment was determined by X-ray diffraction analysis (Bruker, D2PHASER), and the morphologies and element distribution maps of the samples before and after the experiment were observed with a scanning electron microscope (Hitachi High-Tech, TM3030) equipped with energy-dispersive spectroscopy. The alkaline aqueous solution of bubble-trapped bromine compounds formed from non-condensable gases was analyzed by ion chromatography (Shimadzu, HIP-SP). The product gas that passed through the alkaline aqueous solution was recovered by a gas pack and analyzed using a Shim-pack column (150-mm long × 4.6-mm inner diameter) and a CDD-10Asp detector as part of a high-performance liquid chromatography system.

3 Results and Discussion

3.1 Product oil yield

The yields of product oil from the pyrolysis experiments are shown in Figure 2, which varied for different additives. The case using HT showed the highest yield, which was approximately 1.5 times that of the additive-free case. This result indicates the catalytic action of HT [40, 41]. In addition, the gases generated by pyrolysis were all less than 1 L.

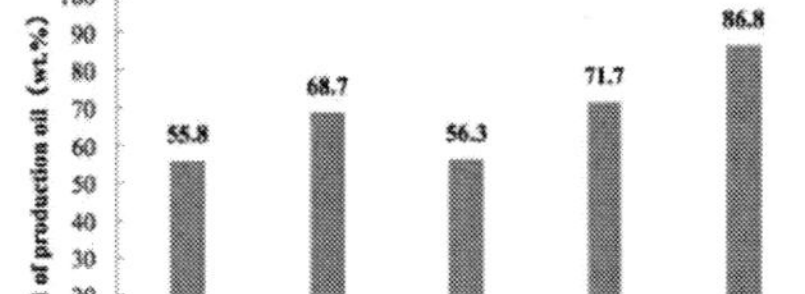

Figure 2. Yield of production oil

3.2 Product oil components

Components of the product oils are given in Table 2. All the oil products contained toluene, ethylbenzene, styrene, and 1-methylethyl benzene. Bromine compounds obtained from the product oil included 2-bromohexane, 3-bromo-1-propenyl benzene, 4,5-dibromodecane, 1-bromomethylbenzene, 3-bromophenol, and 4-bromo-2,6-dimethylbenzaniline.

The results in Table 2 show that bromine compounds were not detected in the oil produced under condition V, which is the plastic containing HT.

Table 2. Components of product oils obtained under each experimental condition.

		Condition				
Compound	Formula	I	II	III	IV	V
Toluene	$C_6H_5CH_3$	o	o	o	o	o
4-Methylheptane	C_6H_{14}	o	o	o	nd	o
2-Bromohexane	$C_6H_{13}Br$	o	o	nd	nd	nd
2,3-Dimethyl-2-heptane	C_9H_{18}	o	nd	nd	o	nd
2,4-Dimethyl-1-heptene	C_9H_{18}	nd	nd	o	nd	nd
2,6-Dimethyl-3-heptene	C_9H_{18}	nd	nd	nd	o	nd
Ethyl benzene	C_8H_{10}	o	o	o	o	o
Styrene	C_8H_8	o	o	o	o	o
2,4,6-Trimethyl-3-heptene	$C_{10}H_{20}$	o	o	nd	nd	o
4,4,5- Trimethyl-2-hexene	C_9H_{18}	nd	nd	nd	o	nd
Cumene	C_9H_{12}	o	nd	nd	nd	nd
1-Methylethylbenzene	C_9H_{12}	o	o	o	o	o
Phenol	C_6H_6	o	nd	nd	o	nd
3-Bromo-1-propenyl benzene	C_9H_9Br	nd	nd	nd	o	nd
4,5-Dibromodecane	$C_{10}H_{20}Br_2$	nd	nd	o	o	nd
1-Bromomethylbenzene	C_8H_9Br	o	nd	nd	nd	nd
3-Bromophenol	C_6H_5BrO	o	nd	o	nd	nd
4-Bromo-2,6-dimethylbenzaniline	$C_8H_{10}BrN$	o	o	nd	nd	nd
1,3-Diphenylpropane	$C_{15}H_{16}$	o	o	o	o	o

o : detected, nd : not detected

3.3 Residue analysis

The surface of each sample was observed by EDS after pyrolysis to analyze any bromine remaining in the residue and the element distribution maps are shown in Figure 3.

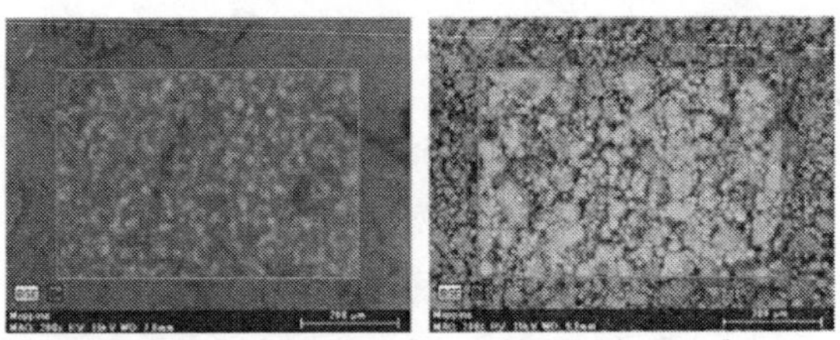

Figure 3. (a). EDS map of condition (b). EDS image of condition V

3.4 Bromine content in the gas

The analysis results of the gas generated by pyrolysis are shown in Figure 4 and Table 3. The gas obtained from thermal decomposition was analyzed for bromine-containing compounds. The bromine included in the plastic from the added TBBA was 2 g, and the bromine in the generated gas was 1843 mg for all the conditions. Most bromine was distributed in the product oil and the residue. The decomposition of the plastic causes gases, such as H_2 and CH_4, to be expelled, but these were low after the addition of HT. This thermal decomposition reaction does not distribute bromine to the gas because of HT, and they are degraded in the product oil and residue.

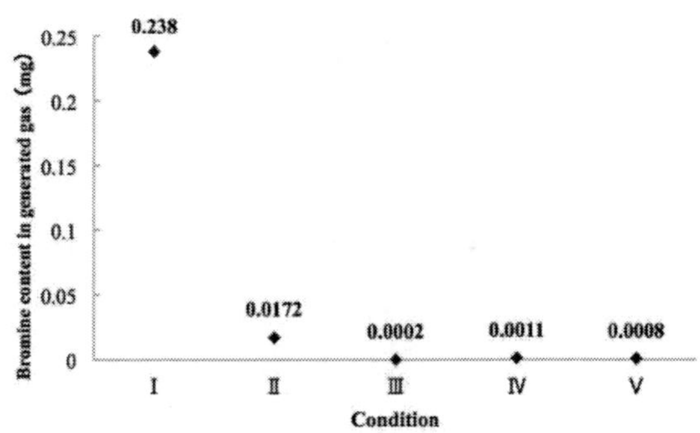

Figure 4. Bromine content in the generated gas

Table 3. Components of the product gas

Condition	H_2	CH_4	C_2H_4	C_2H_6	C_3H_8	C_3H_6
I	○	nd	nd	nd	○	○
II	○	nd	nd	nd	○	○
III	○	nd	nd	nd	nd	nd
IV	○	○	○	○	○	○
V	○	nd	nd	nd	○	○

○ : detected, nd : not detected

4 Conclusion

In this study, we investigated a method of reducing bromine compounds in the product oil obtained by the thermal decomposition of a TBBA-spiked PS and PP mixture using HT. Plastics containing brominated flame retardants can be thermally decomposed to product oils containing no bromine compounds by adding HT. The yield of product oil after adding HT improved. This indicates that the utility of the HT is high, which may be because of a catalytic effect. The residue after adding HT showed more bromine than without additive, suggesting the HT adsorbed the bromine compounds. Thus, the product oil does not contain bromine. The efficacy of plastic thermal decomposition is increased by adding HT, because the bromine compounds are contained in the residue and reduced in product oil.

Acknowlegment

This work was supported by Mr. Kathunori Suzuki of Tokyo Metropolitan Tama High School of Science and Technology, Japan. N. Morita thanks Mr. Masato Ota and Mr. Yuuta Shimohonji for their cooperation in this study.

References

1. Jonathan C. Acomb, Mohamad Anas Nahil, Paul T. Williams, "Thermal processing of plastics from waste electrical and electronic equipment for hydrogen production", *Journal of Analytical and Applied Pyrolysis*, **103**, 320-327, (2013)
2. G. Martinho, A. Pires, L. Saraiva, R. Ribeiro, "Composition of plastics from waste electrical and electronic equipment (WEEE) by direct sampling", *Waste Management*, **32 (6)**, 1213, (2012)
3. United Nations University, "Review of Directive 2002/96 on Waste Electrical and Electronic Equipment (WEEE); final report", (2008)
4. European Union, "EC, Directive 2002/96/EC Waste Electrical and Electronic Equipment", L27/34, (2003)
5. Bhaskar T, Matsui T, Kaneko J, Uddin MA, Muto A, Sakata Y Novel calcium based sorbent (Ca–C) for the dehalogenation (Br, Cl) process during halogenated mixed plastic (PP/PE/PS/PVC and HIPS-Br) pyrolysis. Green Chem, **4**, 372-375, (2002)
6. Grause G, Ishibashi J, Kameda T, Bhaskar T, Yoshioka TKinetic studies of the decomposition of flame retardant containing high-impact polystyrene. Polym Degrad Stab, **95**, 1129-1137, (2010)
7. Onwudili JA, Insura N, Williams PTComposition of products from the pyrolysis of polyethylene and polystyrene in a closed batch reactor: Effects of temperature and residence time. J Anal Appl Pyrolysis, **86**, 293-303, (2009)
8. Hall WJ, Williams P T Fast pyrolysis of halogenated plastics recovered from waste computers. Energ Fuel, **20**, 1536-1549, (2006)
9. Bhaskar T, Uddin MA, Murai K, Kaneko J, Hamano K, Kusaba T, Muto A, Sakata YComparison of thermal degradation products from real municipal waste plastic and model mixed plastics. J Anal Appl Pyrolysis, **70**, 579-587, (2003)
10. Serrano DP, Aguado J, Escola JM, Garagorri E Performance of a continuous screw kiln reactor for the thermal and catalytic conversion of polyethylene-lubricating oil base mixtures. Appl Catal B, **44**, 95-105, (2003)
11. Bhask ar T, Matsui T, Uddin MA, Kaneko J, Muto A, Sakata Y Effect of Sb2O3 in brominated heating impact polystyrene (HIPS-Br) on thermal degradation and debromination by iron oxide carbon composite catalyst (Fe–C). Appl Catal B, **43**, 229-241, (2003)
12. Murata K, Hirano Y, Sakata Y, Uddin M A Basic study on a continuous flow reactor for thermal degradation of polymers. J Anal Appl Pyrolysis, **65**, 71-90, (2002)
13. Blazso′ M, Cze′ge′ny Z, Csoma C Pyrolysis and debromination of flame retarded polymers of electronic scrap studied by analytical pyrolysis. J Anal Appl Pyrolysis, **64**, 249-261, (2002)
14. Bockhorn H, Hornung A, Hornung U, Jakobstro¨er P, Kraus M Dehydrochlorination of plastic mixtures. J Anal Appl Pyrolysis, **49**, 97-106, (1999)
15. Blazso′ M, Jakab E Effect of metals, metal oxides, and carboxylates on the thermal decomposition processes of poly(vinyl chloride). J Anal Appl Pyrolysis, **49**, 125-143, (1999)
16. R. Luijk, H. Haj, and L. Nelissen, "Formation of polybrominated dibenzofurans during extrusion of high-impact polystylene/decabromodiphenyl ether/antimony (III) oxide," *Environ Sci Technol*, **26**, 2191-2198, (1992)
17. Uemichi Y, Makino Y, Kanazuka T Degradation of polyethylene to aromatic hydrocarbons over metal-supported activated carbon catalysts. J Anal Appl Pyrolysis, **14**, 331-344, (1989)
18. A. López, I. de Marco, B.M. Caballero, M.F. Laresgoiti, A. Adrados, A. Aranzabal, "Catalytic pyrolysis of plastic wastes with two different types of catalysts: ZSM-5 zeolite and Red Mud" *Applied Catalysis B: Environmental*, **104**, 211-219, (2011)
19. C. Xie, F. Liu, S. Yu, F. Xie, L. Li, S. Zhang, J. Yang, "Study on catalytic pyrolysis of polystyrene over base modified silicon mesoporous molecular sieve " *Catalysis Communications*, **9**, 1132-1136, (2008)
20. T. Bhaskar *et al.*, "Effect of polyethylene terephthalate (PET) on the pyrolysis of brominated flame retardant-containing high-impact polystyrene (HIPS-Br)," *J Mater Cycles Waste Manage*, **12**, 332-340, (2010)

21. S. Peng *et al.*, "Debromination of flame-retarded TV housing plastic waste,", *J Mater Cycles Waste Manage*, **12**, 103-107, (2010)
22. P. S. Kulkarni, J. G. Crespo, and C. A. M. Afonso, "Dioxins sources and current remediation technologies-a review," *Environ Int*, **34**, 139-153, (2008)
23. C. Vasile, M. A. Brebu, T. Karayildirim, J. Yanik, and H. Darie, "Feedstock recycling from plastic and thermoset fractions of used computers (I): pyrolysis,", *J Mater Cycles Waste Manage*, **8**, 99-108, (2006)
24. R. Luijk, H. Wever, K. Olie, G. Haj, and J. J. Boon, "The influence of the polymer matrix on the formation of polybrominated dibenzo-p-dioxins (PBDDs) and polybrominated dibenzofurans (PBDFs)," *Chemosphere*, **23**, 1172-1183, (1991)
25. X. Yang, L. Sun, J. Xiang, S. Hu, S. Su, "Pyrolysis and dehalogenation of plastics from waste electrical and electronic equipment (WEEE): A review" *Waste Management*, **33**, 462-473, (2013)
26. Research Association For Feedstock Recycling of Plastics Japan, Base and Application of Plastic Recycling, CMC BOOKS Japan, 44ff, (2012)
27. T. Kwon, G. A. Tsigdinos, and T. J. Pinnavaia, "Pillaring of layered double hydroxides (LDH's) by polyoxometalate anions" *J. Am. Chem. Soc.*, **110**, 3653-3654, (1988)
28. N. Morita, T. Wajima, and H. Nakagome, "Reduction in Content of Bromine Compounds in the Product Oil of Pyrolysis Using Synthetic Hydrotalcite," *International Journal of Chemical Engineering and Applications*, **6(4)**, 262-266, (2015)
29. N. Morita, A. T. Saito, T. Wajima, and H. Nakagome, "Halogen Reduction in Pyrolysis Oil from Bromine-containing Plastics Using Hydrotalcite," *Advances in Environment Research* , **87**, 54-61, (2015)
30. N. Morita, M. Nakayasu, A. T. Saito, T. Wajima, and H. Nakagome, "Effect of Hydrotalcite on Bromine Content in Oil Produced from the Pyrolysis of Acrylonitrile-Butadiene-Styrene Plastics," *International Journal of Chemical Engineering and Applications*, **7(4)**, 23-27, (2016)
31. T. Kwon, G. A. Tsigdinos, and T. J. Pinnavaia, "Pillaring of layered double hydroxides (LDH's) by polyoxometalate anions" *J. Am. Chem. Soc.*, **110**, 3653-3654, (1988)
32. S. Miyata, " The syntheses of hydrotalcite-like compounds and their structures and physico-chemical properties I: The systems Mg^{2+}-Al^{3+}-NO_3^-, Mg^{2+}-Al^{3+}-Cl^-, Mg^{2+}-Al^{3+}-ClO_4^-, Ni^{2+}-Al^{3+}-Cl^- and Zn^{2+}-Al^{3+}-Cl^-," *Clays Clay Miner.*, **23**, 369-375, (1975)
33. T. Kameda, M. Nakamura, T. Yoshioka, "Removal of antimonate ions from an aqueous solution by anion exchange with magnesium–aluminum layered double hydroxide and the formation of a brandholzite-like structure." *Journal of Environmental Science and Health, Part A*, **47**, 1146-1151, (2012)
34. T. Kameda, K. Hoshi, T. Yoshioka, "Uptake of Sc^{3+} and La^{3+} from aqueous solution using ethylenediaminetetraacetate-intercalated CuAl layered double hydroxide reconstructed from CuAl oxide." *Solid State Sciences*, **13**, 366-371, (2011)
35. K. Ralla, U. Sohling, K. Suck, F. Sander, C. Kasper, F. Ruf, T. Scheper, "Adsorption and separation of proteins by a synthetic hydrotalcite" *Colloids and Surfaces B, Biointerfaces*, **87**, 217-225, (2011)
36. K. J. Martin, and T. J. Pinnavaia, "Layered double hydroxides as supported anionic reagents. Halide-ion reactivity in zinc chromium hexahydroxide halide hydrates $[Zn_2Cr(OH)_6X.nH_2O]$ (X = Cl, I)," *J. Am. Chem. Soc.*, **108**, 541-542, (1986)
37. T. Kameda, N. Uchiyama, T. Yoshioka, "Removal of HCl, SO_2, and NO by treatment of acid gas with Mg-Al oxide slurry" *Chemosphere*, **82**, 587-591, (2011)
38. E. Suzuki, M. Okamoto, and Y. Ono, "Catalysis by interlayer anions of a synthetic hydrotalcite-like mineral in a halide exchange between organic halides," *Chem. Lett.*, **18**, 1485-1486, (1989)
39. Kazuya Morimoto, Sohtaro Anraku, Jun Hoshino, Tetsuro Yoneda, Tsutomu Sato, "Surface complexation reactions of inorganic anions on hydrotalcite-like compounds" Journal of Colloid and Interface Science, **384**, 99-104, (2012)
40. Lin Deng, Zhou Shi, "Synthesis and characterization of a novel Mg-Al hydrotalcite-loaded kaolin clay and its adsorption properties for phosphate in aqueous solution", Journal of Alloys and Compounds, **637**, 188-196, 2015.
41. Ganapati D. Yadav, Payal A. Chandan, "A green process for glycerol valorization to glycerol carbonate over heterogeneous hydrotalcite catalyst", Catalysis Today, **237**, 47-53, 2014.

Implementation of Cleaner Production in a Natural Dye *Batik* Industry SME: A way to Enhance Biodegradability of *Batik* Wastewater?

Elzavira Felaza[1,a] and Cindy Rianti Priadi[1]

[1]*Environmental Engineering, Faculty of Engineering Universitas Indonesia, Indonesia*

Abstract. Utilization of natural dye in *batik* industry is a preventive solution taken to avoid environmental problems. Natural dyes is used to substitute synthetic dyes, naphtol and indigosol which has COD of 10.000-20.000 mg/L. Although the solution has been taken, the wastewater quality still exceed the limits stated on Regulation of the Ministry of the Environment No.5/2014. The research aim is to generate cleaner production solution, specifically material substitution and process modification to increase biodegradability. The implementation is conducted six times. The research process is pre-assessment of natural dye and *batik* wastewater, implementation of material substitution in mordanting, washing material and process modification. Natural dyes used has BOD of 203-975 mg/L, COD 1.316-2.453 mg/L and BOD/COD 0,1-0,4. Through statistical analysis with 95% level of confidence, the results show no significant changes to BOD, COD, BOD/COD as well as colour reduction. The results of options implemented is BOD value of 99-450 mg/L, COD 402-1.102 mg/L, TSS 105-540 mg/L, colour 291-2.408 Pt-Co and BOD/COD 0,2-0,4. Hence, wastewater treatment is needed and unit designed for wastewater flow of 0,09 m^3/day is anaerobic baffled reactor with the size of 0,6 m x 0,45 m x 0,5 m, HRT of 36,4 hour and SRT of 6 days.

1 Introduction

Being Indonesia's cultural heritage, *batik* is famous of its contribution in culture and economy. There are now 50.000 *batik* industries growing rapidly in Indonesia (Ministry of Industries, 2014) [1]. Aside from the positive outcomes that these industries has brought, there are negative impacts as well. The *batik* processing generates waste that has low biodegradability and it pollute the environment (Subki, 2011) [2]. The environmental pollution occurred in region with centralized *batik* industries (Mratihatani, 2013) [3]. Preventive solution has been taken by several small and medium *batik* industries to solve the problem which is by the use of natural dyes. This solution is categorized as material substitution in types of cleaner production implementation. The natural dyes is derived from various types of plants (Purnomo, 2004) [4], hence it is assumed to be more biodegradable than the synthetic dyes.

The natural dyes is easily degraded through biological process or biodegradable with the ration of BOD_5/COD approximately 0,5 (Chan et.al, 2000) [5]. The high level of biodegradability indicated that the wastewater can degrade the organic substances without the need for physical and chemical process, hence it can be treated biologically. One of the cause of the non-biodegradable nature is that the *batik* processing still requires chemical substances in mordanting, washing and fixating process (Supraptro, 2000) [6]. The examples of chemical substances used is alum which contains Al and turkey red oil which is the detergent used for washing process. This is presumed to be the cause why the accumulation of natural dye batik wastewater has low biodegradability, hence the preventive solution of cleaner production is needed. The types of cleaner production options which can be used are input substitution and process modification (Van Berkel, 2000) [7].

The aim of this study is to discover which process and material caused the low level of biodegradability in natural dye batik wastewater, know the right process modification and material substitution to support the cleaner production, learn the impact of cleaner production to the biodegradability as well as recommend wastewater treatment needed after the preventive actions are taken. The results of this research will be a recommendation for further implementation of cleaner production in natural dye *batik* industry to support the use of natural dye and the development of eco-industries.

2 Theoretical Overview

There are several types of *batik* processing in the *batik* industry. The process can be divided in to traditional process and fabrication process with modern technology. Overall, the *batik* processing consists of pre-treatment, waxing, dyeing, washing and drying. The *batik* processing in natural dye *batik* industry is shown in Figure 1.

[a] Corresponding author: efelaza@yahoo.com

© The Authors, published by EDP Sciences. This is an open access article distributed under the terms of the Creative Commons Attribution License 4.0 (http://creativecommons.org/licenses/by/4.0/).

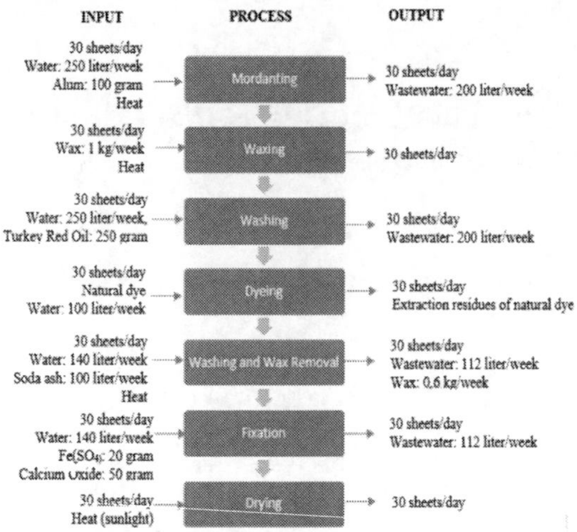

Figure 1. Batik processing flow diagram

The input used for dyeing in this case is natural dyes. Indonesia has more than 150 types of plant which can be used as the source of natural dyes, identified by the Center of Batik Industrial Research and Development Yogyakarta (Supraptro, 2000) [6]. The natural dyes is divided into several categories, which are mordant, direct, acid/base and vessel. Each of these category produces different colors and endurance which varied based on the media used. Natural dyes has BOD_5/COD value of approximately 0,5 (Chan P. et.al, 2000) [5]. The difference of the BOD_5/COD ratio or biodegradability between natural dyes and synthetic dyes is shown in Table 1.

Table 1. Characteristics of Natural Dyes and Synthetic Dyes

Parameter	Natural dyes (with mordant)	Natural dyes (without mordant)	Synthetic Dyes
BOD_5 (mg/L)	799– 2700,0	1000-4800	-
COD (mg/L)	1700- 4800,4	1700-4600	-
BOD_5/COD (mg/L)	0,2 – 0,6	0.5-1,1	0,4 – 1,1

Source: Chan, P. et.al, 2000 [5]

The natural dyes is used in the case study location, however the characteristic of the accumulation of wastewater still exceed the quality standard stated on the Regulation of the Ministry of Environment number 5 year 2014. The results of the wastewater accumulation from each step of batik processing is shown in Table 2.

Table 2. Preliminary Results of the Accumulation of Natural Dye Batik Wastewater

Parameter	Test Methodology	Unit	Acc. Sample 1	Acc. Sample 2	Washing with TRO	Fixation
BOD_5	IKM/5.4.5/BLK-Y	mg/L	81,74	18,10	282,3	40,5
COD	APHA 5220-C,2005	mg/L	1320,96	1198,08	668,9	86,9
BOD_5/COD	-	-	0,062	0,015	0,422	0,466
Oil/Grease	SNI 06-6898.10-2004	mg/L	9	7	5	3
Color	SNI 06-2413-1991	Pt Co	692,2	880,1	60,5	195,5
Turbidity	SNI 06-6989.25-2005	NTU	89,76	22,9	60,5	20,4

The wastewater characteristic results in Table 2 shown that the wastewater still exceed the quality standard. There are two types of approach in tackling pollutants in the wastewater, which are end-of-pipe and pollution prevention. Cleaner production is a sustainable application of integrated preventive strategy to process, produce, increase efficiency and minimize the risk for human and the environment (UNEP, 1995) [8]. The concept of cleaner production is focused on material and energy conservation, prevention of hazardous substances, minimize the level and amount of emission and waste (Nurdalia, 2006) [9]. The types of cleaner production consist of product modification, input substitution, technology modification, better housekeeping and on-site recycling (Van Berkel, 2000) [7]. Material substitution in batik industries is used to support the cleaner production effort. One of the example is the use of *lerak* or *Sapindus mukorossi* as detergent (Ernawati, 2013) [10]. Material substitution of dyes to the less hazardous substance can be a way to increase biodegradability of textile wastewater (Ozturk, 2009) [11].

3 Research Methodology

The methodology used in this study is implementation of cleaner production, specifically input substitution and process modification. The steps of this study consist of preliminary test of natural dyes and accumulated wastewater characteristic in each process, implementation of washing input substitution, implementation of mordanting input substitution and process modification. Washing material substitution alternative used is *lerak,* lemon and baking soda, while the alternative for the mordant substitute is *tunjung*. The volume of each material used is 1 liter for *lerak*, 750 mL for lemon and 750 mL for baking soda. *Tunjung* which is used for mordant alternative is 100 gram. Process modification used is filtration of wastewater from the results of washing and wax removal using 0,1 mm filter.

In the preliminary test of wastewater characteristics, sampling is conducted through composite sampling where the samples is collected by mixing the samples of wastewater produced from each process, while the sampling for wastewater from each process is collected using grab sampling technique which represents the condition of wastewater at the time of sampling from each stage of *batik* processing. The total samples for sampling and implementation of cleaner production is 40 samples.

The research variables of this study is divided to independent and dependent variables, where the independent variable is the material of substitution. Whereas the dependent variables are COD, BOD, TSS and color. The wastewater test conducted is adapted to the National Standard of Indonesia which is shown in Table 3.

Table 3. Methodology for Wastewater Characteristic Test

Parameter	Unit	Methodology
COD	mg/L	SNI 6989.73:2009
BOD	mg/L	SNI 6989.72:2009
TSS	mg/L	SNI 06 – 6989.27 – 2005
Color	mg/L	SNI 06- 6989.80:2011

The data of this study consist of primary and secondary data. The primary data is collected through survey and test of wastewater quality. The collection of primary data is using direct observation in case study location, interview with the SME staff and testing of wastewater characteristics. Whereas the secondary data for this study is collected through the literature of batik processing, wastewater quality standard and cleaner production alternatives.

The analysis of the result of cleaner production is based on the Guidance Manual: How to Establish and Operate Cleaner Production Centers (UNIDO and UNEP), where the steps used for this study is planning element which consist of goals setting and program scope as well as identification of pollutant source, pre-assessment element, which consist of information collection, basic information preparation and raw materials and material balance, along with assessment element which consist of detailed information collection, material balance and identification and evaluation of cleaner production recommendation.

In order to validate the analysis results from wastewater samples, statistic test named independent t-test is conducted (Berthouex, 2002) [12]. The aim of this method is to compare two variable which are randomly distributed. The final result of this test is conclusion on statistic basis of the results of implementation which can be concluded whether the cleaner production made a significant difference or not.

4 Results and Discussion

On this study, natural dyes which were used in the case study location is tested, which are *soga, indigofera, jalawe* and *tingi*. The comparison between the natural dyes used in the case study location and the ones in the literature as well as synthetic dyes is shown in Table 4.

Table 4. Comparison of Dyes Quality in the Case Study Location and Literature

Name of Dyes	BOD (mg/L)	COD (mg/L)	BOD/ COD	TSS (mg/L)	Color (Pt-Co)
Soga	203	1.316	0,15	164	14,217
Indigofera	260	1.734	0,15	98	480
Jalawe	984	2.453	0,4	615	27.128
Tingi	475	1.665	0,29	54	14.889
Naphtol (2014)	3.200	13.800	0,23	-	-
Napthol (2010)	5.400	19.921	0,27	-	-
Indigosol	3.050	10.230	0,29	-	-
Natural Dyes (2000)	796-4.755	1.759-4.790	0,16-0,6	-	-

Source: Zuhria, 2014 [13]; Lestari, 2010 [14] dan Chan, 2000 [5]

The results of the quality of the natural dyes are then tested statistically through independent t-test. The results of the statistical estimation is shown in Table 5.

Table 5. Results of Independent t-test of the comparison of BOD, COD and BOD/COD of natural and synthetic dyes

Parame-ter	BOD (mg/L)		COD (mg/L)		BOD/COD	
	S*	N*	S	N	S	N
Sample 1	3.200	203,4	13.800	1.316	0,232	0,155
Sample 2	5.400	260,2	19.921	1.734	0,271	0,150
Sample 3	3.050	984,6	10.230	2.453	0,298	0,401
Sample 4	-	475,8	-	1.665	-	0,286
Average	3.883,3	481,0	14.650,3	1.792	0,267	0,248
Variance	1.730.833,3	189.709,2	24.021.170,3	227.633,3	0,001	0,014
$t_{5,0.025)}$*$s(y_1-y_2$	1.721,1		6.115,6		0,155	
Interval	5.123	1.681	18.970	6.740	0,17	-0,14

*S = Synthetic, N= Natural

The results stated in Table 5, shown that the BOD and COD of natural dyes are statistically proven to be lower than synthetic dyes. However, in terms of BOD/COD, there are no conclusion can be taken statistically. By analyzing the results, it is proven that there is still a need for improvement of cleaner production implementation in natural dyes *batik* industry, because the wastewater quality still exceed the limit stated by the regulation. The results of the cleaner production implemented in the study location is as shown in Figure 2.

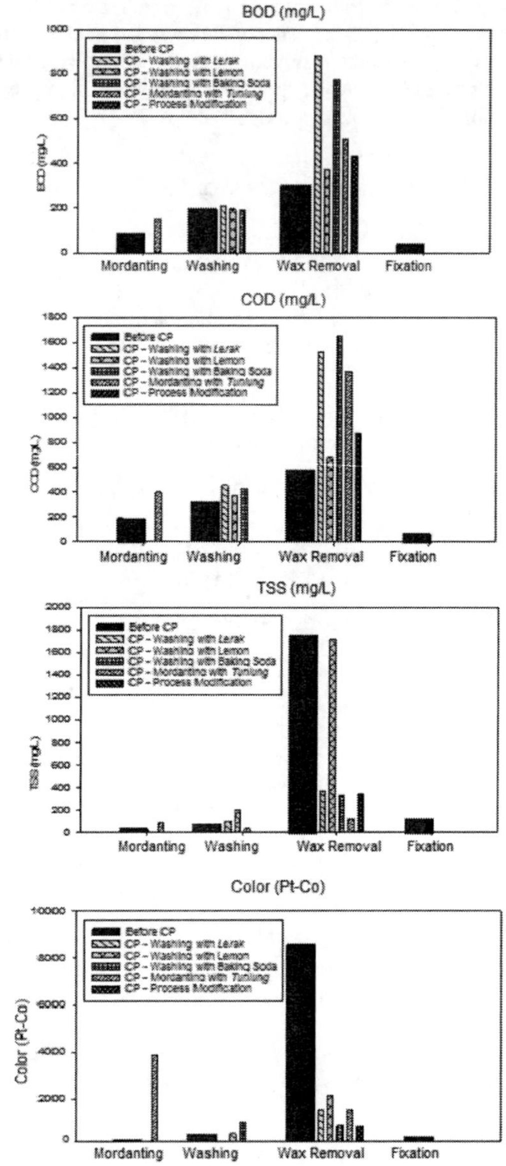

Figure 2. Wastewater Quality after Cleaner Production

The results of implementation of cleaner production conducted in this study in Figure 2, shown that the substitution of *mordant* from alum to *tunjung,* results in higher BOD, COD, TSS and color. Hence, the best option of material for *mordanting* process is alum. By the results of implementation in washing process, in terms of BOD and COD there is no significant decrease by the options implemented. In the wax removal process, it is shown that the process modification results in decrease of TSS and color in the wastewater.

The option of cleaner production chosen for further analysis is *mordanting* with alum, washing with lemon and process modification of wax removal. The accumulation of wastewater before and after the implementation of the chosen option is tested to obtain the wastewater quality. The results of the implementation is shown in Table 6.

Table 6. Results of Wastewater Accumulation before and after Cleaner Production

Name of Industry	BOD (mg/L)	COD (mg/L)	BOD/COD
Before CP	394	651	0,6
After CP	230	646	0,36
Textile Industry (2000)	393	950	0,4
Batik C (2006)	326	22.076	0,014
Batik F (2006)	407	1.337	0,3
Batik I	652	22.613	0,028

Source: Nurdalia, 2006 [9]

The results shown that the BOD and COD is decreased due to the cleaner production option applied. However, in terms of BOD/COD, the chosen cleaner production option did not increase the biodegradability of the wastewater. The ratio of BOD/COD is decreased from 0,6 to 0,36. The BOD/COD should not be the main parameter for the improvement of wastewater quality, because although the biodegradability decreased, if the BOD and COD are seen separately, both of these parameters decreased.

Statistical analysis of independent t-test is conducted to assure the difference of the wastewater quality before and after the implementation. The results of the test proven that with 95% level of confidence, there are no conclusion which can be taken statistically that there is a significant difference before and after the implementation of cleaner production, due to the large number of variant and low number of samples taken.

The quality of wastewater after cleaner production still exceed the limit stated in the regulation. Hence, wastewater treatment unit is needed for this *batik* industry. The unit suggested for treating the wastewater is anaerobic baffled reactor with the volume of 0,6 m x 0,45 m x 0,5 m with HRT of 36,4 hour and SRT of 6 days.

5 Conclusion

The conclusions of this study are of the following:

1. Accumulation of natural dye batik wastewater from the whole *batik* processing resulted in average biodegradability of 0,6.
2. The natural dye used in this study has biodegradability which varied between 0,15-0,4. In terms of quality, natural dye is proven to be better than the synthetic dyes
3. Material substitution and process modification with the best results are alum for mordanting process, washing with lemon and process modification of washing and wax removal. Statistically, with significance level of 95%, the cleaner production option used in this study does not bring a significant change to the wastewater quality. The fluctuation of wastewater characteristics caused insignificant change of the result of material substitution and process modification to the wastewater quality, in terms of biodegradability
4. The accumulation results after the chosen cleaner production option are wastewater with BOD value of 99-450 mg/L, COD value of 401-1.102 mg/L, TSS value of

104-540 mg/L and color value of 201-2.408 Pt-Co. The biodegradability of wastewater after the cleaner production option is 0,2-0,4.

The recommendation given for the similar study is to increase the number of samples collected for research in home industries, conduct modification process or material substitution for the extraction of natural dyes for better biodegradability of natural dye and conduct further study using different methodology of biodegradability test, such as respirometer for oxygen uptake, biological methane potential and anaerobic toxicity assay.

References

1. Indonesian Min. Industries. (2014)
2. N.S. Subki, H. Rohasliney. Intl. Conf. Chem, Bio, Env.Sci. (2011)
3. A. Mratihatani. Prgm. Srjn. Fak. Eko. Bsn. Unv. Dip. (2013).
4. M.A. Purnomo. Jrnl. Sn. Rp. STSI. Srkt. **1**, 2 (2004)
5. P. Chan. Intst.Txtls.Clth.Hngkg.Plytch.Uni. (2000)
6. H. Supratpro. Bl. Bsr. Pngmbgn. Indstr. Krjn. Btk (2000)
7. R. Van Berkel. J. Clnr. Prdctn. **15**, 8-9 (2000)
8. UNEP. Glb. Env. Otlk (1995)
9. I. Nurdalia. Unv. Dipngr. Smrg. (2006)
10. Ernawati. Intrnl. Cnf. TIMEE. (2013)
11. E. Ozturk. J. Clnr. Prdctn. Trky (2009)
12. P.M. Berthouex. *Statistics for Environmental Engineers* (Lewis Publishers, 2002)
13. F. Zuhria. Unv. Gdjh. Md. (2014)
14. K. Lestari. Unv. Gdjh. Md. (2010)

Research on Catalytic Oxidation Pretreatment of Organic Pesticide Wastewater with High Concentration

Jing Jiang[1,a]

[1]*Department of architectural and environmental Engineering, Chengdu Technological University, Chengdu 610000*

Abstract: Pesticide wastewater has the characteristics of high organic pollutant, high concentration, deep color and high toxicity, which has become difficult to treat the organic wastewater with high concentration at home and abroad. This article uses three methods of US(ultrasonic), Fenton(Fe^{2+}&H_2O_2) and combination of US&Fenton were used in the comparative research on the treatment of organic pesticide wastewater with high concentration. **Experimental conditions:** time of 130 min, ultrasonic power of 280W, frequency of 418kHz, pH value of 3.5, H_2O_2 concentration of 0.3mol/l; dosing mode: two thirds was added at 0 min, the rest one third was added at 65 min. The results showed that the treatment effect of the combination of US&Fenton was significantly better than that of independent US and independent Fenton; after the treatment on the organic pesticide wastewater with high concentration, the degradation rate of COD reached 85%, the chromaticity degradation rate reached 99%; the ratio of COD/BOD was about 1.4, with better biodegradability, which has provided a good condition for the subsequent biochemical treatment. Solved the problem of pesticide wastewater was difficult and expensive to degradation.

1. Introduction

With the wide application of pesticide, pesticide wastewater discharged in the process of pesticide production has become a potential environmental hazard. This kind of industrial wastewater has become one of the pollutions which are the most seriously and the most difficultly treated in China. Advanced oxidation technology usually refers to the treatment technology which oxidizes and degrades organic pollutants by producing high-reactivity hydroxyl radicals (HO) under ambient temperature and pressure, including the photocatalytic oxidation, O_3 oxidation, Fenton reagent oxidation as well as the combinations of oxidation technology. The application of advanced oxidation technology to treat landfill leachate, pesticide wastewater and dye wastewater has been reported in the literature [1-5]. However, due to the characteristics of pesticide wastewater, the application of advanced oxidation technology is not yet common in practice.

The composition of pesticide wastewater is complex, containing a large number of toxic and harmful substances, resulting in its poor biodegradability; the use of traditional biochemical treatment method is generally difficult to get good treatment results, so it is needed to strengthen the pretreatment of wastewater [6, 7].

The combination of Fenton method and ultrasonic method was used to pre-treat the organic pesticide wastewater with high concentration, and the effect of the process conditions on US&Fenton treatment system was discussed.

2. Experiment

2.1 Experimental materials

Wastewater used in the experiments is from a chemical plant in Chengdu, the wastewater is reddish brown, with strong pungent odor, and poor biodegradability. According to the report of wastewater detection, the water mainly contains dimethyl phthalate, tributyl phosphate, etc.

Experiment reagents:$FeSO_4{\cdot}7H_2O$ （analytical reagent AR）, 30%H_2O_2（analytical reagent AR）, $K_2Cr_2O_7$（analytical reagent AR）

Experimental apparatus: Visible spectrophotometer 721G (Shanghai Precision Scientific Instrument Co., Ltd.), JY92 - II ultrasonic instrument (Ningbo scientz Biotechnology Co., Ltd.), PHS-3B digital precision pH meter (Shanghai Leici Instrument Factory).

2.2 Experimental methods

Put wastewater in the reactor, add a certain amount of ferrous sulfate and hydrogen peroxide, turn on the ultrasound instrument, and add H_2O_2 according to a certain way under the condition of aeration rate of 150 L/h. After a period of time of reaction, adjust the pH

[a] Corresponding author: 25859260@qq.com

© The Authors, published by EDP Sciences. This is an open access article distributed under the terms of the Creative Commons Attribution License 4.0 (http://creativecommons.org/licenses/by/4.0/).

value to 10, and then wait 1h, finally take the supernatant for the index measurement.

The initial concentration of water sample COD was 2253 mg/L, the ultrasonic frequency was418kHz, the power was 280W, the pH value was 3.5, the concentration of Fe^{2+} was 25mmol/L, the concentration of H_2O_2 was 0.3mol/L, and the three kinds of treatment methods of independent Fenton, independent US and combination of US&Fenton were used to treat wastewater for 130min.

2.3 Analysis method

Potassium dichromate method was used to determine wastewater COD [8]; Fe^{2+} concentration and total iron concentration were measured according to HJ/T345-2007 "*Water Quality Determination of the Iron Phenanthroline Spectrophotometric Method*" [9].

3 Results and Discussion

3.1 Degradation of COD with different methods and color contrast

It can be known from Fig 1 that the degradation effect of independent US on water sample COD was not obvious, it was only 15%, the maximum value was reached at 90min. The degradation effect of independent Fenton on water sample COD was 65%, the maximum value was reached at about 90min. The degradation effect of combination of US&Fenton on water sample COD was up to 85%, which was more than the degradation sum of independent US and independent Fenton, and it reached the maximum value at 130 min, and tended to be stable, proving US and Fenton have synergistic effect, and its degradation effect on water sample COD was obvious. The COD value of water sample reduced from 2253 to 338, the ratio of COD/BOD was about 1.4, the biodegradability was better, which have achieved the pretreatment effect of pesticide wastewaters with high concentration.

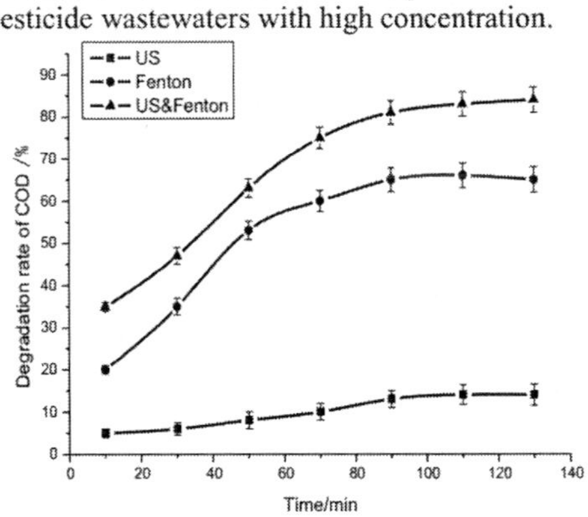

Figure 1. COD degradation with different methods

Figure 2 shows the degradation rate contrast of 3 methods for the water samples in the experiments. The degradation rate of US on experimental water sample was 30%, and the highest value was reached at 90min. The degradation rate of Fenten was 74%, and the highest value was reached at 90min. The degradation rate of US&Fenten on the chromaticity in experimental water sample reached 99%, the highest value was reached in 110 min; the reaction process was more durable, the treatment effect was significantly higher than those of the US and Fenten. The pretreatment effect of the combination of US&Fenten on organic pesticide wastewater with high concentration was obvious, achieved the goal of the wastewater pretreatment, which has provided favorable conditions for follow-up biochemical treatment.

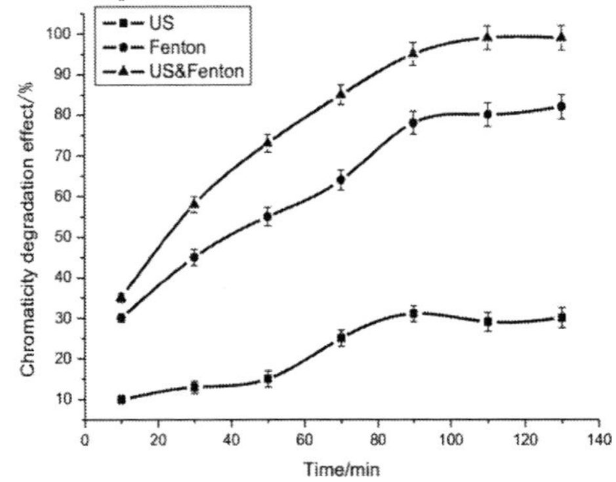

Figure 2. Chromaticity degradation effect with 3 different methods

3.2 Effect of pH value on degradation rate

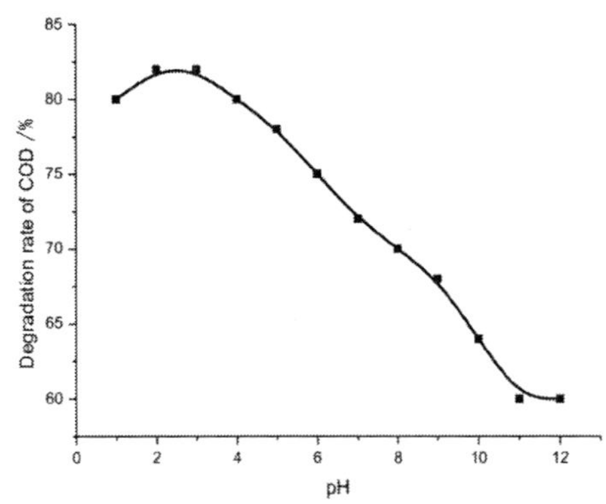

Figure 3. Effect of pH value on COD degradation rate

With the combination treatment method of US&Fenten, it can be seen from Figure 3 that the experimental conditions were as follows: the initial concentration of experimental water samples COD was 2253 mg/L, ultrasonic frequency was 418 kHZ, the power was 280W, the concentration of Fe^{2+} was 25 mmol/L, the concentration of H_2O_2 was 0.3 mol/L. it can be seen from Fig 4 that when the pH value is 2 ~ 5, the degradation efficiency of COD was higher; when the pH value was 3, the degradation rate of COD had its peak value, and the degradation efficiency decreased with the increase of pH value. Therefore, it is concluded that the optimum condition is the pH value of 3.5. This is because when the pH value is less than 3, $[Fe(II)(H_2O)_6]^{2+}$ will be produced in the system, the reaction rate of the compound with H_2O_2 is lower than that of $[Fe(II)(OH)(H_2O)_5]^+$, thus, reducing the production of OH, affecting the effect of catalytic oxidation reaction [10]. And when the pH value is much higher, it not only inhibits the production of OH, but also makes Fe^{2+} and Fe^{3+} in the solution subside in the form of hydroxide precipitation and lose catalytic ability; at the same time, the higher pH value will make the H_2O_2 produce invalid decomposition, reducing the utilization rate of H_2O_2 [11].

3.3 Effect of ultrasonic parameters on the degradation rate

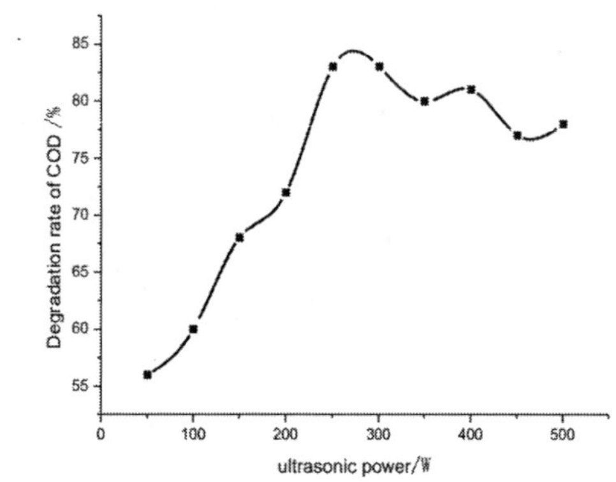

Figure 4. Effect of ultrasonic power on the COD degradation rate

With the combination treatment method of US&Fenten, the experimental conditions were as follows: the initial concentration of experimental water samples COD was 2253 mg/L, the pH value was 3.5, the concentration of Fe^{2+} was 25 mmol/L, the concentration of H_2O_2 was 0.3 mol/L. it can be seen from Figure 4 that degradation efficiency of COD increased with the increase of the power of ultrasonic power, peaked in 200 ~ 300 w; when the power is greater than 300, the degradation rate of COD decreased. Generally speaking, when the ultrasonic frequency is constant, the degradation effect should increase with the increase of power, the phenomenon in the experiment is speculated that too great power will result in the barrier effect caused by cavitation bubbles in ultrasonic larger negative phase, resulting in instable energy and attenuation, thereby reducing the utilization rate of ultrasonic energy [12, 13]. As a result, too great ultrasonic power may cause energy waste, from the perspectives of economy and treatment effect, it is determined that the best ultrasonic power is 280 W.

3.4 Effect of H_2O_2 dosage on degradation rate

With the combination treatment method of US&Fenten, the experimental conditions were as follows: the initial concentration of experimental water samples COD was 2253 mg/L, the ultrasonic power was 280W, the pH value was 3.5, the concentration of Fe^{2+} was 25 mmol/L. H_2O_2 with different concentrations was added; the organic pesticide wastewater with high concentration was treated with the combination method of US&Fenton for catalytic oxidation degradation for 130 min, the result is shown in Figure 5.

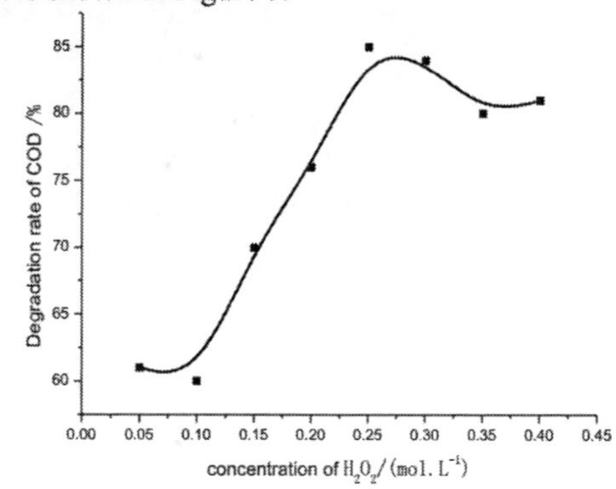

Figure 5. Effect of H_2O_2 concentration on COD degradation rate

It can be seen from the Figure, when the concentration of H_2O_2 in the solution was less than 0.25 mol/L, it increased with the concentration of H_2O_2, the degradation rate of COD in experimental sample water was increased. When the concentration of H_2O_2 in the solution increased to 0.25 ~ 0.3 mol/L, the degradation rate reached its peak; afterwards, the degradation rate of COD in experimental sample water decreased with the increase in the concentration of H_2O_2. Therefore, it is established that the best experimental concentration condition of H_2O_2 is between 0.25 ~ 0.3 mol/L.

4 Conclusions

Under the experimental conditions of 418 kHZ ultrasonic frequency, the 280W power, 3.5 pH value, 25mmol/L concentration of Fe^{2+}, 0.3mol/L concentration of H_2O_2, three methods were used to treat organic pesticide wastewater with high concentration for 130 min; the results proved that the effect of combination treatment method of US&Fenton was the best, COD degradation rate reached 85%, chromaticity degradation rate reached 99%, the ratio of COD/BOD was about 1.4, with better biodegradability, which has provided a good condition for the subsequent biochemical treatment.

References

1. Zhang Yongpeng, Zhou Junying, et al. Research Progress of Application of Several Advanced Oxidation Technology in Pesticide Wastewater Treatment [J] Journal of Pesticide Science, 2007, 9(2):103-109.
2. Zhang Zhisheng, Li Guangming et al. Research Progress of Advanced Oxidation Technology for Treatment of Landfill Leachate [J] Sichuan Environment, 2005, 24(5):72-78.
3. Li Wenshu, Li Yongmei, et al. Application of Advanced Oxidation Technology in the Treatment

of Persistent Organic Pollutants[J]Industrial Water Treatment, 2004, 24(11):9-12.

4. Zhang Yi, Hu Bing, et al. Research Progress on Advanced Oxidation Technology for Degradation of Organic Ohosphorus Pesticides in Water [J] Environmental Pollution & Control, 2006, 28(5):361-3 64.
5. Liu Yan, Sun Dezhi, et al. Research Progress on Advanced Oxidation Technology for Treatment of Dye Wastewater J] Industrial Water Treatment, 2006, 26(6):1-5.
6. Ge Xiangfeng, Xu Mingfang, et al. Analysis of Influencing factors of the Photocatalytic Degradation of Trichlorfon Pesticide Wastewater [J] Ecological Science, 2004, 23(2):1 24-127.
7. Wu Qimo, Wu Jianzhong, et al. Treatment of Pesticide Wastewater by Oxidation Process [J] Environmental Engineering, 2005, 23(5):88-90.
8. Editorial board of National Environmental Protection Bureau *"Detection and Analysis Methods of Water and Wastewater"* [M] Fourth Edition, Beijing: China Environmental Science Press, 2002: 210-211,
9. China's Environmental Monitoring Station. HJ/T345—2007 *"Water Quality Determination of the Iron Phenanthroline Spectrophotometric Method"* [S] Beijing: China environmental science press, 2007.
10. He Dewen, Qin Yan et al. Experimental Study on Treatment of Ultrasonic Oxidation for Printing Ink Wastewater [J] Journal of Central South University, 2009, 40(6):1483-1487.
11. Chen Chuanhao, Xie Bo, et al. Mechanism of the Influence Factors in Fenton Reagents for the Treatment of Wastewater [J] Environmental Science, 2000, 21(3):93-96.
12. Yoshio N , Michiyo N , Hiroshi O.Sonochemicaldegradation of chlorophernols in water[J].Ultrason Sonochem, 2000, 7(3): 115-120.
13. Su Chunyan, Guan Jie, et al. Study on Catalytic Oxidation Degradation of Nitrobenzene in Pharmaceutical Wastewater [J]. Journal of Changchun University of Technology, 2008, 29(5): 575-578.

MATEC Web of Conferences 62, 06001 (2016) DOI: 10.1051/matecconf/20166206001
ICCFE 2016

A Comparative Study on Equilibrium Adsorption of Dyes on Adsorbents Prepared from Coastal Plant, *C. equisetifolia* Seeds via Green Modification and Activated Carbon

Jazulhafiz Jefri Jaafar[1], Mohamad Awang[1, a], Anis Ayuni Aman Zuki[1], Mohammad Hussin Zain[2], Asmadi Ali @ Mahmud[1]

[1]*School of Ocean Engineering, Universiti Malaysia Terengganu, 21030 Kuala Terengganu, Terengganu, Malaysia*
[2]*School of Fundamental and Liberal Studies, Universiti Malaysia Terengganu, 21030 Kuala Terengganu, Terengganu, Malaysia*

Abstract. Eco-friendly adsorbents were prepared from coastal plant, *C. equisetifolia* seeds treated with β-cyclodextrin (Rhusorbent) for the removal of dyes. The study was carried out to determine the effectiveness of Rhusorbent to remove common dyes such as malachite green (MG) and neutral red (NR) dyes from aqueous solutions in batch experiments. The commercial activated carbon (AC) was also used in this equilibrium absorption study which includes the effect of adsorbent dosage, initial dye concentration and contact time. The percentage of MG dye removal of about 92.4% and 88.4% was achieved for both Rhusorbent and AC respectively at concentration of 60mg/L and 0.4g absorbent. The equilibrium adsorption of about 91.1% and 79.1% was achieved for both Rhusorbent and AC respectively involving NR dye. The scanning electron micrographs of Rhusorbent show uneven and rough surface prompting evidence for the effective adsorption of MG and NR dyes on the surface of adsorbents. For both adsorption studies, Langmuir and Freundlich isotherm models fitted well the adsorption data involving both Rhusorbent and AC. C. equisetifolia seeds are a potential source for bio-inspired adsorbents for the removal of MG and NR dyes.

1 Introduction

Water has become increasingly important for both agricultural and industrial sectors as well as for daily use. But today, the point and non-point sources of water pollution are contaminating our valuable water resources. Water pollution is a global problem that has led to interests in pollution control. On the positive side, dyes can give beautiful color to various products, and therefore they are widely used in many fields such as textiles, paper, plastic, food, painting, and medicine [1]. However, the release of effluent dyes into the ecosystem is a dramatic source of aesthetic pollution and of perturbations in aquatic life. Most synthetic dyes are toxic and can bring about serious water pollution, destroy community structure of aquatic organisms, and further become a hazard to all mankind. Since the removal of dyes from wastewater is considered an environmental challenge and government legislation requires textile wastewater to be treated, therefore there is a constant need to have an effective process that can efficiently remove these dyes [2].

Adsorption being a physical process, in-expensive and less time consuming, is widely accepted [3]. Adsorption onto activated carbon is widely practiced for removal of dissolved dyes from wastewater. However, adsorbent-grade activated carbon is cost-prohibitive. So it is necessary to seek some alternative low-cost adsorbents to activated carbon. The plant waste from C. equisetifolia plant seeds has been chosen as an adsorbent due to its vast availability since its seeds is abundant in nature, plus the fact that it is non-toxic, environmental friendly, biodegradable and cheap raw materials.

2 Methodology

2.1 Preparation of Adsorbent

C. equisetifolia plant seeds were washed with distilled water to eliminate dust, sand impurities and other unwanted chemicals. Then, the seeds were dried in an oven at the temperature of 70 C for a period of 24 hours. The seeds were ground with a crusher and sieved into small particles with an average size ranging between 2.0 and 3.0 mm. Then ground seeds were treated with 0.01M β -cyclodextrin for 24 hour. Then the β -cyclodextrin-treated seeds were washed with water and dried in oven at temperature of 70 C for a period of 24 hours to produce Rhusorbent. Activated carbon (AC) with similar size range was heated at 70 C for 5 hours prior to adsorption test. Surface morphology of RhuSorbent and AC was investigated using scanning electron microscope.

2.2 Batch Adsorption Studies

[a] Corresponding author: mohamada@umt.edu.my

© The Authors, published by EDP Sciences. This is an open access article distributed under the terms of the Creative Commons Attribution License 4.0 (http://creativecommons.org/licenses/by/4.0/).

In each batch, adsorption test was carried out in a 250 ml conical flask in which the MG and NR dyes initial concentrations were fixed at 100 ml with different concentrations. After that, Rhusorbent of 0.4g was added into the solution. The effect of contact time on the adsorption process was examined.

The mixed solution was stirred on the rotary shaker at rotation speed of 250 rpm at the room temperature. Then, the residual concentration was determined using absorbance values measured before and after the adsorption by UV-Visible spectrophotometer (UV-Vis) at wavelength of the dye solutions. UV-Vis scans were taken at 30, 60, 90, 120, 150, 180, 210 and 240 min. The MG and NR dyes removal percentage was calculated based on the following equation:

$$\text{Dye Removal } \% = (C_0 - C_i)/C_0 \times 100 \quad (1)$$

where, C_0 and Ci (mg/L) are the initial and final dye concentrations respectively. The average dye removal percentage was calculated based on three replicates.

3 Results and Discussion

3.1 Morphological characterization of adsorbent

Figs. 1(a) and 1(b) show the scanning electron micrographs of Rhusorbent and AC. While the surface morphology of the AC does show rough surface, the micrograph of Rhusorbent also indicates rough surface and caves-like and more uneven surface. The presence of rough and irregular surfaces of β -cyclodextrin-treated Rhusorbent in this case may be associated with the breaking of the covalent bond and the hydroxyl group present in the cellulose or lignin of fibers [4]. The rough surface may also be formed due to the removal of hemicelluloses, lignin and cellulosic constituents after treatment [5, 6]. As a result, more reactive hydroxyl group exposed on the fibers surface contributing to rough surface. The presence of cracks on the plant seed surface may be due to the reaction between acetyl groups with the fibers hydroxyl groups and thus the molecular orientation of the cellulose backbone was changed. As a result, the fiber becomes rougher and stiffer in character. This reduces the hydrophilicity of the fiber [7].

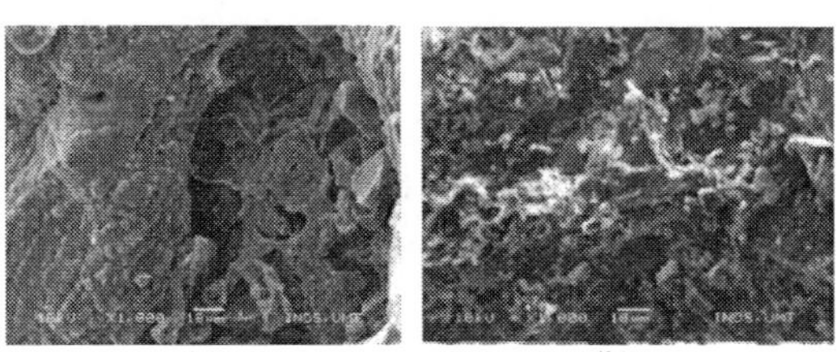

(a) (b)

Figure 1. Morphology of (a) Rhusorbent and (b) activated carbon

3.2. Batch Adsorption Studies

Table 1 and Table 2 show the percentage of MG dye removal at different initial dye concentrations for Rhusorbent and AC. The equilibrium adsorption of MG onto both Rhusorbent and AC was achieved within 240 min at various initial MG dye concentrations. Table 3 and 4 are the results of NR dye removal at different initial dye concentration versus time shows a similar trend to that of MG dye removal. Both percentage of MG and NR dyes removal decreased as the initial dye concentration increased as higher dye concentration lowers adsorption due to the saturation of adsorption sites.

Table 1. Percentage of MG removal for Rhusorbent treated with β-cyclodextrin

Time (min)	Percentage of MG Removal (%)				
	60 mg/L	100 mg/L	200 mg/L	300 mg/L	400 mg/L
30	56.62	53.44	21.29	20.90	21.13
60	68.21	55.11	27.08	22.01	28.53
90	71.39	58.08	33.43	22.92	33.29
120	76.15	68.99	39.70	24.70	33.80
150	79.47	72.42	53.66	38.43	34.39
180	89.09	80.63	69.27	41.69	34.82
210	91.41	84.99	69.82	42.93	34.98
240	92.39	85.58	70.29	43.91	35.57

Table 2. Percentage of MG removal for activated carbon

Time (min)	Percentage of MG Removal (%)				
	60 mg/L	100 mg/L	200 mg/L	300 mg/L	400 mg/L
30	64.26	66.44	28.96	26.96	13.61
60	72.72	67.46	24.31	30.48	14.08
90	75.78	78.31	39.44	30.97	16.42
120	77.79	82.36	45.52	41.63	19.11
150	83.56	86.41	54.42	45.37	19.91
180	85.27	85.91	71.59	45.18	20.09
210	87.62	86.16	73.13	45.35	19.96
240	88.38	86.59	73.29	45.09	20.03

Table 3. Percentage of NR removal for Rhusorbent treated with β-cyclodextrin

Time (min)	Percentage of NR Removal (%)				
	60 mg/L	100 mg/L	200 mg/L	300 mg/L	400 mg/L
30	51.18	37.45	20.56	20.90	19.26
60	64.00	45.82	24.42	22.01	23.11
90	81.76	72.19	30.41	22.92	27.08
120	88.81	83.27	35.28	24.70	29.64
150	90.29	85.48	38.29	38.43	31.55
180	90.31	85.51	41.81	41.69	35.94
210	90.79	86.23	42.47	42.93	36.54
240	91.10	86.69	42.67	43.91	37.09

At the lowest initial dye concentration (60 mg/L), it is observed that the percentage of MG dye removal for Rhusorbent and AC at 240 min was about 92.4% and 88.4% respectively. At the highest initial dye concentration (400 mg/L), the percentage of MG dye removal using Rhusorbent at 240 min was higher than that of AC. The equilibrium adsorption of MG dye for Rhusorbent and AC was 35.6% and 20.0% respectively. The removal percentage of NR dye using RhuSorbent was better that that of AC at all initial dye concentrations. At the lowest initial dye concentration (60 mg/L), the percentage of NR dye removal using Rhusorbent and AC at 240 min was 91.1% and 79.1% respectively. Meanwhile, at the highest initial dye concentration (400 mg/L), the percentage of NR dye removal using Rhusorbent and AC at 240 min was 37.1% and 24.6%, respectively. The adsorption for both MG and NR dyes at different dye concentrations was rapid at the initial stages and gradually decreased with the progress of adsorption until the equilibrium was reached.

Table 4. Percentage of NR removal for activated carbon

Time (min)	Percentage of NR Removal (%)				
	60 mg/L	100 mg/L	200 mg/L	300 mg/L	400 mg/L
30	47.02	36.44	15.34	16.15	13.16
60	53.65	42.10	22.31	19.86	14.17
90	63.1	45.95	26.11	26.16	15.60
120	77.39	58.37	26.36	28.83	17.10
150	83.56	66.47	30.2	31.03	19.69
180	78.34	77.14	34.87	32.82	21.69
210	79.36	76.30	35.49	33.26	24.17
240	79.11	77.24	36.13	33.36	24.60

3.3 Adsorption Isotherm

3.3.1 Langmuir model

The Langmuir equation has been applied for the monolayer sorption onto a surface of a finite number of identical sites with negligible interaction between of molecule [8]. It is also observed that the initial removal of dye is fast and with the rise in concentration the percentage uptake gradually decreases. The form of the Langmuir isotherm can be represented by the following equation:

$$q_e = q_m\,[(K_LC)/(1+K_LCe)] \quad (2)$$

where q_e was the amount of dye adsorbed per gram of adsorbent (mg/g), C denotes the equilibrium concentration of dye in solution (mg/L); K_L represents the Langmuir constant (L/mg) that related to the affinity of binding sites and q_m was a theoretical limiting adsorption capacity when the monolayer surface was fully covered with dye molecules and assists in the comparison of adsorption performance (mg/g). Linear plots obtained for $1/q_e$ against $1/C_e$ with R^2 values in Figures 2 and 3 indicate that the Langmuir isotherm holds good for the MG-Rhusorbent and NR-Rhusorbent systems.

3.3.2 Freundlich model

The Freundlich isotherm model is based on the assumption that adsorption occurs on the heterogeneous adsorption surface having unequally available sites with different energies of adsorption [9]. The Freundlich isotherms model had the following form:

$$q_e = K_FC_1/n \quad (3)$$

where q_e was the amount of dyes adsorbed per gram of adsorbent (mg/g); C was the equilibrium dye concentration in solution (mg/L); K_F and n were the Freundlich constants, which represented the adsorption capacity and the adsorption strength, respectively. The magnitude of 1/n quantified the favorability of adsorption and the degree of heterogeneity of the adsorbent surface [6].

The Freundlich isotherm is represented by equation:

$$\log q_e = \log K_F + 1/n \log C \quad (4)$$

The Freundlich adsorption isotherm plot is made between log C_e and log q_e as shown in Figures 4 and 5. The isotherm data for RhuSorbent fitted the Freundlich model with 0.99 R^2.

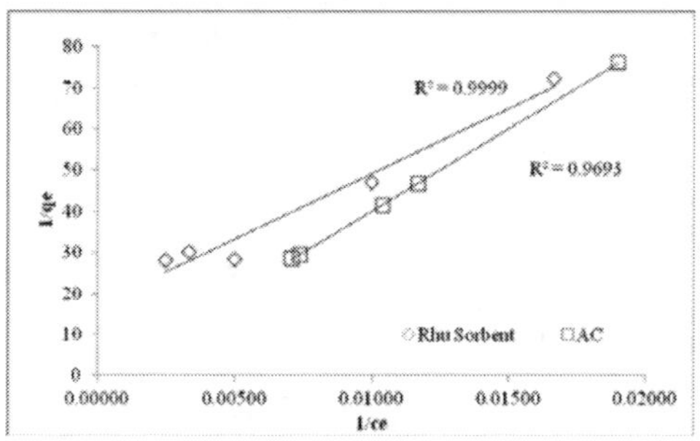

Figure 2. Langmuir isotherm for MG dye adsorption onto Rhusorbent and AC

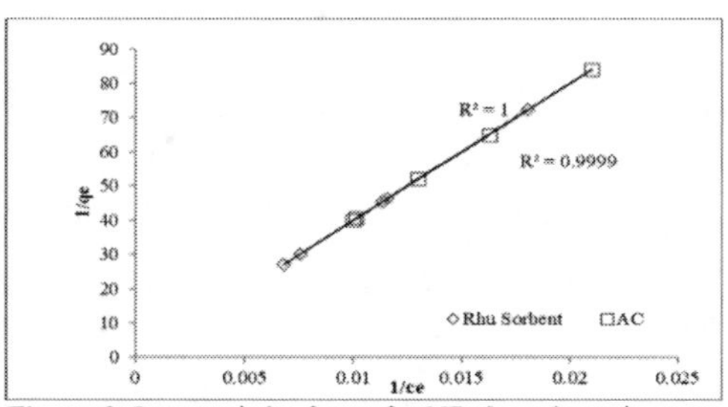

Figure 3. Langmuir isotherm for NR dye adsorption onto RhuSorbent treated and AC

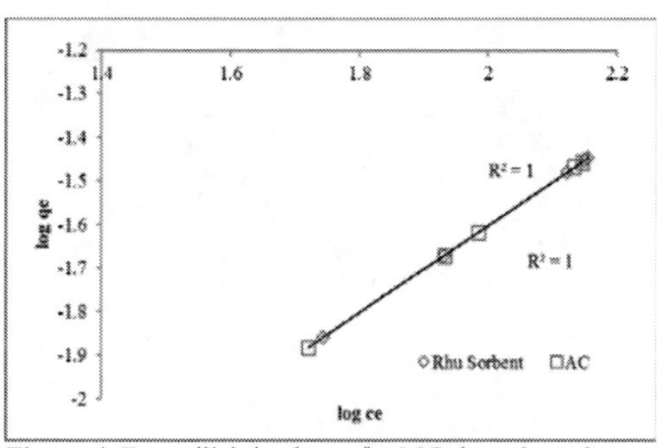

Figure 4. Freundlich isotherm for MG dye adsorption onto Rhusorbent treated and AC

DOI: 10.1051/matecconf/20166206001

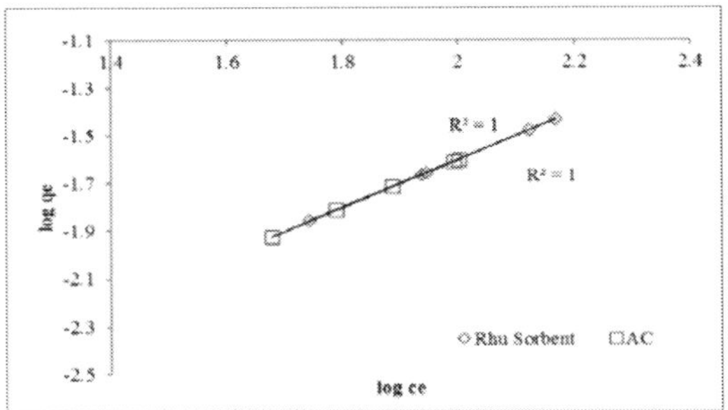

Figure 5. Freundlich isotherm for NR dye adsorption onto Rhusorbent treated and AC

4 Conclusions

C. equisetifolia seeds are a potential source for bio-inspired adsorbents for the removal of MG and NR dyes. The effectiveness of green modification of adsorbents prepared from C. equisetifolia seeds using β-cyclodextrin (Rhusorbent) in dye removal may add another eco-friendly adsorbent to the present list and hence protect the environment. The future works may include the kinetic study on the adsorption of the same dyes as well as the effectiveness of the green modification using other bio-inspired adsorbents.

Acknowledgement

The authors wish to thank the Ministry of Higher Education Malaysia (VOT 59274) and the School of Ocean Engineering, Universiti Malaysia Terengganu for the financial support.

References

1. J. Tan, X. Zhang, X. Wei, L. Wang, Bioresources **7**, 3 (2012)
2. M.A. Mohd Salleh, D.K. Mahmoud, W.A. Wan Abdul Karim, A. Idris, Desalination **280** (2011)
3. N. Sharma, D.P. Tiwari, S.K. Singh, *International Journal of Scientific & Engineering Research* **3**, 2 (2012)
4. I.O. Bakare, F.E. Okiemen, C. Pavithran, H.P.S Abdul Khalil, M. Brahma Kumar, Material Design **3** (2010)
5. J. Rout, M. Misra, S.S. Triphathy, S.K. Nayak, A.K. Mohanty, Comp. Sci. and Tech. **61**, 9 (2001)
6. E. Sinha, S.K. Rout, Bulletin of Materials Science **32**, 1 (2009)
7. V.P. Cyras, C. Vallo, J.M. Kenny, A. Vasquez, Journal Composite Materials **38**, 16 (2004)
8. T.C. Chandra, M.M. Mirna, Y. Sudaryanto, S. Ismadji, Chemicals Engineering Journal **127** (2007)
9. B.H. Hameed, D.K. Mahmoud, A.L. Ahmad, Journal of Hazardous Materials **158** (2008)

A Review on Enhancement of Biogas Yield by Pre-treatment and addition of Additives

Dr. Prashant Baredar[1,a], Dr.S Suresh[2] ,Dr. Anil Kumar[3] and Krishnakumar P[4]

[1]*Associate Professor, Energy Centre, MANIT, Bhopal, India*
[2]*Assistant Professor, Chemical Department, MANIT, Bhopal, India*
[3]*Assistant Professor, Energy Centre, MANIT, Bhopal, India*
[4]*Reasearch scholar, Energy Centre, MANIT, Bhopal, India*

Abstract. Energy from biomass has been in use from historical days itself but biomass as energy source is not as popular as that of other renewable energy resources. Anaerobic digestion is commonly used to convert organic biomass into biogas. The main drawback of this technology is the long retention period and low efficiency. To tackle this problem pre-treatment and adding additives are best suitable options available. This paper reviews the different pre-treatment techniques and additives that can be used to improve the biogas (mainly methane) yield and to reduce the retention period of the process.

1 Introduction

Biogas is considered as a renewable energy can be produced from a variety of raw materials. Anaerobic digestion and fermentation can be used to generate biogas from biodegradable materials such as organic wastes, sewage, municipal waste, manure, plant residues, vegetable wastes and energy crops. Large amount of activated sludge wastes, containing organic and mineral substances, are produced by municipal and industrial plants. Waste sludge handling represents a bottleneck in industrial plants, due to environmental, social, economic and legal issues. Organic wastes are of natural origin that possesses biochemical features ensuring quick microbial decomposition at comparatively normal operating conditions.

Basically organic wastes generated today originate from industrial, agricultural and municipal sector. Municipal wastes are generated due to human biological and social activities contain large portion of organic materials. Agricultural wastes are common in livestock and food production which can be utilised for biogas production. Industrial wastes are the most difficult type of waste for biological treatment, depending of its content and origin. Many industries uses chemicals in their manufacturing processes to achieve better product quality and these chemicals are difficult to treat. Recently, organic waste has had a lot of attention, due to possibilities of energy that can be recovered from these wastes and to prevent their adverse effect on environment. Most of the organic wastes are in solid form which has moisture contain up to 90%, so thermo chemical treatment of these wastes are difficult. Biodegradable organic waste can be treated in the presence or absence of air [1]. Aerobic process is called composting and anaerobic process is called digestion. Digestion is complicated, slow and expensive process, applicable only for selected raw materials. Recently, anaerobic digestion has become a widespread choice for biodegradable organic waste treatment around the world. It is well suited for organic wastes of high moisture content (over 80%), yielding methane rich biogas.

[a] Corresponding author: prashant.baredar@gmail.com

2 Anaerobic digestion

Anaerobic digestion is a series of steps in which microorganisms break down biodegradable organic waste materials into biogas and slurry in the absence of free oxygen. Biogas is a mixture of different gases mainly methane, carbon dioxide and small traces of H2S, NH3 and H2.The main features of digestion process are waste mass and volume reduction, biogas and nutrient-rich fertilizer productions. Table 1 shows typical solid and organic substance contents and biogas yields for most common organic wastes, treated with anaerobic digestion. The digestion process is divided into hydrolysis, acidogenesis, acetogenesis and methanogenesis.

Anaerobic digestion has a great future amongst the different renewable energy resources as well as in waste management technologies. However, the low overall bio digestion efficiency of the process and long retention periods (20-30 days) results in low efficiencies. In anaerobic digestion, the hydrolysis process is the rate-restricting step in which soluble organic materials are converted to biogas are generated. Consequently, the biogas yield depends mainly on the biodegradability and hydrolysis rate of the digestion. Biogas production can be

© The Authors, published by EDP Sciences. This is an open access article distributed under the terms of the Creative Commons Attribution License 4.0 (http://creativecommons.org/licenses/by/4.0/).

improved by pre-treatment of the organic wastes and adding additives to the slurry. During pre-treatment, both solubilisation of particulate matter and biological decomposition of organic long chain polymers to small chain monomers takes place ie. cell walls of the organic wastes are destroyed and extracellular polymeric substances are digested resulting in the conversion of organic material to a suitable form for easy digestion by acidogenic micro-organisms. Additive increases the microbial activity increasing the rate digestion process.

Table 1. Types of organic wastes and their biogas yield [3].

Organic waste	Total Solid[%]	Volatile Solid [%]	Biogas yield [m^3kg^{-1} of VS]
Municipal organic waste	15-30	80-95	0.5-0.8
Municipal wastewater sludge	3-5	75-85	0,3-0,5
Brewery spent grain	20-26	80-95	0.5-1.1
Yeast	10-18	90-95	0.5-0.7
Fermentation residues	4-8	90-98	0.4-0.7
Rumen content (untreated)	12-16	85-88	0.3-0.6
Vegetable wastes	5-20	76-90	0.3-0.4
Grass cuttings (from lawns)	20-37	86-93	0.7-0.8
Grass silage	21-40	87-93	0.6-0.8
Corn silage	20-40	94-97	0.6-0.7
Straw from cereals	~86	89-94	0.2-0.5
Cattle manure (liquid)	6-11	68-85	0.1-0.8
Cattle excreta	25-30	75-85	0.6-0.8
Pig manure (liquid)	2-13	77-85	0.3-0.8
Pig excreta	20-25	75-80	0.2-0.5
Chicken excreta	10-29	67-77	0.3-0.8
Sheep excreta	18-25	80-85	0.3-0.4
Horse excreta	25-30	70-80	0.4-0.6
Whey	4-6	80-92	0.5-0.9

Table 2. Reviews on different pre-treatment methods to enhance AD using various substrates.

Substrate	Pretreatment	Conclusions
OFMSW	All pretreatment methods	Physical pretreatments are widely applied for OFMSW, whereas other methods are not spread at industrial level
All organic substrates	All pretreatment methods	The most popular pretreatment methods are thermal and ultrasonic for WWTP sludge, chemicalfor lignocellulosic substrates, and mechanical for OFMSW
Lignocellulosic substrates	Thermal, thermo-chemical, chemical	Thermal pretreatments as well as lime and ammonia based chemical methods are more effectivein improving the digestibility of lignocellulosic substrates
Pulp & paper sludge	Thermal, thermo-chemical, chemical	Pretreatments could result in reduced HRT, increased methane production, and reduced sludge size
WWTP sludge	Ultrasound, chemical, and thermal	Pretreatments result in enhanced biogas production (30–50%)
WWTP sludge	Thermal, thermo-chemical, and chemical	The effect of pretreatment methods depends on the characteristics of sludge and the intensity ofthe method. Pretreatments could yield a better digestate with high recoverable nutrients

3 Pre-treatment

Different pre-treatment techniques such as thermal, chemical, biological, mechanical and their combinations are reviewed in this paper. Pre-treatment techniques are used to make the organic matter become more accessible to anaerobic microorganisms. This process improves the overall digestion efficiency and the degree of digestion, reducing retention time and increasing methane production rates [3]. The paper will provide a sequence of literature reviews about optimum conditions to obtain enhanced biogas production in various pre-treatments techniques. Table 2 shows the important findings of different researches on pre-treatment of various biomasses.

3.1. Thermal pre-treatment

It is well known for many years that thermal pre-treatment can improve the degradability of organic wastes. While the lipids and the carbohydrates of the wastes are easily degradable, the proteins are protected by cell walls from hydrolysis. Heat supplied during thermal pre-treatment breaks the chemical bonds of the cell wall, thus makes the protein content easily accessible for biological digestion. Maximum bio digestibility, in percentage, means the maximum percentage of substrate chemical oxygen demand (COD) that can be converted to methane, was calculated by El-Mashad et al [4]. Effect of thermal pre-treatment is studied in a wide range of temperatures ranging from 60to 270 ºC. In practice, the optimum temperature for the pre-treatment process is 160-180° C and treatment time between 30 to 60min. Pressure associated with it may vary from 600 to 2500kPa. Various researches on thermal pre-treatment have been conducted to verify this conclusion. Bougrier et al compared the thermal pre-treatment (130° C, pH=10, 150ºC and 170ºC during 30min) performance of waste activated sludge obtained from urban wastewater plants with untreated sludge. The results suggested that there was positive effect on solubilisation rates and methanization rate when thermal pre-treatment was performed [5].

Haug et al. worked on thermal treatment at low temperatures in order to improve digestibility of the waste and it also helped them to avoid the problems that occurred with high temperature thermal pre-treatment.

Their experiments showed that the effect on digestibility of activated sludge was high at 175ºC. The anaerobic digestion of the thermally pre-treated biomass at this temperature resulted in an increased methane production of 60-70% and higher temperatures led to decrease in methane production [6].

3.2 Chemical pre-treatment

Chemical pre-treatment is an efficient and cost-effective method to hydrolyze the cell walls and thus increase solubility of the organic matter contained within the cells. According to different principles, chemical treatment methods can be classified into acid and alkaline oxidation. The most frequent oxidative methods are ozonation and peroxidation.

3.3 Colour illustrations

You are free to use colour illustrations for the online version of the proceedings but any print version will be printed in black and white unless special arrangements have been made with the conference organiser. Please check with the conference organiser whether or not this is the case. If the print version will be black and white only, you should check your figure captions carefully and remove any reference to colour in the illustration and text. In addition, some colour figures will degrade or suffer loss of information when converted to black and white, and this should be taken into account when preparing them.

3.3.1 Ozonation

Ozone is a strong cell-lytic agent, which can destroy the microorganisms present in activated sludge and oxidize the organic substances released from the cells [7]. Among the techniques to disintegrate sludge, ozonation of sludge is one of the effectual way and yields the utmost degree of disintegration. Ozonation process changes the characteristics of the sludge greatly. The sludge biodegradation is affected by the amount of ozone. Several researchers have investigated the impact of ozone amount on sludge biodegradation. Ozonation treatment has two counter effects: degradation of molecules and cell structures which is responsible for the increase in biogas production and oxidation of organic molecules which is responsible for the decreases in biogas production [10]. According to Weemaes et al. the biogas yield increased by 80% with ozone treatment of 0.1 g O3/g COD; the effect was not evident at higher concentration of ozone [7]. A review of studies on the combination of ozonation with activated sludge process was proposed by Chu et al [8].

3.3.2 Peroxidation

Several peroxidation techniques, including the popular Fenton peroxidation and novel reactions of peroxymonosulphate (POMS) and dimethyldioxirane (DMDO), can achieve a transformation of available COD into soluble BOD enhancing the biogas production. Fenton pre-treatment disintegrates extracellular polymeric substances and breaks cell walls, releasing intracellular water. Hence increasing the soluble COD and BOD in the sludge water. Low pH value required is the major disadvantage of this method. Dewil et al. studied biogas generation by treating sludge with Fenton peroxidation, DMDO and POMS techniques on laboratory scale. The results showed a maximum increase of 75% with Fenton, while the biogas production became twice by PMOS and by DMDO method biogas production was increased by a factor of 2.5[9].

3.4 Thermo-chemical pre-treatment

Alkali treatment combined with thermal treatment is called thermochemical treatment. There is no agreement on the efficiency of this method. Kim et al. showed that the order of effectiveness in sludge solubilization was NaOH>KOH>Mg(OH)2 and Ca(OH)2, whereas Penaud et al. demonstrated that pre-treatment with KOH was more efficient than using NaOH and observed no effect on the biodegradability [10]. Haug et al. determined a decrease in biodegradability of 60%[6]. Tanaka et al. showed that thermo chemical pre-treatment significantly raised the biodegradability upto 230%. It's noted that thermo chemical method gives the best result compared with thermal, chemical, ultrasonic methods under same conditions. Valo et al. tested the COD removal rates and biogas production under two conditions (thermal treatment at 170° C and thermo-chemical at 130 ° C with PH = 10 for 30 min). They found that the COD removal rates were significantly increased compared with the untreated raw sludge, being 71% and 60% of raw sludge COD, respectively, while the biogas generations were increased by 54% and 74% [11].

3.5 Mechanical pre-treatment

Mechanical pre-treatment favours solubilisation of particulate matters in liquid state which in turn improves the yield of the biomass. In general, in mechanical pre-treatment the most often techniques used are ultrasonic pre-treatment, grinding and high pressure homogenization. By these methods, the aim is to increase the degradability of organic matters by disrupting the flocs and/or lysing the bacterial cells.

3.5.1 Ultrasonic pre – treatment

Ultrasonication is an effective and promising method to improve the biodegradability of the sludge. This method has several inherent advantages such as efficient sludge disintegration (>95%), improvement in biodegradability and bio solids quality, increase in methane production, reduction in retention time, sludge reduction and energy recovery [12]. This method enhances the sludge digestibility by disrupting the chemical, physical and biological properties of the sludge. Ultrasonic treatment accelerates the hydrolysis reactions by disrupting cell walls. The bacterial cells are divided by pressure waves and cavitations produced by ultrasonic generator leading to extraction of intracellular organic elements. The disruption of sludge particles obtained from ultrasonication treatment will boost subsequent acidogenesis, acetogenesis and methanogenesis reactions leads to an enhanced methane production and reduction of sludge volume [13]. Wang et al. showed that the methane generation was increased by12%, 31%, 64% and 69% with corresponding ultrasonic pre-treatment of 10minutes, 20 minutes, 30 minutes and 40 minutes, respectively. The optimum pre-treatment for enhancing the methane generation should be approximately 30 min.

3.5.2 Grinding

One dominant pre-treatment technique is the wet milling, which is more of a grinding method. Wet milling uses small beads to break cell walls, the size of the beads used are thus critical for maximal sludge disintegration. Of several milling devices, the ball mill using small diameter (0.2-0.275 mm) balls has the best performance [13]. During grinding sludge is pressed through a conical or cylindrical space with shear-stresses high enough to break the bacterial cell walls.

3.5.3 High pressure homogenizer

One of the most frequently used methods on large-scale is high pressure homogenization, compressing the sludge to a pressure of 60 MPa [1]. The compressed suspension is then depressurized using a valve and projected against an impaction ring at very high speed. The cells are hereby subjected to turbulence, cavitations and shear stresses, resulting in cell disintegration. Lack of available literatures on mechanical pre-treatment, the methods mentioned above is not comprehensive. However, it's seen that their efficiency of improving anaerobic digestion of waste activated sludge is rather lower than other methods [14].

4 Additives

Organic as well as inorganic chemicals can be added to improve methane gas production. Additives can increase microbial activity under different environmental conditions. Some organic substances are naturally available [15]. Usually additives are classified into organic and inorganic additives.

4.1. Inorganic Additives

There are many inorganic additives for example, iron salt that enhances gas production. Different salts were added to the slurry at different concentrations; FeSO4 was added at 50 mM, FeCl3 at 70μM. The addition of magnesium and calcium salts increased methane generation and avoided slurry foaming [15]. Malik et al. obtained an increased biogas production by 8–11% adding urea and diammonium phosphate to the sludge. Kumar et al.(1987) experiment on commercial charcoal Darco G-60 resulted in 17% and 34.7% increase in biogas yield in batch and semi continuous digesters, respectively. Nickel ions increase the methane production by 54%. Nickel was found to enhance the production up to 5 ppm. The optimum range appeared to be at 2.5 ppm in a water hyacinth-bovine waste substrate. The increase in biogas is due to the activity of nickel dependent metallo-enzymes involved in digestion [16]. Eosin blue dye at a concentration of 0.1 μM increased the biogas by 25-35% when mixed to manure slurry [17].

4.2. Organic Additives

Organic additives can maintain favourable conditions for increased gas production in anaerobic digestion. These conditions include: inhibition of acetogenesis and methanogenesis, pH, etc. Powdered leaves of some plants such as gulmohar, leucacenaleucocephala, acacia auriculiformis, dalbergiasisoo and eucalyptus tereticonius have found to improve biogas yield by18-40% [15]. Enhanced biogas production due to organic additives is due to the adsorption of substrate on the surface of the additive. Plant residues treated with 1% NaOH for 7 days combined with slurry manure at a weight ratio of 1:1 showed a twice increase in biogas production [17].

Tomato plant wastes when added to rabbit wastes improved methane production when the amount of tomato waste is more than 40% [18]. Plant waste like rice straw, corn stalks, cotton stalks and wheat straw when mixed with cattle manure increased gas production by 10-80%. Addition of bean seeds, guar, black gram and guar gum at a rate of 1500 ppm in the biodegradation of mango processing wastes increased methane production significantly. The addition of organic and inorganic substances to slurry waste increases the efficiency of anaerobic digestion and methane gas production. Some additives can increase the gas production even by 80%. Most of these options are easy and economically viable.

5 Conclusions

This paper reviewed various methods which can be used to minimize these drawbacks of anaerobic digestion. Pre-treatment of the waste and adding additives to the sludge are effective methods reduces the retention period and increases the digestion efficiency. Effectiveness of different techniques is discussed in the paper. The change in biogas production is different for different techniques. The increase in the biogas production in percentage by various techniques is summarized in the Table 3. It can be seen that mechanical pre-treatment methods requires high capital investment and are energy intensive. Thermal and thermo chemical methods require high temperatures to achieve better results. Cost of oxidative methods is high and requires low pH. However it appears that taking into account the increased financial requirements of sludge disposal, the operational cost and the investments required for sludge, ozonation may be the best option available due to their decreased operational costs for sludge treatment and disposal.

Table 3. Percent increase in biogas production by different methods.

Method			Percentage increase in methane production (%)
Pre-treatment	Thermal		60-70
	Chemical	Ozonation	80
		Peroxidation	75-150
	Thermo-chemical		75-230
	Ultrasonic		12-69
Addition of additives	Organic		10-80
	Inorganic		8-150

References

1. S. Umesh, S. Swati, B. Kansingh and V. Mayank, *Activated Sludge Treatment Process to Increase Bio Gas Production-A Need of Present Energy Crisis Scenario*, (2013)
2. Zupančič, D. Gregor, and V. Grilc, *Management of Organic Waste*, 1-28 (2012)
3. Ariunbaatar, Javkhlan, A. Panico, G. Esposito, F. Pirozzi, and Piet NL Lens, *Applied energy* **123**, 143-156 (2014)
4. El-Mashad, M. Hamed, G. Zeeman, Wilko KP van Loon, Gerard PA Bot, and G. Lettinga, *Bioresource technology* **95**, 191-201 (2004)
5. Bougrier, Claire, J-P. Delgenes, and H. Carrere. *Process Safety and Environmental Protection* **84**, 280-284 (2006)
6. Haug, T. Roger, D. C. Stuckey, J. M. Gossett, and P. L. McCarty, *Journal (Water Pollution Control Federation)* 73-85 (1978)
7. S. Chu, Libing, S. Yan, X. Hui Xing, X. lin Sun, and B. Jurcik, *Water Research* **43**, 1811-1822 (2009)
8. Yeom, I.T., Lee, K.R., Lee, Y. H., Ahn, K. H. & Lee, S. H., *Water Science and Technology* **46**, 421–425(2000)
9. Kim, Jeongsik, C. Park, Tak-Hyun Kim, M. Lee, S. Kim, S. Wook Kim, and J. Lee, *Journal of bioscience and bioengineering* **95**, 271-275 (2003)
10. Penaud, V., J. P. Delgenes, and R. Moletta, *Enzyme and microbial technology* **25**, 258-263 (1999)
11. Tanaka, Shuzo, Toshio Kobayashi, Ken-ichiKamiyama, and Ma Lolita N. SigneyBildan. *Water Science and Technology* **35**, 209-215 (1997)
12. Pilli, Sridhar, PuspenduBhunia, Song Yan, R. J. LeBlanc, R. D. Tyagi, and R. Y. Surampalli. *Ultrasonics sonochemistry* **18**, 1-18 (2011)
13. W. Qunhui, M. Kuninobu, K. Kakimoto, I. Hiroaki, and Y. Kato, *Bioresource technology* **68**, 309-313 (1999)
14. Baier, Urs, and P. Schmidheiny, *Water Science and Technology* **36**, 137-143 (1997)
15. A. Yadvika, A. Santosh, T. R. Sreekrishnan, KohliSangeeta, and R. Vineet, *Bioresource Technology*, (2004)
16. Geeta, G. S., K. SlJagadeesh, and T. K. R. Reddy, *Biomass* **21**, 157-161 (1990)
17. Dar, Gh Hassan, and S. M. Tandon. *Biological wastes* **21**, 75-83 (1987)
18. D. Trujillo, J. F. Perez, and F. J. Cebreros. *Bioresource technology* **45**, 81-83 (1993)

Development of NaY zeolite derived from biomass and environmental assessment of carbon dioxide reduction

Patcharin Worathanakul[1,2] and Patchaya Tobarameekul [1]

[1,2]*Department of Chemical Engineering, Faculty of Engineering, King Mongkut's University of Technology North Bangkok, 10800, Thailand*
[1]*Center of Eco-material and Cleaner Technology, King Mongkut's University of Technology North Bangkok, 10800, Thailand*

Abstract. Carbon dioxide is one of greenhouse gases. The carbon dioxide caused by the industry activities and impact to the global warming. The objectives of this research were to synthesize NaY zeolite from bagasse ash as silica source and loaded with different weight percentage of Cu(II) for carbon dioxide reduction. The carbon footprint of Cu/Y zeolite for carbon dioxide reduction was calculated. The synthesized NaY zeolite from bagasse ash can be easily formed at Si/Al ratio of 0.75 with the additional heat after crystallization 70 °C for 1 hour. The crystal size of NaY zeolite was approximately 0.22-0.37 μm diameter. The results of carbon dioxide adsorption were increased when the flow rate of carbon dioxide decreased. Finally, the carbon footprint value was shown that synthesis step was shown the highest of greenhouse gas emission. This research can increase the value of wastes and reduce pollution emission.

1 Introduction

Global warming caused by greenhouse gases is a major environmental problem, and many efforts have been made to reduce their emission. The increase effects atmospheric CO_2 is one of the greenhouse gases. Many researchers have suggested that CO_2 emissions could be reduced by capturing using solvents [1], membranes [2] and solid absorbents [3]. However, liquid absorption method suffers from several drawbacks such as high regeneration energy, large equipment size, solvent degradation, and equipment corrosion [4].

To avoid the disadvantages of liquid absorption, solid adsorbent is one of alternative ways. Zeolite is a suitable candidate for CO_2 adsorption because of its high surface area, thermal and mechanical stabilities. It is a hydrated aluminosilicate material that exhibits crystallinity and has a porous structure with large surface area.

There are many tools to address global warming problem and measure impact of each product on the environment. One of them is called "Carbon footprint assessment", carbon footprint assessment can also be information to determine an ability to manage the environmental impact of individual entrepreneurs, by using life cycle assessment method (LCA) of the product. The measurement and calculation of greenhouse gas emissions have been stated, that there are many activities such as electricity and water supply consumption, quantities of wastewater and garbage, and the amount of fossil fuels used causing greenhouse gas emission. Carbon footprint is one of tools to quantify and subsequently to GHGs emissions reduction [5]. It expressed as CO_2 equivalents (CO_2-eq) but also carried out the hot spot of environmental impact of products cycles and can be used for production process improvement [6].

Generally sugar industry produced sugar only but the current sugar industry has become a combination of sugar mill, power plant and distillery, etc. which produce sugar, electricity and ethanol. In sugar mill, sugar is produced through several processes which are given below sugar milling, juice clarification and treatment, crystallization, centrifugal separation and drying [7].
Bagasse is by-product of sugar industry obtained after the extraction process. This by-product was used in different processes such as fuel resources [8] which sugarcane bagasse is burnt in the boiler, producing steam that is expanded in turbines coupled with electric generators [9] and silica source [10]. Bagasse ash is a general term describing all forms of the ash produced from burning bagasse. From large quantities of bagasse ash from industry lead to many researches to enrich its value for example, using bagasse ash to synthesize silica or to synthesize catalysts.

With good physico-chemical properties of zeolite for CO_2 adsorption, zeolite Y or NaY is a highly versatile molecular sieve from the faujasite family of zeolites whose 7.4 Å, three-dimensional pore structure and solid acidity make it useful as a catalyst, ion exchanger, adsorbent, etc. Zeolite Y is commonly prepared with a high aluminum content, but in most cases it is employed in a silicon-enriched form [11].

The objective of this research was to develope porous materials at the nanometer scale for NaY zeolite synthesis using seeding method for CO_2 reduction. Therefore, the production of NaY from bagasse ash was used Si/Al ratio of 0.75 with different temperatures, times and weight percentages of ion exchange. The CO_2 reduction of NaY from bagasse ash and carbon footprint assessment of products was evaluated the greenhouse gas emission throughout life cycle of products. Finally, ways to reduce energy consumption in the process for minimizing the environmental impact was studied.

2 Methodology

2.1 Materials

[a] Corresponding author: patcharin.w@eng.kmutnb.ac.th, patcharin_wr@yahoo.com

© The Authors, published by EDP Sciences. This is an open access article distributed under the terms of the Creative Commons Attribution License 4.0 (http://creativecommons.org/licenses/by/4.0/).

Silica and NaY zeolite adsorbent was prepared by reacting the following: bagasse ash obtained from a bagasse field in Thailand contained 96 wt.% SiO_2, 0.98 wt.% Al_2O_3, 0.77 wt.% K_2O and 0.13 wt.% Fe_2O_3 (Khonburi Sugar Public Company Limited). Other materials used were; hydrochloric acid (37 wt.% HCl, Merck), sodium aluminate (Al_2O_3 50-56 wt.% Fe_2O_3 0.05 wt.% and Na_2O 40-45 wt.%, Sigma Aldrich), sodium hydroxide (99 wt.% NaOH, Merck), Sodium siligate (Na_2O 18 wt.% and SiO_2 63 wt.%, Sigma Aldrich), copper nitrate (99.5 wt.% $Cu(NO_3)_2{\bullet}3H_2O$, Merck), distilled water, deionized water and carbon dioxide (Carbon dioxide : HP Grade, Linde)

2.2 Silica extraction

The impurities in bagasse ash can be reduced to using hydrochloric acid treated bagasse ash. Direct extraction of silica can be performed by stirring the sodium hydroxide solution and bagasse ash. During this process, silica is extracted in the form of sodium silicate together with other organic moieties. The sodium silicate obtained is converted to silica by solvent extraction method. The silica from bagasse ash in this sample ash was studied the effect of acid treatment.

2.3 NaY zeolite preparation

The seed gel was prepared first by adding the sodium silicate solution prepared from bagasse ash or silica colloid into sodium aluminate solution under stirring until homogenous. The feedstock gel was also prepared in the same way. The seed gel was then mixed into the feedstock gel. Synthesis of NaY zeolite was prepared with different temperatures and time. Ion exchange of the zeolite was then performed in different wt.% of $Cu(NO_3)_2.3H_2O$.

2.4 Characterization analysis

Chemical composition of bagasse ash was examined by X-ray fluorescence spectroscopy (XRF: Bruker AXS, Germany). The crystallinity was measured by X-ray diffraction analysis (XRD: Model D_8 Advance, Bruker AXS, Germany). The surface areas of the zeolites were determined with Brunauer–Emmett–Teller (BET: (Autosorb 1C, Quantachrome, USA) analysis. A thermogravimetric analyzer (TGA: Netzsch 449F3) was used to measure the CO_2 adsorption capacity and Scanning Electron Microscopy (SEM: JSM-7001F) were determined morphology and characteristics crystalline phase of the silica and NaY zeolite.

2.5 Carbon footprint of NaY for CO_2 reduction

The Carbon footprint assessment consists of four main steps according to life cycle assessment method (LCA)

2.5.1 Goals and scope

The research boundary began from raw materials acquisition, NaY preparation, distribution, CO_2 reduction and disposal step according to Business-to-Customer scope (Fig. 1). Input and output were mainly collected from laboratory data. It was assumed that there is no emission at the distribution stage. At the disposal stage, waste was assumed to be assigned to landfills. The functional unit was 0.3 g of NaY with CO_2 reduction.

2.5.2 Inventory analysis

Inventory analysis is the methodological step where an overview is given of the environmental interventions caused by or required for the processes within the boundaries of the studied system. To collect and gather the information of an input-output substances quantity and energy used in each procedures of the CO_2 reduction process in accordance with the target and the scope that were defined.

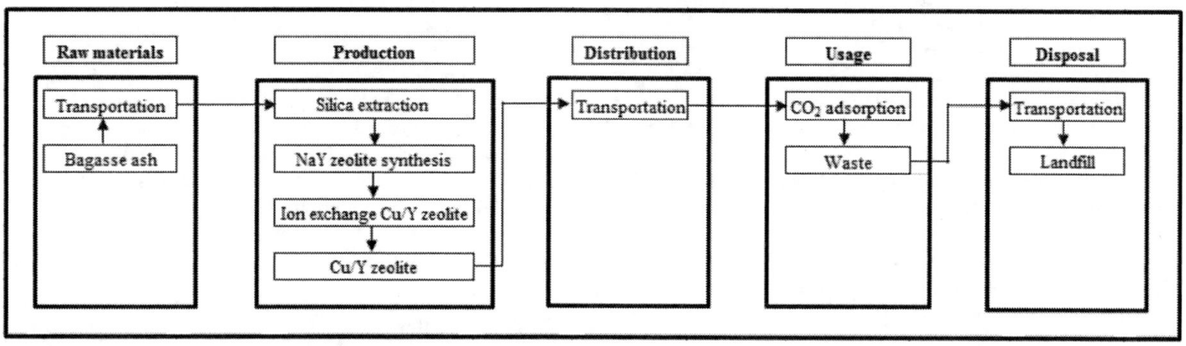

Figure 1. Boundary of NaY for CO2 reduction.

There are five steps including; acquisition of raw materials, NaY preparation, distribution, CO_2 reduction and disposal steps. All materials were transported to the laboratory area. There is no transportation to distributors

DOI: 10.1051/matecconf/20166206003

because of usage only in the laboratory. In the disposal stage, all waste was assumed to be transported to a landfill site.

2.5.3 Impact Assessment

Impact assessment examines the environmental burdens of the emissions and resource use quantified in the inventory analysis. The amount of greenhouse gases emissions for NaY production for CO_2 reduction and its application was calculated in kgCO_2-eq unit.

2.5.4 Interpretation

Interpretation was affected to environment concerns and gave information with the hot spot step. This step can be to reduce energy consumption in the process for minimizing the environmental impact. The identification of significant issues, conclusions and recommendations are made in this step, which can be changed and improved to the appropriate process in the future.

3 Results and Discussion

3.1 Characterization of silica

3.1.1 Chemical composition of silica

The composition of minerals leached from Bagasse ash was shown in Table 1. Found that silica production was composed of silicon dioxide the 92.64 wt.% and other impurities by gross weight. However, the silica purity of the bagasse ash from acid-leached bagasse ash in this study was higher than directly from bagasse ash without leaching.

Table 1. Chemical composition of silica from bagasse ash.

Chemical compositions	Wt.%	
	Directly	Acid-leached
SiO_2	77.81	92.64
Al_2O_3	14.40	0.81
P_2O_5	194.64	1.26
Fe_2O_3	191.38	2.34
K_2O	1.15	2.07
Others	0.62	0.88

3.1.2 X-ray diffraction analysis (XRD) of silica

Silica from bagasse ash in Fig. 2 was observed that, only a broad peak with 2Θ at 22 degree corresponds to the research of Mohamed, et al [11].

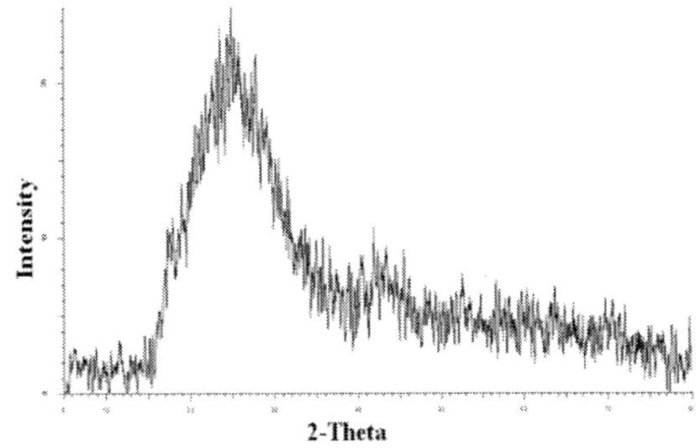

Figure 2. XRD patterns of silica from bagasse ash.

3.1.3 Scanning Electron Microscopy (SEM) of silica

From the SEM image showed the physical structure of silica in Fig. 3. It was found that amorphous shape. It was observed a characteristic of amorphous silica.

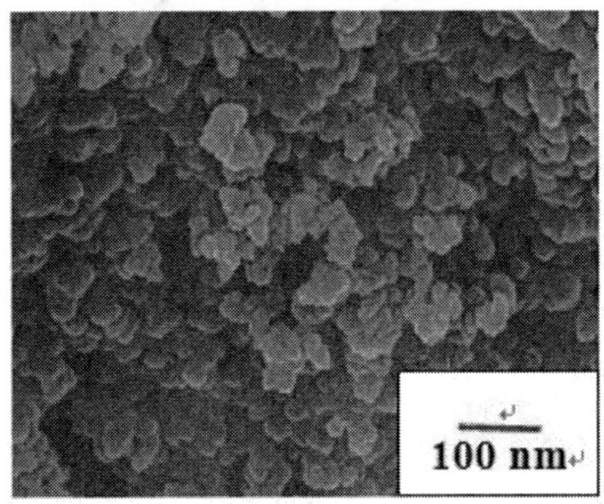

Figure 3. SEM of silica from bagasse ash.

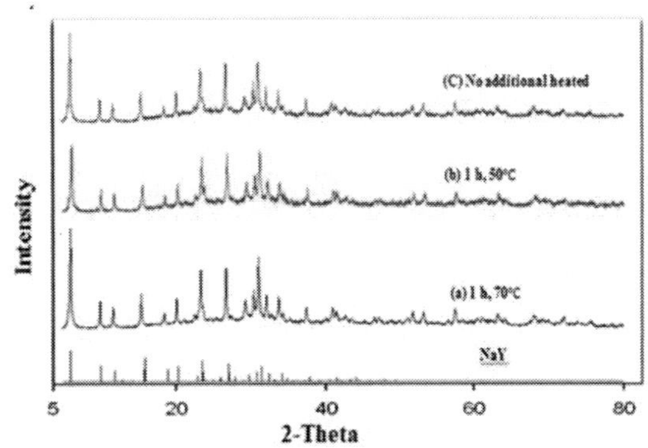

Figure 4. XRD patterns of NaY zeolite.

3.2 Characterization of NaY zeolite

3.2.1 X-ray diffraction analysis (XRD) of NaY zeolite

The synthesized NaY zeolite in Fig. 4 showed peaks corresponded to peaks standards [12]. In addition, the NaY zeolite was characterized by XRD compared with

DOI: 10.1051/matecconf/20166206003

the patterns effect of temperature and time after crystallization. When the crystallization time was reduced from 24 h to 3 h, it was found that the height of peaks was decreased. While there is no heating, the samples are amorphous structure due to decreases of crystallization rate.

3.2.2 Scanning Electron Microscopy (SEM) of NaY zeolite

Fig. 5 shows SEM image of NaY at 0.75 Si/Al with spherical shape. The crystal size of NaY was approximately 0.22-0.37 μm diameter corresponds to the research of Rakpasert (2012) [13].

Figure 5. SEM of NaY with Si/Al ratios of 0.75 with additional heat 1 h at 70 °C.

3.2.3 Brunauer–Emmett–Teller (BET) of NaY zeolite

BET surface area of 5.5 wt% of Cu loading to NaY was 52.33 m^2/g, total pore volume was 0.204 cm^3/g and pore size was 18.2 Å. The increased surface area and total pore volume of NaY results in an increase in the number of adsorption sites for CO_2 adsorption. The volume for trapping CO_2 was determined by pore volume. The large amount of adsorption sites and the trapping volume led to the high CO_2 capacity. In terms of pore, the micropore is more effective than the mesopore and macropore.

3.3 Carbon footprint of zeolite for CO2 reduction

Finally, the life cycle assessment of Cu/Y zeolite was calculated for carbon dioxide reduction. The carbon footprint value was 14.84 kg CO_2-eq. Figure 6 shows percentage of carbon footprint in each stage which synthesis process of production stage was shown the highest of greenhouse gas emission. It came from energy consumption of incubator which was used in ion exchange procedure and Cu/Y synthesis procedure. Secondary step is usage for CO_2 reduction has GHG emission ratio equal to 26.9%, followed by raw materials extraction and waste disposal respectively.

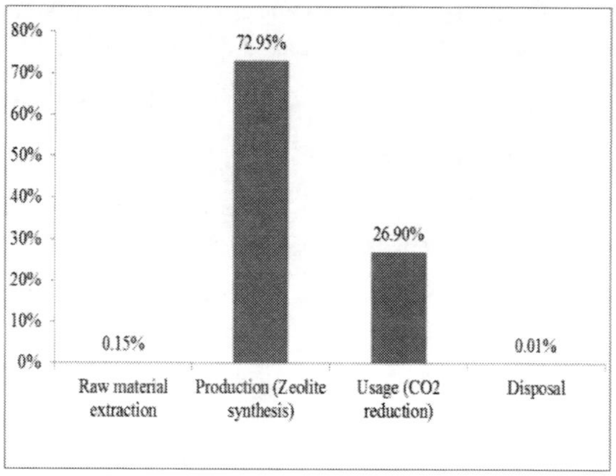

Figure 6. Percentage of carbon footprint in each stage.

4 Conclusions

The bagasse ash is well raw material in NaY zeolite particles synthesis. Physicochemical properties of the produced NaY zeolite after optimizing the various synthesis conditions were characterized. The optimum crystallization time for the synthesis of NaY zeolite with bagasse ash was 1 h at 70 °C. The crystal size of NaY zeolite was approximately 0.22-0.37 μm in diameter. Metals loading on NaY zeolite didn't alter the NaY structure. Therefore, the NaY with the large surface area, the total pore volume, and the micropore was the most favorable in the physical adsorption of CO_2. The carbon footprint of 0.3 g NaY with CO_2 reduction was 14.84 kgCO2-eq. Synthesis process of production stage was shown the highest of greenhouse gas emission. There are three ways to reduce environmental impact from the process which do not affect to the physical properties of NaY zeolites. First, decreasing the incubate time of NaY zeolites in NaY zeolites synthesis step. Second, select the appropriate size of electrical equipment. Last, choose the energy-efficient appliances in the process. All of these will lead to less GHG emission from this CO_2 reduction process.

Acknowledgments

This research was funded by King Mongkut's University of Technology North Bangkok Contract no. KMUTNB-NRU-58-21.

References

1. P. N. Sutar, A. Jha, P. D. Vaidya, E.Y. Kenig. Chem. Eng. J. **207**, 718-724 (2012)
2. W. N. W. Salleh, A. F. Ismail, T. Matsuura, M.S. Abdullah, Separation & Purification Reviews **40**, 261-311 (2011)
3. S. Y. Lee, S. J. Park. J. Ind. Eng. Chem. **23**, 1-11 (2015)

4. C. Chen, J. Kim, D.A. Yang, W.S. Ahn, Chem. Eng. J. **168**, 1134–1139 (2011)
5. T. Wiedmann, J. Minx, *ISA UK Research Report.* (Centre for Integrated Sustainability Analysis, Durham, UK, 2007)
6. S. Bolwig, P. Gibbon, *Risø-R-Report 1719 (EN).* (Risø National Laboratory for Sustainable Energy, Technical University of Denmark (DTU), 2009)
7. M. K. Chauhan, V.S. Chaudhary, S.K. Samar. Renewable and Sustainable Energy Reviews. **15**, 3445 – 3453 (2011)
8. N. Tippayawong, K. Nakpan, *International Conference on Green and Sustainable Innovation*, (Thailand, 2006)
9. M.O.D.S. Dias, R. M. Filho, P. E. Mantelatto, O. Cavalett, C.E.V. Rossell, A. Bonomi, M.R.L.V. Leal. Environ. Dev. **15**, 35–51 (2015)
10. P. Worathanakul, S. Kittipalarak, K. Anusarn. Adv. Mat. Res. **383**, 4038-4042 (2011)
11. R. M. Mohamed, I.A. Mkhalid, M.A. Barakat. Arabian Journal of Chemistry **8**, 48–53 (2015)
12. D. M. Ginter, A.T. Bell, C.J. Radke, M. L. Occelli, H. E. Robson, *Synthesis of Microporous Materials, Molecular Sieves.* (Van Nostrand Reinhold. New York, 1992)
13. N. Rakpasert. *Preparation of Different Metals Loading on FAU Zeolite for NO Reduction* (King Mongkut's University of Technology North Bangkok, Thailand, 2012)

Analysis of the Effect of a Marine Energy Farm to Protect a Biosphere Reserve

Eugen Rusu[1,a]

[1]*Department of Mechanical Engineering, 'Dunarea de Jos' University of Galati, Romania*

Abstract. Sacalin Peninsula in the Black Sea is a new land located south of the Saint George branch of the Danube. Since 1938 this area became a biosphere reserve since many rare species of animals and plants are to be found there. The generation of this new peninsula is due to the sedimentary process induced by the Danube River outflow and it was started more than 150 years ago. In the winter of 2013 this environment was seriously affected by some very strong storms putting in real danger this ecosystem. From this perspective, the objective of the present work is to evaluate the protection that might be offered to this area by a marine energy farm that would be deployed in front of the peninsula. In order to assess the coastal protection offered by the proposed solution, simulations with the SWAN (Simulating Waves Nearshore) wave model have been performed for the most relevant storm patterns. The results show that a marine energy farm can provide a real sheltering effect to the ecological reserve. Such approach seems to be also economically viable since this coastal environment represents a real hot spot in the Black Sea from the point of view of marine energy resources.

1 Introduction

Sacalin is a newly created peninsula in the Black Sea, close to the southern branch (Saint George) of the Danube River. The process of the land generation started more than 150 years ago (at about 1850) due to the sediments transported by the river outflow. It was first developed as an island system, composed by two nearshore islands, Great Sacalin (Sacalinu Mare) and Small Sacalin (Sacalinu Mic). However, in the last decades, due to the sedimentary process the islands were linked together with the mainland generating in this way a peninsula.

Nowadays, the Sacalin Peninsula has a length of about 18.3 km. At this point, it has to be highlighted that a measurement performed in 1977 gave a length of only about 14 km, fact that demonstrates that this is a very dynamically changing coastal environment. As regards the width of the island this is variable. In the north, decreases from about 830 meters to about 350 meters. However it is really hard to estimate the width in the northern part since the island joined the mainland in a peninsula. For the rest of the peninsula, the width varies between 100–150 meters. Finally we can mention also that now the surface of the peninsula is over 21000 hectares. The position of the Sacalin Peninsula in relationship with the Romanian nearshore is illustrated in Figure 1. In the right upper side of this figure, the bathymetric map of the target area is illustrated suggesting also the longshore currents induced by the Danube River outflow.

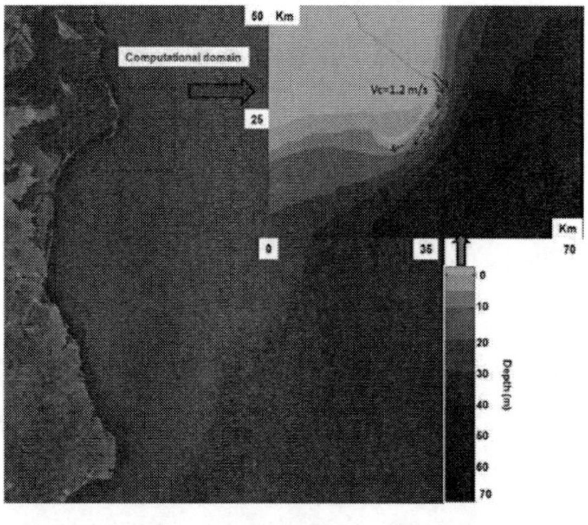

Figure 1. The Romanian nearshore and the target area of the present study (the Sacalin Peninsula). In the upper side the bathymetric map of the target area is illustrated representing also the longshore currents induced by the Danube River outflow

During the time, various rare species of animals have been settled in this environment. Thus, besides about 230 species of birds, which it was discovered that are nesting for the summer in that area, rare species of reptiles and mammals are also to be found there together with some rare plants, such as cabbage and marsh canary. That is why in 1938 this area was declared an ecological reserve and the access of the tourists is nowadays not allowed.

[a] Corresponding author: erusu@ugal.ro

© The Authors, published by EDP Sciences. This is an open access article distributed under the terms of the Creative Commons Attribution License 4.0 (http://creativecommons.org/licenses/by/4.0/).

Despite the fact that there is an obvious natural tendency to enlarge this costal environment due to the sediment transport patterns, in the winter of 2013 a sequence of very strong storms practically broke the peninsula putting in real danger all the rare species of animals and plants living there. From this perspective, in order to provide a realistic explanation related to the causes that generated these damages to the Sacalin biosphere reserve, Rusu and Butunoiu [1] performed some high resolution simulations with numerical wave models considering the most relevant wave propagation patterns. They used for this the spectral phase averaged model SWAN (Simulating Waves Nearshore) [2]. Their results put in evidence the fact that there is a dominant process of driving the sediments brought by the Danube towards the south.

This was called the *'constructive process'* and corresponds to the most relevant pattern of the wave propagation, that is when the direction of the incoming waves is from north, northeast. Nevertheless, when the direction of the incoming waves is from the south, the results show that the wave forces are considerably higher and in the case of the strong storms, the waves may penetrate the peninsula as it was happened in the winter of 2013. Since such phenomena can happen again sometime in the future, as a continuation of [1], the present work explores the possibility of deploying a marine energy farm that may have as an effect also the coastal protection of the Sacalin ecological reserve.

2 Wave modelling and relevant environmental patterns in the Black Sea

The wave and wind climate in the Black Sea became in the last decades an issue of increasing importance. This is due to the continuous enhancement of the human nearshore and offshore activities, including also a significant increase of the maritime traffic.

In order to provide reliable information concerning the wave conditions in the Black Sea, a multi-level wave prediction system, SWAN based was implemented [3, 4]. SWAN is a spectral phase averaged wave model, which integrates the spectral action balance equation in time, geographical and spectral spaces. The system comprises a large computational domain that covers the entire Black Sea basin and the Sea of Azov, which is subsequently focused towards various coastal areas. The advantage is that one single model covers the full scale of the modelling process, although the physical processes considered in each computational domain might be rather different from one computational level to another. Moreover, in order to improve the reliability of the wave predictions, some data assimilation technique have been also implemented, considering either satellite data [3], or in situ measurements [5].

A very important area in the Black Sea, from the point of view of the maritime traffic, is that located in front of the Danube Delta. This is also the principal southern gate in the seventh Pan European transportation corridor. On the other hand, the interactions that occur between the relatively strong waves and the currents produced by the Danube River outflow induce a wide variety of processes and phenomena, sometimes rather unusual. For this reason, several studies were focused on this coastal environment as for example [6, 7, 8].

For the present work, the characteristics of the computational domain, the input fields considered and the physical processes activated are briefly presented in Table 1.

Table 1. Characteristics of the computational domain defined and the main physical processes activated.

Computational grid characteristics		**Input or process**	
Coordinates	Cartesians	Waves	X
Δ x (m)	200	Wind	X
Δ y (m)	200	Current	X
Δ θ (°)	5	Gen	Westh.
No points in X-direction	351	Interactions	Quad. Triads
No points in Y-direction	251	Whitecapp.	X
No of frequencies	35	Refraction	X
No of directions	36	Diffraction	X
Mode	Stationary	Nearshore processes	Friction Breaking Set up

As regards the possibility of deploying a marine energy farm in the coastal environment of the Black Sea, it has to be highlighted that although the wave conditions are less energetic than in the ocean, the idea of the hybrid farms seems however to be viable. Moreover, various studies [9, 10] showed that the wind energy potential in the Black Sea is in line with that from other locations where offshore wind farms already operate successfully. On the other hand, other previous studies [11, 12] indicated that besides renewable energy, such hybrid farms may provide also an effective coastal protection.

3 Analysis of four scenarios

Previous analyses performed in the Black Sea [3, 13] put in evidence that the western part of the sea is its most energetic part. Thus, every winter at least one or two storms occur characterised by significant wave heights greater than 7 meters. Such events can be called trivial strong storms.

Nevertheless, extreme storms may also occur in a time widow usually smaller than a decade. In such cases significant wave heights even greater than 10 meters can be noticed. From a statistical perspective there can be defined two geographic regions where the maximum waves are encountered. One is in the south-western part of the sea, offshore the border between the Bulgarian and the Turkish coastal environments and the other in the north-western part of the sea, offshore the mouth of the Danube.

DOI: 10.1051/matecconf/20166206004

Based on these data, four different environmental scenarios from the point of view of the directions of the wave propagation have been defined and they are presented in Table 2. Corresponding to each scenario and energy case, model simulations have been carried out for the present situation when no farm or other coastal protection are present and also considering various configurations of marine energy farms in front of the Sacalin Peninsula and also considering various transmission ratios in terms of significant wave height (Hs). Figures 2, 3, 4 and 5 present the results for the four scenarios and energy cases considering a transmission ratio of 60%. Figure 6, presents in more detail the location in the target area of the marine farm that was found more appropriate for the coastal protection and the positions of 8 pairs of reference points corresponding to water depths of about 5 (nearshore points - NP) and 8 meters (offshore points - OP), respectively.

Table 2. Description of the four scenarios considered.

Scenario	Direction on boundary in Nautical convention (°)	Case A High waves		Case B Extreme waves	
S1	30	Hs 5m	Tp 10s	Hs 8m	Tp 12s
S2	90				
S3	150				
S4	200				

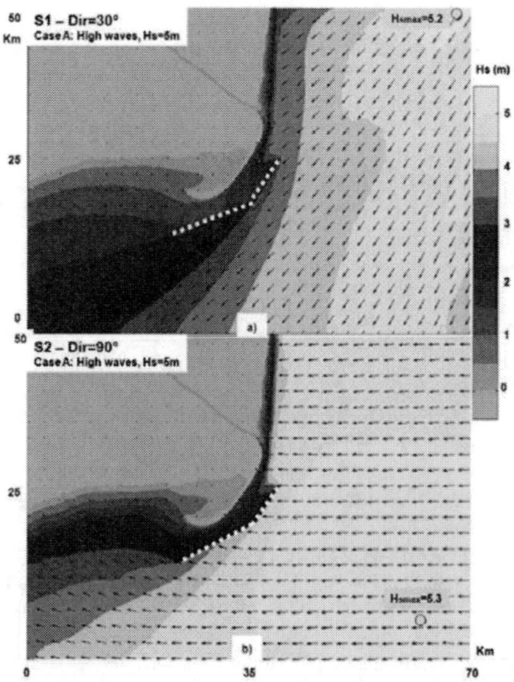

Figure 2. Case A - high waves, significant wave heights fields and wave vectors, a) S1 - wave direction on the boundary 30° in Nautical convention; b) S2 - wave direction on the boundary 30° in Nautical convention.

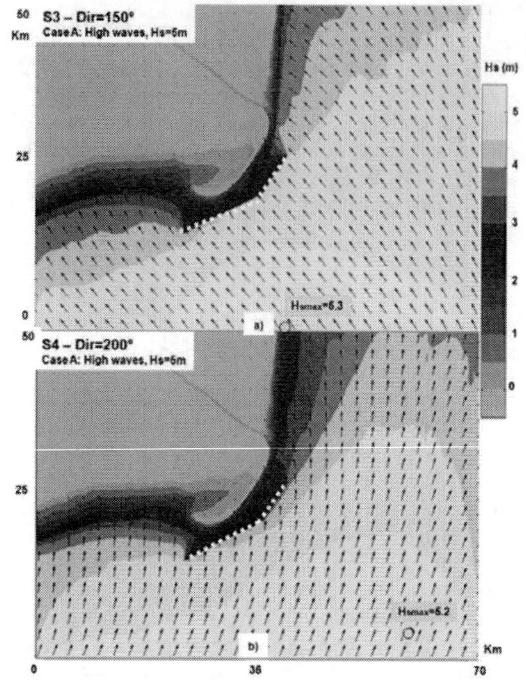

Figure 3. Case A - high waves, significant wave heights fields and wave vectors, a) S3 - wave direction on the boundary 150° in Nautical convention; b) S4 - wave direction on the boundary 200° in Nautical convention.

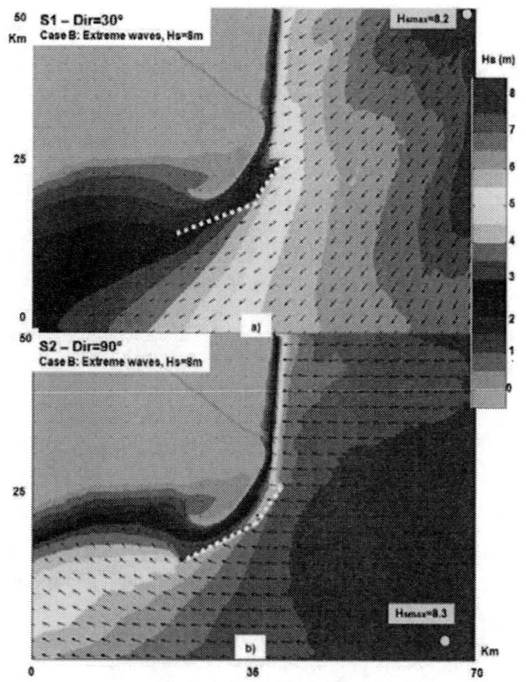

Figure 4. Case B - extreme waves, significant wave heights fields and wave vectors, a) S1 - wave direction on the boundary 30° in Nautical convention; b) S2 - wave direction on the boundary 30° in Nautical convention.

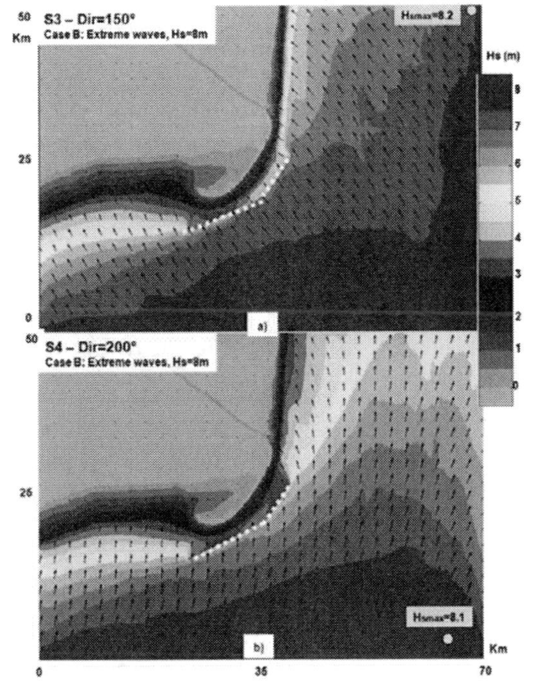

Figure 5. Case B - extreme waves, significant wave heights fields and wave vectors, a) S3 - wave direction on the boundary 150° in Nautical convention; b) S4 - wave direction on the boundary 200° in Nautical convention.

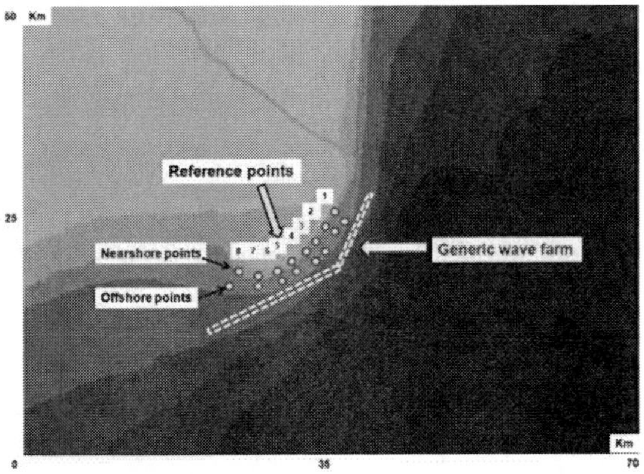

Figure 6. Location in the target area of the marine farm and of the 8 pairs of reference points corresponding to water depths of about 5 and 8 meters, respectively.

4 Discussions

A first issue to be discussed at this point is related to the efficiency of such a marine energy farm in the nearshore of the Black Sea. As it was explained before, although less energetic than the coasts of the oceans, the target area is one of the most energetic in the Black Sea form the point of view of the waves. As regards the wind power this is in line and sometimes even superior than in places where wind farms already successfully operate. From this perspective, the idea of a hybrid marine energy farm appears to realistic enough.

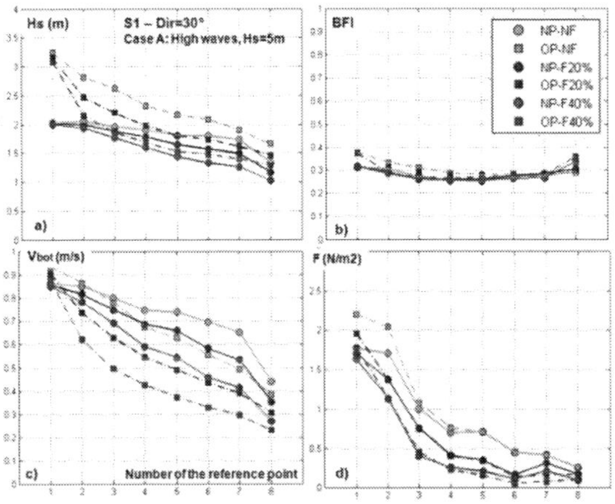

Figure 7. Case A - high waves, S1 - wave direction on the boundary 30°; variations along the 8 reference points: a) Hs; b) BFI; c) Vbot; d) F.

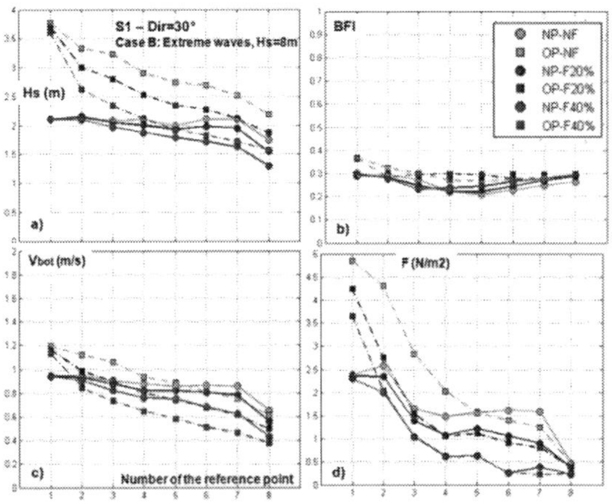

Figure 8. Case B - extreme waves, S1 - wave direction on the boundary 30°; variations along the 8 reference points: a) Hs; b) BFI; c) Vbot; d) F.

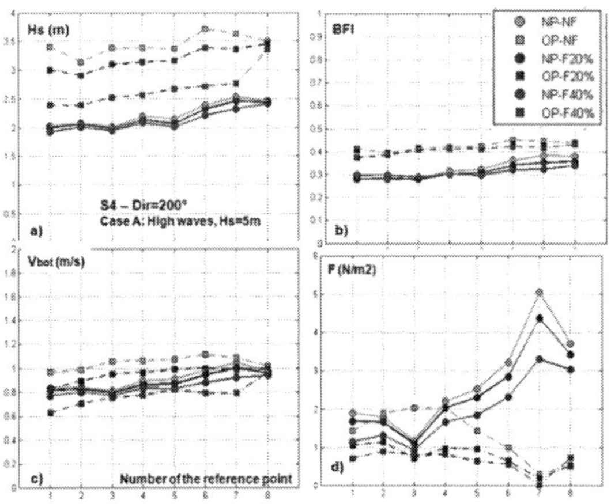

Figure 9. Case A - high waves, S4 - wave direction on the boundary 200°; variations along the 8 reference points: a) Hs; b) BFI; c) Vbot; d) F.

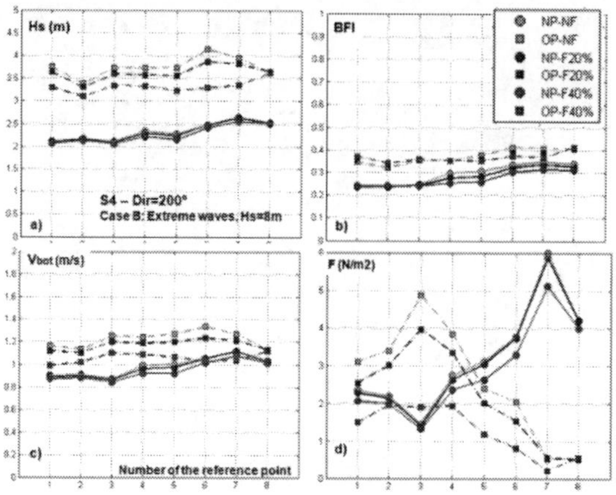

Figure 10. Case B - extreme waves, S4 - wave direction on the boundary 200°; variations along the 8 reference points: a) Hs; b) BFI; c) Vbot; d) F.

As regards the coastal protection, various studies (see for example [14]) showed that the coastal works are very expensive. Thus, a marine farm might represent an option for such cases. Figures 7, 8, 9 and 10 present the values in the reference points of the parameters Hs (significant wave height), BFI (Benjamin Feir Index that indicates if there is a higher probability of the apparition of the extreme waves in relationship with the standard Rayleigh distribution), Vbot (orbital velocity at the bottom) and F (wave forces) corresponding to three relevant situations: NF - no farm, F20% - farm with 20% absorption (including reflections) and F40% - farm with 40% absorption. Thus, the results presented in Figures 7, 8, 9 and 10 demonstrate that the presence of the marine farm may provide a protection to the Sacalin Peninsula.

5 Conclusions

The present work explores the possibility to implement a marine energy farm in order to protect an important and unique biosphere reserve in the Black Sea. Besides the fact that a hybrid marine energy farm might be economically viable, the results presented show that such a project might play an important role in protecting the coastal area targeted.

Thus, when the waves come from the south, the wave forces become considerably higher, almost double for the high waves than in case of the waves coming from the north, north-east. Such situations of high and extreme waves coming from the south are not very likely to be encountered, but they are possible and in fact this was the case that almost destroyed the peninsula in the winter of 2013.

Acknowledgements

This work was supported by a grant of the Romanian Ministry of National Education, CNCS – UEFISCDI, project number PN-II-here ID-PCE-2012-4-0089 (project DAMWAVE).

References

1. L. Rusu, D. Butunoiu, Journal of Marine Science and Technology – Taiwan **23**, 669-677(2015)
2. N. Booij, R.C. Ris, L. H. Holthuijsen, J. Geophys, Res. **104**, 7649–7666 (1999)
3. L. Rusu, Energies 2015 **8**, 10370-10388 (2015)
4. D. Butunoiu, E. Rusu. Journal of Environmental Protection and Ecology **13**, 1332-1349 (2012)
5. D. Butunoiu, E. Rusu. *Proceedings of the Second International Conference on Traffic and Transport Engineering ICTTE*, Belgrade, 180-187 (2014)
6. A. Ivan, C. Gasparotti, E. Rusu, Journal of Environmental Protection and Ecology **13**, 1673-1682 (2012).
7. A. Ivan, L. Rusu, Environmental Engineering and Management Journal **14**, 751-761 (2015)
8. E. Rusu, S. Macuta, Environmental Engineering and Management Journal **8**, 147-151 (2009)
9. F. Onea, E. Rusu, International Journal of Green Energy **11**, 465-487 (2014)
10. F. Onea, E. Rusu, Meteorological Applications **21**, 316-329 (2014)
11. A. Zanopol, F. Onea, E. Rusu, Energy **72**, 652-670 (2014)
12. S. Diaconu, E. Rusu, The Scientific World Journal, Article Number: 498013, (2013)
13. L. Rusu, D. Butunoiu, E. Rusu, Journal of Environmental Protection and Ecology **15**, 445-454 (2014)
14. I. Omer, R. Mateescu, L. Rusu, D. Niculescu, E. Vlasceanu, Journal of Environmental Protection and Ecology**16**, 424-433 (2015)

Feasibility Study on the Promotion and Application of Straw Bale House in Northern Rural Areas in China

Xiaoming Li[a]

Shandong Urban Construction Vocational College, Jinan, China

Abstract. With the accelerated pace of new rural areas construction in northern China, people have a higher requirements in terms of houses' quality and function. At the same time, the backward construction methods and developing with the wind blindly has also led to a rapid growth in resource consumption. In order to better meet the requirements of hot comfort, energy saving and environmental protection properties of rural houses in northern areas. Based on the research methods including investigation, mathematical statistics, documentation, and logical analysis, combining with the residential problems in northern rural areas, and the superiority of straw bale house, and the favorable conditions for straw bale buildings. This article identified the feasibility of straw bale housing construction in the northern countryside, and draw the conclusion that the excellent economy, simple construction methods and good energy efficiency of straw bale house will be accepted by the majority of farmers friend.

1 Introduction

The pace of studying on ecological building materials is faster and faster. Although the new environmental protection materials are energy saving and good quality, it can also not be accepted by the people in northern rural areas of China. Because their backward construction methods and poor economic conditions. Comparatively speaking, straw materials are easy to obtained in rural areas. And people have no need to spend very great investment and a lot of effort on the study of straw baling technology. Furthermore, it is an urgent need to promote the innovation of wall materials and protect arable land. It is obvious that recycling straw resources is an feasible way to protect the environment and reduce the atmospheric fog and haze. So the straw bale houses have be widely constructed in Europe and America countrysides (Fig. 1). It has been fully proved that straw bale have superior performance as a kind of building material [1]. Today, innovation of better wall materials and promotion of energy-efficient buildings is an urgent need for arable land protecting and energy conservation. Recycling straw resources is an important way to protect the environment and reduce air pollution. And the straw bale building can meet these needs perfectly. For a Inadequate resources China, it is a important initiative to improve the energy efficiency of rural houses, who are 75% of the nationwide building area [2].

The date of investigation on part of the northern rural areas (Some research data is shown in Fig. 2) indicate that there have no housing insulation measures in 57% of the northern rural areas. Some of adobe houses and stone houses have able to meet the families' insulation requirement. But most of brick houses and concrete block houses are unsatisfactory [3]. Northern summer temperatures are not too high. But the winter are usually long and cold. Coal always be used as the primary fuel for heating and cooking. According to the statistics, wall area accounted for the largest proportion in building envelope.Through the wall, there will loss about 20% of the total heat capacity of the building in winter. And there will absorbed about 35% of the total absorption in summer. The heat loss of door and window accounted for 30% in the total building energy consumption. So it is important to design for the energy-saving of exterior walls, windows and doors. But which is always be overlook by residential construction in countryside. According to statistics, the coal consumption of a 100m^2 house in winter is about is 5 tons. Over the past five years, the price of coal has risen from 200 yuan to 500 yuan per ton. While the average annual income of northern farmers is only 2,500 yuan or less. So the fired spend occupies a large part compared with other household income [2]. In addition, the energy consumption for the adobe Kang heating can not be underestimated in most of the regions. In conclusion, for a house without enough insulation measures, room rate may not be much. But the residents have to pay for the subsequent energy consumption.

[a] Corresponding author: xiaomingqwe@163.com

© The Authors, published by EDP Sciences. This is an open access article distributed under the terms of the Creative Commons Attribution License 4.0 (http://creativecommons.org/licenses/by/4.0/).

DOI: 10.1051/matecconf/20166207001

Figure 1. Large numbers of straw bale buildings are constructed in foreign countries.

2 The Problems of Houses in Northern Rural Areas

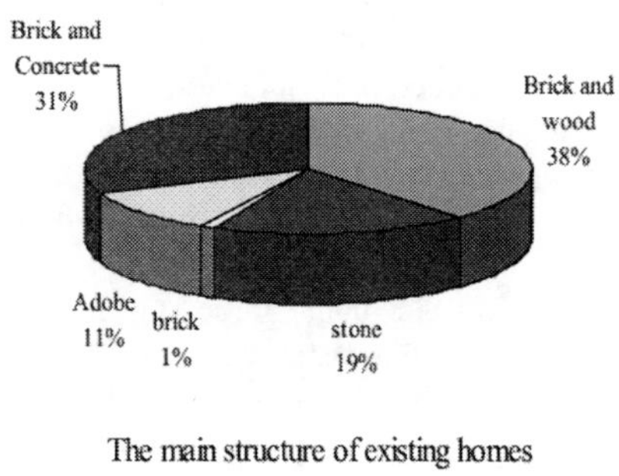

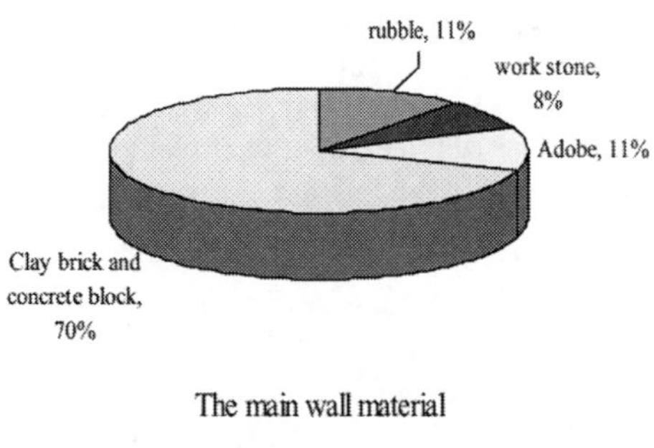

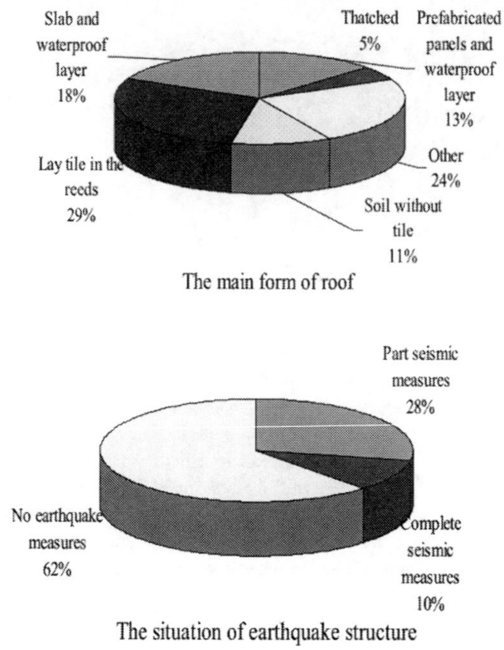

Figure 2. Part of survey data of the northern rural areas.

Owing to the fact that northern rural residents lack of experience in terms of earthquake awareness and sophisticated construction technology. Most rural houses' construction quality are poor. And their structure function can not meet the earthquake requirements. Taking into account that the local brick and concrete block are cheap and easy to access. Northern rural housing construction materials are used under the premise of a huge energy consumption. Which can not meet the requirements in saving resources. Moreover, northern rural housing lack of insulation measures. Resulting in a high energy consumption and high heating costs. Due to the lack of appropriate professional guidance and unified settlement planning, residential construction standards in rural areas is uneven, the workers have a poor build technical capacity, and residents are built blindly. Which increasingly unable to meet the needs of new rural construction. Therefore, it is a priority to explore a new materials as a substitute for clay brick for the northern rural areas.

3 Energy-Saving and Cost Advantages of Straw Bale House

American architect Jim Hanford has analyzed the energy performance of five kind of walls including adobe houses, fiber concrete, wood frame, foam blocks and straw bale in the US Department of Energy (DOE) laboratory. The ultimate computer analysis showed that straw bale structure can save more than 60% of energy, which has the best energy-saving effect [4]. Moreover, it is revealed by the contrastive study of thermal engineering and economic behavior between straw bale house and brick house that, wall heat consumption is very alarming in an ordinary brick house. At the same time, huge amount of coal consumption in winter increased the burden of atmosphere environmental. Straw bale buildings can relieve the pressure faced by rural ecological environment (Table 1). And it will exploit sustainable local resources effectively. If the combustion of one kilogram of coal will exhaust 1.5 to 2 kg CO^2 , about 44.686 more million tons CO^2 will be released to air by producing 10 million solid clay brick than the same volume of straw bale. Furthermore, whether manufacturing energy consumption, transport energy consumption or environmental performance, straw bale is better than solid clay brick. According to the statistics, the production of 10 million volumes straw bale will save clay 19,747,800 tons ,land 1750875m², coal 167,634,500 tons, and will reduce the amount of soot emissions from coal-fired 61,312.5 tons. Straw bale house have low heat consumption, high coal saving rate, and short payback period. Those aspects are all better than brick house [5].

Table 1. Comparison of energy consumption for making burned clay brik and straw bale(kJ/kg)

Project	Clay brick		Straw bale's average value
	Average value	Max value	
Production process-related heat loss	2615~2930	3560~3770	
Energy consumption associated with material properties	96~180	340~500	120
Basic power consumption	171~189	216~234	
Total energy consumption	2984~3197	4134~4486	

4 The Favorable Conditions to Build Straw Bale House in Northern Rural Areas

In recent years, the northern rural government regard building ecological civilization village as the strategic emphasis of Balancing Urban and Rural Development and Building Socialist New Countryside. Which aim is improving the rural environment and building a better home. Enclosing the theme of "Constructing Ecological Environment, Developing Ecological Economy, Fostering Ecological Civilization". Working from actual conditions, A group of civilized ecological characteristics villages were constructed based on the mountain specialties, local conditions, overall planning, classified guidance and overall progress. These measures enable the rural landscape has changed dramatically. At the same time,a good ecological, economic and social benefits were created.

4.1 Climatic conditions advantages

In recent years, numerous exemplary housing are built combined with active technology and passive technology, like solar thermal storage technology, photovoltaic technology and ground-source heat pump heating technology. Which have received a good social and economic benefits. In this trend, comprehensive design program combing active and passive technology will have a broader applicability.

Most of the northern rural areas have a dry climate, long cold winter and lower temperatures. There are rich of wheat straw resources in northern China's rural areas. So the raw materials of straw bales are easily to access and have a low price. In addition, these local natural resources are renewable, degradable and recyclable, which is the typical properties of green buildings. Moreover, most of the local materials have a simple processing method. And it has a low process energy consumption.The straw bale house have advantages in energy efficient, low cost, good insulation, Etc. Therefore, it has strategic significance for the promotion of straw bale house in underdeveloped mountainous countryside. Furthermore, the straw bale house which has the advantage of eco-efficiency, can provide a practical and effective way for comfort and savings to spend the long winter in mountain areas.

4.2 Material supply advantages

Establishment of a modern building system makes concrete, steel and cement spent a lot of land, water, energy and materials. It led to the outstanding problems of air pollution, water pollution, greenhouse effect and solid waste emissions. Which increased a heavy burden of the environment. And it seriously restricts the sustainable development of our country.

Figure 3. The idle wheat straw in northern rural areas.

Northern rural areas in China are abound of corn and wheat (Fig. 3). And the building can use the local materials without transportation.So, it is enough and convenience for straw bale manufacturing. Moreover, if fully utilized instead of straw burning, air pollution and the greenhouse effect will be less and less. This also reduces the phenomenon that remaining straw occupation of limited land. The most important thing is the accidental or deliberate arson fire can be avoid. Making full use of local materials in residential construction in villages, and combined with the ecological construction technology can not only reduce residential energy consumption and environmental stress. It can also reduce the construction costs and improve people' s living comfort. So it has a real significance in the transition of rural residence construction.

4.3 Cost advantages

There are a lot of poor people in northern rural areas. At the same time, we are accelerating the pace of new rural construction. And a large number of rural tenements have to be rebuilt. In the past years, crop straw is usually dealt as waste. Which caused waste of resource and the straw' s oversupply. The construction price of straw bale house is lower than ordinary brick house. And its excellent thermal insulation properties can help farmers save a lot of energy consumption costs in their subsequent life. Furthermore, straw bale house have a simple construction technique [6]. Even the people with no construction experience can design and construct it easily. Which will save a lot of labor costs. China's poor population are distributed in the mountainous areas. From the point of farmers' view, economic and practical straw bale house is one of the effective ways to help the poor. In northern rural areas, most of wealthy families build brick houses to symbolize their wealth and success. The aim of lots of hard work for a life time by the farmers is to own a brick house. So they put most of the income into the new house construction. Building costs become their greatest concern. Northern rural residents per capita household income were lower than the national average. To build an inexpensive straw bale house is definitely a good choice for most of the farmers, especially for the elderly people.

4.4 Straw Bale Houses Should be Vigorously Promoted in the Northern Rural Areas

Straw bale material makes full use of the renewable plant straw and reduces the pollution of the atmosphere caused by straw burning. Instead of brick, the application of straw bale meet the policy about phasing out of clay. Which can protect arable land and ecological environment and produce a obvious environmental effects [7]. In addition to the widely planting of corn and wheat, straw bale will have a good developing prospect and a strong vitality in northern rural areas.

Nowadays, housing construction in rural northern have two major trends: Firstly, the so-called standards becoming increasingly high. People always pursue luxury exterior and interior. Secondly, due to the lack of architectural concept and scientific technical standards. Rural housing is still stay in the spontaneous construction phase [8]. The comfort, construction function is difficult to improve. Resulting in uncontrolled energy consumption of rural building. As a long time, there are lack of wall materials alternatives to clay brick. Even clay will consume huge amounts of energy, still become the first choice of rural buildings. Thus, considering the global resources and energy shortage, it has a very great practical significance to increase focus on the environmental performance of rural housing. It's imminent to improve the quality of the physical environment of rural housing and reduce energy consumption.

5 Summary

Straw bale house is an energy efficient building, who has the benefits of social, economic and environment. And it will has a bright future in the northern rural areas. Straw bale wall has many good performances that many other new environmentally friendly building materials do not. These performances make the straw bale house outstanding in terms of saving energy for heating in winter, saving total investment, ease air pollution and other aspects. The northern rural areas have a sufficient foundations in climate, local materials, and access to energy savings. In the overall context of Accelerating rural ecosystem construction and Promoting vigorously the new socialistic rural building, governments should mobilize the whole society to work together to accelerate the introduction and promotion of straw bale house.

References

1. P. Xu, "The Study of Taking Raw Material without Wood Plant Fiber as the Construction Wood-based Panels," *New Building Materials*, vol. 50, pp. 19-24, Nov. 1996.
2. China *Building Energy Saving Annual Development Report, Building Energy Research Center of Tsinghua University*, 2014.
3. B. Liu, "The Energy Analysis and Application Prospect of Straw Bale Building in the Northwest Territories," *Industrial Construction*, vol. 3, pp. 17-19 , May 2007.
4. Y. S. Tao, "The energy saving effect analysis of aerated concrete building," *New Building Materials*, vol. 1, pp. 55-57, July 2005.
5. M. Li, "The Eco-construction Technical Research on Rural Straw Bale House in Yimeng Mountain Areas," M.E. dissertation, Dept. Elect. Eng., Shandong Agricultural Univ., Taian, China, 2014.
6. X. G. Li, J. Ma, "Research and Application of Straw Bale," *Development Guide to Building Materials*, vol. 3, pp. 57-60 ,Sep. 2006.
7. L. Ahasteen, D. Bainbridge, L. "Denestone. House of straw. U.S.Department of Energy Efficiency and Renewable Energy," DOE/GO 10094-01, 1995.
8. X. M. Li, Q. D. Li, H. J. Liu, X. D. Xu. "Eco-Renovation of the Existing Rural Housing in Yimeng Mountain Area," *Applied Mechanics and Materials*, vol. 368, pp. 186-189 , Nov. 2013.

AUTHOR INDEX

AUTHOR INDEX

EDP Sciences – Web of Conferences
17, Avenue du Hoggar
Parc d'Activité de Courtabouf
BP 112
F-91944 Les Ulis Cedex A
France

CURRAN ASSOCIATES INC.
proceedings.com

ISBN 978-1-5108-2600-7
9 781510 826007

10th International Conference on Energy Materials and Environment Engineering (ICEMEE 2024)

E3S Web of Conferences Volume 560 (2024)

Kuala Lumpur, Malaysia
14-16 June 2024

edp sciences